Biology and Criminology

Routledge Advances in Criminology

Biology and Criminology

The Biosocial Synthesis

Anthony Walsh

Routledge
Taylor & Francis Group

New York   London

First published 2009
by Routledge
711 Third Avenue, New York, NY 10017

Simultaneously published in the UK
by Routledge
2 Park Square, Milton Park, Abingdon, Oxon OX14 4RN

First issued in paperback 2012
Routledge is an imprint of the Taylor & Francis Group, an informa business

Library of Congress Cataloging in Publication Data
Walsh, Anthony, 1941-
  Biology and criminology : the biosocial synthesis / by Anthony Walsh. — 1st ed.
    p. cm. — (Routledge advances in criminology ; 7)
  Includes bibliographical references and index.
  1. Criminal behavior. 2. Criminal behavior — Genetic aspects. 3. Sociobiology. 4. Criminology. I. Title.
  HV6115.W34 2009
  364.2'4 — dc22
  2008053719

ISBN13: 978-0-415-65366-4 (pbk)
ISBN13: 978-0-415-80192-8 (hbk)
ISBN13: 978-0-203-87584-1 (ebk)

This book is dedicated to my wife Grace, my parents Lawrence and Winifred, my sons Robert and Michael, my stepdaughters Heidi and Kasey, and my grandchildren Robbie, Ryan, Mikey, Randy, Christopher, Stevie, Ashlyn, and Morgan.

Contents

Boxes

Figures

Tables

Tables

Foreword

I've known the author of this book, Anthony Walsh, for many years. There-fore, I will dispense with the formality of calling him by anything other than his nickname.

Tony and I first came to know one another after I published an article many years ago entitled "The Decline and Fall of Sociology" (Ellis 1977). The theme of the article was that sociology was heading to scientific oblivion if sociologists continued to deny the importance of biological variables for comprehending human behavior. Sadly, this denial persists and readers can decide for themselves whether the fate that I envisioned has come to pass.

Fortunately, unlike sociology, criminology has made significant strides toward incorporating biological concepts into many of its recent theories and research agendas. Criminology's shift away from pure environmental-ism is very evident in a forthcoming book that two coauthors and I have written to summarize what is currently known about variables that are and are not associated with criminality (Ellis, Beaver, and Wright 2009). Sur-prisingly, one of the longest of the book's nine chapters is devoted to links between criminality and biological variables.

While Tony and I are in total agreement about the importance of biology for understanding criminal behavior, the routes each of us took to arrive at this view were quite different. So, before I address the importance of Tony's new book, let me sketch how each of us came to believe that the role of biology in crime causation is quite significant.

TONY'S DEVELOPMENT

Tony began his career as a cop, only to begin college at the age of 30. His initial goal in college was to become a physical therapist, which led him to major in biology. However, an undergraduate sociology professor radical-ized Tony into a Marxist way of thinking and convinced Tony to pursue a master's degree in sociology rather than working toward physical therapy certification. He became a recovering Marxist after about two years, but maintained his interest in sociology, especially along the lines of criminology. Consequently, Tony entered a PhD program and returned to the field of criminal justice as a probation and parole officer.

By the time Tony had finished his PhD, he had read my article, "Decline and Fall of Sociology." This article almost instantly resonated with Tony's biological background, and apparently helped him to complete his transition from radical sociology to biosocial sociology.

MY DEVELOPMENT

In academic terms, my career followed a fairly straight sociological trajectory. This included having taken my first criminology course using Southerland and Cressey's very popular (but extremely antibiological) text of the 1960s. I understood and could parrot back the environmentalist mantra espoused in all of the sociology courses of the time, but I remember never being completely convinced that biology had no bearing on human behavior. Especially in criminology, I recall thinking that crime might result from more than just social experiences and learning. Also, I purposely took some biology and biologically oriented psychology courses that inspired me to think about how the brain controls all thought and behavior. By the time I graduated from college, I recall already thinking about human behavior outside of a purely environmentalist box.

Looking back, I think there was another reason my sociological thinking becoming increasingly "biologized" as time went by. It came from having watched the behavior of horses and cattle while growing up on a Kansas farm. I recall being impressed by the behavioral differences between male and female farm animals despite their never having been acculturated into their proper sex roles.

Especially regarding aggression, it seemed to me that if sex differences could be produced without culture or social training, then similar human differences might also be "hardwired". I related this to the fact that most violent crimes are committed by men. Increasingly, I became interested in how sex hormones might alter the brains of males beginning around the time of puberty to cause aggression and violent crime.

Despite the continued unpopularity of a biosocial perspective in sociology, I am sure that one day, this discipline will be transformed. And I now believe that criminology will help it to do so.

ABOUT THIS BOOK

For me, the main lesson to be learned from both Tony's and my experiences is that for those who are serious about becoming a competent criminologist, take some courses in biology. Do so even if your sociology or criminal justice professors assure that these courses are a waste of time. They're not. Basic biology is a must, but you should also take courses in neurology, endocrinology, and evolutionary theory, if possible.

Tony's book provides readers with the culmination of much of what he has learned and synthesized from the growing field of biosocial criminology. Readers will see that, in two ways, the book cuts through the fuzzy reasoning of environmentalism and brings to the forefront the power that comes from meshing biological and social learning concepts.

One way involves discussing the growing influence of evolutionary approaches to comprehending criminality. In particular, violent sexual crimes can be understood in terms of reproductive advantages afforded offenders, especially when the criminal justice system is being poorly equipped to identify and severely punish them.

The second way biosocial criminology is flexing its muscle is by identifying some of the underlying biochemistry and neurology associated with the increased probabilities of offending. Especially promising have been findings surrounding the role of the sex hormone testosterone and the neurotransmitter serotonin. These, in addition to numerous other promising lines of investigation, are breathing new life into a discipline that struggled for decades with strictly environmental theories of criminality that made minimal progress toward understanding criminality.

—Lee Ellis

Preface

Former president of the American Society of Criminology (ASC), C. Ray Jeffery, wrote an editorial in the ASC's flagship journal entitled *Criminology—whither or wither?*" He was asking where his beloved criminology was going and warning that it had better go in the right direction or else it would dry up and wither. He argued that criminology should have dropped "anomie, opportunity theory, differential association, social learning theory, conflict theory, and labeling theory," 20 (now more than 50) years ago (1977:284). For a president of the ASC to argue that his discipline should drop its major theories is peculiar to say the least, but Jeffery has long campaigned for a more biologically informed criminology, arguing that criminology must meet the biological challenge "or sink into the mire" (1977:285). Although I agree completely with Jeffery's point that criminology desperately needs biology, I believe that he was too radical in his blanket condemnation of existing criminological theories. The longevity enjoyed by these theories suggests that there are conceptual cores to each of them with sufficient logical appeal and empirical support to have captured the attention of generations of criminologists. These theories are not wrong; they are merely incomplete, encompassing the first stage (demographics and their correlates) in the search for causes. Criminology is an inherently interdisciplinary science, and the causes of criminal behavior can be sought at many levels as long as each level is part of a coherent and mutually reinforcing whole.

Most criminologists would probably agree in the abstract that criminology is inherently interdisciplinary, but in practice they tend to ignore any perspective other than the one they were trained in, and with which they have become comfortable. Criminologists with different disciplinary orientations conduct their research and formulate their theories apparently oblivious to what is going on in other camps. Biological factors do not operate in an environmental vacuum, nor do environmental factors operate in a biological vacuum, and we must cease formulating our theories as if they do. I argue that scientific progress is only possible when immature sciences are vertically integrated with the more mature and fundamental sciences. Vertical integration has been observed across all mature scientific disciplines, although many practitioners in the early history of those disciplines stoutly resisted it.

Although this book agues for applying insights from evolutionary biology, genetics, and the neurosciences to criminological theorizing, it is not a call for a "biological" criminology; such a criminology is not possible. Nor is it a book introducing a new criminological theory, for we already suffer an embarrassment of riches in this area. It is a call for more *biologically informed* criminological theories. More precisely, it is a work that examines how *existing* criminology theories can be better understood, expanded, and revised by incorporating relevant methodological, conceptual, and theoretical insights from the biological sciences. This should be welcomed by criminologists who favor environmental explanations because, as many biosocial scientists have pointed out, insights from the biological sciences provide us with a deeper understanding and appreciation of the vital role of the *environment* than strictly environmentalist theories ever could.

I make the argument for a biosocial criminology in the context of traditional criminological concepts that have long served as explanations of criminal behavior such as socioeconomic status, age, race, gender, conflict, the family, and child abuse and neglect.

Chapter 1 makes the argument for vertical integration with biology. It briefly explores the history of the animus between sociology and biology, and why this animus has retarded the progress of sociology (and by extension, criminology) as a science. It does this primarily by examining how other sciences in their youth resolved the same kinds of tensions between them and the more mature science adjacent to them. I also explore the deep ideological barriers that separate criminologists, the resistance to biological explanations among them and the reasons for it, and rebuttals to arguments frequently used to reject biological thinking in the human sciences. The chapter also explains why criminology is ripe for a paradigm shift and the barriers set against such a shift from the perspective of Thomas Kuhn's work on paradigms.

Chapter 2 introduces the theories, concepts, and methods of behavior genetics, molecular genetics, and epigenetics. The issue today is not whether genes affect behavior, but rather *how* they do. We explore twin and adoption study designs that allow behavior geneticists to separate genetic from environmental affects and to calculate heritability coefficients. The genetic influences on criminal behavior are explored via the concepts of gene x environment interaction and gene/environment correlation. The changing affects of genes on personality and behavior over the life span, and how genes are differentially expressed in different environments, are also explored. Molecular genetics is also getting into the game with its focus on the effects of various genetic polymorphisms on behavior and personality. Epigenetics is the final branch we explore. This branch of genetics explores how environmental events are captured by the genome in somatic time without altering the DNA sequence. Genetic research of all kinds offers an unprecedented opportunity for criminologists to assess the shifting balance of genetic and environmental factors over time and place.

Chapter 3 examines the evolutionary origins of behaviors that modern societies have come to define as criminal. The basic concepts and logic of evolutionary psychology are examined by discussing some of the many misunderstandings of evolutionary logic, such as "organisms are designed to be directly concerned with maximizing their fitness," or "what is natural is good." I then discuss altruism, empathy, and cooperation, how these behaviors create a niche for cheats (criminals), the mechanisms that have evolved for detecting cheats, and discuss whether the co-evolutionary "arms race" between cooperators and cheats may have resulted in a psychopathic genotype that makes cheating an obligatory strategy. The relationship between violence, status, and reproductive success is also discussed in this chapter. The chapter concludes with brief introductions to four specific evolutionary theories of criminal behavior.

Chapter 4 looks at the contributions to criminological theory from the neurosciences. Genes have surrendered much of their control of human behavior to a much more plastic organ we call the brain. The brain physically "captures" the environment as it wires itself in response to environmental experience. However, the brain has many assumptions built into it over eons of evolutionary time; it is no *tabula rasa*. The evolutionary importance of attachment is explored in terms of the neurological consequences of non-attachment. The affects of child abuse and neglect on various neurological structures that are important to pro- and antisocial behavior are also discussed, as is the role of the autonomic nervous system in the acquisition of a conscience. The chapter concludes with a brief introduction to neurologically specific criminological theories (reward dominance and prefrontal cortex dysfunction) and how they help to illuminate the consequences of abuse and neglect.

The remaining six chapters attempt to show the relevance of the proceeding material to traditional criminological theories in the context of major theoretical concepts contained in those theories.

Chapter 5 examines the anomie/strain tradition and socioeconomic status. After briefly exploring the evolution of this tradition from Durkheim to Agnew, I examine what the biosocial sciences have to tell us about the central concepts of this tradition; e.g., determinants of occupational success and coping with strain. These determinants are intelligence and conscientiousness. The importance of status striving and the chemistry that facilitates (not "causes") it are also discussed.

Chapter 6 explores the differential association/social learning tradition, beginning a brief discussion of these theories and their attempts to explain the age-crime curve. The peculiarities of the endocrinology and neurology of adolescence are examined in an attempt to understand the age effect, which some criminologists such as Gottfredson and Hirschi have asserted is essentially impervious to explanation. Delinquency cannot be understood without understanding the neurohormonal changes of puberty. I then examine syndromes (ADHD, ODD, and CD) that are important to understanding more serious forms of delinquency and end with an examination of the differences between persistent offenders and those who limit their offending to adolescence.

Chapter 7 examines the control tradition in criminology which does not ask why some people commit crimes, but rather why most of us do not. The assumption is that antisocial behavior is the default option that emerges in the absence of social and/or self-control. The concepts of concern to control theories are the mechanisms of social control, which primarily means the family. Our first task, then, is to examine the evolutionary roots of the family, why the mother/father/child triad is the "expected environment of rearing," and the neurobiology of mother/infant attachment and male/female attraction and attachment.

The impact of rampant illegitimacy on antisocial behavior of all kinds and why illegitimacy is so prevalent in some environments are discussed, with emphasis on the role of the sex ratio. Also addressed is the biological basis (primarily serotonin levels and prefrontal cortex functioning) of self-control, a trait thought by traditional self-control theorists to be solely a function of socialization.

Chapter 8 examines the ecological/social disorganization tradition with race being the central concern. After examining and critiquing the basic assumptions of this tradition, I examine race differences in criminal behavior. This tradition is preeminently about how the neighborhood cultural milieu affects the behavior of those living there. Inner-city neighborhoods are wracked with crime; thus we look at the nature of these neighborhoods. I first examine the impact of slavery, how blacks adapted to it, and the cultural consequences of these adaptations. The extremely low sex ratio in the black community is seen as a structural reason for the almost 70% illegitimacy rate in the inner city. The neurological and behavioral effects of viewing violence and being victimized by it as children are examined, as well as other concerns such as the low rate of breastfeeding (here's where epigenetic modification of DNA may be very important) and high rates of in utero exposure to teratogenic compounds such as alcohol and childhood exposure to lead are also addressed.

Chapter 9 looks at the critical tradition in criminology as exemplified by the revolutionary Karl Marx and the evolutionary Max Weber. I begin by examining Marx's (and Marxists in general) thoughts on human nature, alienation, equality, and the class struggle as they pertain to crime. The biggest mistakes Marx made relevant to criminologists (although Marx wrote very little about crime per se) were his assumptions about human nature and his assumption that conflict (and thus crime) would be eliminated under communism.

Max Weber, although less influential than Marx, was more sophisticated about both human nature and conflict. He addressed many of the themes that Marx did, but often arrived at opposite conclusions. He believed that conflict was inherent in human sociality and could not be eliminated. Additionally, he viewed it as necessary and functional because without it the only option for social change was revolution. Crime is simply the result of the conflict of interests among different groups, some with the power to criminalize acts contrary to their interests.

Chapter 10 focuses on the central issue of feminist criminology: "Why do females everywhere and always commit far less crime than males?"

Feminist criminology attempts to answer this question by appeals to differential socialization because they assume that gender is a social construct.
We look at typical arguments along these lines before venturing into the
neurohormonal basis for gender (the process of "sexing" the brain) illustrated by the various pseudohermaphrodite types.

We then look at the ultimate (evolutionary) reasons behind the proximate
neurohormonal differences between the sexes, examining Campbell's "staying alive" hypothesis and Taylor's "tending and befriending" hypothesis. The
research on the traits (fear, empathy, nurturing, aggression, dominance, etc.) of
importance to these hypotheses is discussed. The chapter concludes with a brief
biosocial discussion of rape, a crime of deep concern to feminist criminologists.

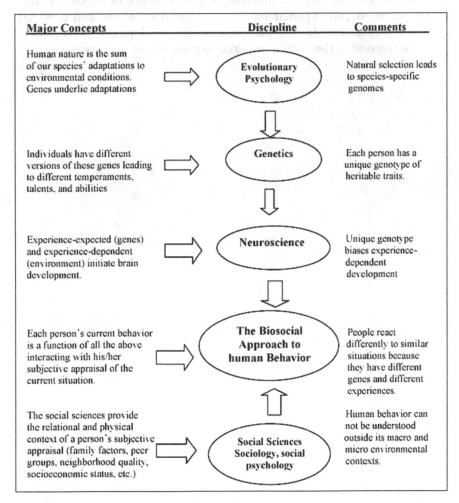

Figure P.1 The biosocial approach to behavior in a nutshell. Source: A. Walsh & K.
M. Beaver (2009). Biosocial Criminology: New Directions in Theory and Research.
Source: A. Walsh & K. M. Beaver (2009). *Biosocial Criminology: New Directions
in Theory and Research*

Figure P. 1 provides a "nutshell" model of the biosocial approach to human behavior as presented in this book. It names the disciplines that contribute to the approach and identifies major concepts from them. The figure makes it clear that we must look at all levels of explanation—from the social to the genetic, and from the distal to the most proximate—if we wish to understand human behavior. Our evolutionary history has provided us all with the same species-specific human nature specified by the human genome. However, these genes come in different versions that provide each person with a unique genotype which leads to individual differences in temperaments and to trait variation. Genotypic differences are modified by developmental histories as nature and nurture combine to produce personality, variations in which lead individuals to seek and/or be exposed to different situations and experiences. Environmental situations and personal experiences further mold each person's ongoing development as they interpret and respond to these situations and experiences in their own way.

Acknowledgments

We would first of all like to thank senior editor Benjamin Holtzman for his faith in this project from the beginning. Thanks also for the commitment of his very able assistant Jennifer Morrow. This tireless team kept up a most useful dialog between authors, publisher, and a number of excellent reviewers. The copy editor, James Barbee, spotted every errant comma, dangling participle, and missing reference and misspelled word in the manuscript, for which we are truly thankful, and our production editor Michael Watters made sure everything went quickly and smoothly thereafter. Thank you one and all.

I am also most grateful for the reviewers who spent considerable time providing me with the benefit of their expertise during the writing/rewriting phase of the book's production. Their input and encouragement has undoubtedly made the book better than it would otherwise have been. These expert criminologists were: Matt DeLisi, John Paul Wright, and John Eagle Shutt.

Most of all I would like to acknowledge the love and support of my most wonderful and drop-dead gorgeous wife, Grace Jean; a.k.a. "Grace the face." Grace's love and support has sustained me for so long that I cannot imagine life without her; she is a real treasure and the center of my universe.

1 Why Criminology Needs Biology

Francis T. Cullen, Distinguished Research Professor of sociology and criminal justice at the University of Cincinnati, describes himself as a proud member of the sociological criminology paradigm. He has contributed numerous books and countless articles to this paradigm and is a tireless worker for criminal rehabilitation and social justice. Notwithstanding all this effort over his long and distinguished career, he is a true scientist who follows the data to wherever it leads him and is thus "persuaded that sociological criminology has exhausted itself as a guide for the future study on the origins of crime. It is a paradigm for the previous century, not the current one" (2009:xvi).

Cullen believes the paradigm that will take the place of sociology as the guiding paradigm for 21st-century criminology will be biosocial; "a broader and more powerful paradigm," but he adds that biosocial criminologists must educate their sociological colleagues by "introducing biosocial science in ways that are accessible and understandable" (2009:xvii). This may be the easiest task for biosocial criminologists to accomplish; more difficult will be getting their sociological colleagues interested enough to listen. The sociological tradition in criminology has become so strong that Pierre van den Berghe (1990:177) characterizes sociologists (and most criminologists by extension) as not only oblivious to biology; but "militantly and proudly ignorant." This is a pity, because as Matt Robinson (also a sociologically trained criminologist) has opined: "the biological sciences have made more progress in advancing our understanding about behavior in the last 10 years than sociology has made in the past 50 years" (2004:x). But neither Robinson nor I suppose that there could be such a creature as a biological theory of criminal behavior. Criminal behavior, indeed all behavior, is *always* the result of a blend of the biological and the social, nature and nurture, genes and environment. A purely biological approach would be just as limited as a purely environmental approach; thus this book is about *biosocial* criminology.

There are three broad biosocial approaches to the study of criminal behavior: genetic, evolutionary, and neurobiological. While they employ different theories and methods and work with different levels of analysis, their principles are conceptually consistent across all three levels.

Furthermore, all three approaches are so environmentally friendly that they may be called "biologically informed environmental approaches," although "biosocial approaches" slides easier across the tongue.

Why should criminologists concern themselves with what biology has to offer? After all, biology as applied to human behavior is still associated with illiberal politics by many social scientists. The long answer is provided throughout the book; the short answer is that a review of the behavior genetic literature led the reviewer to remark that behavior genetics studies often reach the same conclusions about environmental solutions to social problems that "left-leaning sociologists" do (Herbert, 1997:80). If this is so, why should we burden ourselves by becoming familiar with another body of literature telling us the same thing? The short answer is again supplied by Herbert when he points out that the conclusions arrived at by behavior geneticists were arrived at using "infinitely more sophisticated tools." These "infinitely more sophisticated tools" (theories, models, concepts, instruments, method-ologies) developed by behavior geneticists (as well as molecular geneticists, evolutionary biologists, and neuroscientists) can be brought to bear on the concepts and assumptions of traditional criminological theories as quality control devices functioning to separate the wheat from the chaff.

There is a lot of chaff in traditional criminological theories, but there is also a lot of wheat; criminology's problem is that it is blind to the dif-ference. Much of the discipline's inability to move forward and ally itself with more robust sciences lies in the ideological barriers it has constructed which prevent those hiding behind them from acknowledging that biology can help to illuminate criminology (Wright et al., 2009). Moir and Jessel (1995:10) put it well when they wrote: "the evidence that biology is a cen-tral factor in crime, interacting with cultural, social, and economic factors, is so strong . . . that to ignore it is perverse." Yet it is all too often ignored, and few criminologists consider themselves perverse for doing so.

SOCIOLOGY CONTRA BIOLOGY

The origin of criminology's historical reluctance to embrace biology is probably its heritage as a child of sociology, which has long been openly hostile to biology. This attitude has led to the current sad state of sociol-ogy. The decade of the 1990s saw an increasing number of social scientists commenting on sociology's decline and on its stubborn refusal to consider what the biological sciences can tell us about human behavior (e.g., Crip-pen 1994; Ellis 1996; Horowitz 1993; Lopreato and Crippen 1999; Nielsen 1994; Udry 1995; Walsh 1995a). Tooby and Cosmides (1992:23) offer one of the most stinging of these criticisms:

> After more than a century, the social sciences are still adrift, with an enormous mass of half-digested observations, a not inconsiderable body of empirical generalizations, and a contradictory stew of ungrounded,

middle-level theories expressed in a babble of incommensurate technical lexicons. This is accompanied by a growing malaise, so that the single largest trend is toward rejecting the scientific enterprise as it applies to humans.

The assault continued into the new millennium with an edited book (Cole 2001) containing articles from some of the brightest stars in sociology such as Howard Becker, Peter Berger, Randall Collins, Richard Felson, and Seymour Lipset, who collectively bemoan the state of their discipline for almost 400 pages. Common themes were that sociology is saturated with left-wing ideology, the mundane nature of its studies, its methodological faddism, and its lack of scientific rigor.

These and many other critics urge the social sciences to vertically integrate with biology. Having observed the natural sciences develop coherent and florescent theories from which have flowed a cascade of testable propositions and hypotheses resulting in robust explanations of biological phenomena, these critics see no reason why the social sciences cannot follow suit. To do this, social science must make itself consistent with what is known in the natural sciences so that there is a harmonious interlocking of causal explanations running from biology to psychology to sociology. There is no defensible *scientific* reason why sociology should not be continuous with biology in the same way that biology is continuous with chemistry, and chemistry with physics.

The schism between biology and sociology can be traced to the work of Lester Ward, the first president of the American Sociological Society (wisely changed to the American Sociological Association after acronyms became trendy). Ward, a biologist turned sociologist, saw sociology as a guide to creating a better society. He was vehemently opposed to the misuse of Darwinism by social Darwinists to justify exploitation and other social ills. Ward's reformist campaigns set the stage for sociology to become the self-appointed conscience of science, ever on the lookout to expose and oppose any idea or theory that it believed threatened to promote insidious social policies.

This is a noble and valuable mission, and sociology is to be applauded for staking out science's moral high ground, but advocacy must not be confused with science. Sociology's agenda has led to it becoming "so enmeshed in the politics of advocacy and the ideology of self-righteousness that it is simply unaware of, much less able to respond to, new conditions in the scientific as well as social environments in which it finds itself" (Horowitz 1993:5). The scientific ignorance and ideological blinders Horowitz refers to often lead to misidentifying the causes of the problems sociologists seek to meliorate. This misidentification leads to interventions that do not work, which eventually leads policymakers and scientists in other disciplines to mistrust and discount anything further these well-meaning reformers may have to say. Ward himself persisted in his belief in Lamarckism despite its categorical rejection by biologists for no other reason than he felt that the idea of the inheritance of acquired characteristics was more optimistic and progressive than Darwinian natural selection (Degler 1991:22).

For much of the 20th century sociology seemed to have achieved its goal of becoming a respected and autonomous science. Although there were always sociologists who yearned to link their discipline to biology, sociology's view of human nature as socially constructed had become more or less the view accepted by the liberal democracies of the Western world following their clash with the racist dogmas of Nazism. A major challenge to sociology's autonomy emerged with the publication of Edward O. Wilson's *Sociobiology* (1975), a 697-page book which, with the exception of the "infamous" Chapter 27, dealt with the social behavior of non-human animals. The opening paragraph of the chapter, entitled "Man: From Sociobiology to Sociology," seemed to relegate sociology to a minor branch of biology:

> Let us now consider man in the free spirit of natural history, as though we were zoologists from another planet completing a catalog of social species on earth. In this macroscopic view the humanities and social sciences shrink to specialized branches of biology; history, biography, and fiction are the research protocols of human ethology; and anthropology and sociology together constitute the sociobiology of a single species. (1975:547)

Such a bold statement was guaranteed to raise the hackles of sociologists already hostile to "biologizing." There followed numerous attacks on sociobiology as well as on Wilson by such groups as Science for the People (1978), which saw sociobiology as "biological determinism" legitimizing all the alleged evils of the status quo. Yet Stephen Jay Gould, a leading figure in Science for the People, later argued that it is wrong to view sociobiology as motivated by a political agenda, and stated that if social scientists find the theory wanting: "They must find and use a more adequate evolutionary biology, not reject proffered aid and genuine partnership" (1990:51).

Reductionism

As the story has it, Sir Isaac Newton's explanation of rainbows in cold scientific terms supposedly prompted the poet William Wordsworth to pen his objections in the form of his famous poem *The Tables Turned*. In it he lamented the reduction of the rainbow's mysterious beauty to cold science thusly:

> Our meddling intellect
> Misshapes the beauteous forms of things
> We murder to dissect.

Applying biological science to the sociological rainbow is likewise sure to be met with similar but less eloquent objections, for reductionism is one of sociology's favorite boo words. Wordsworth eventually came to realize that rainbows are no less beautiful when understood as refracted light, and that reductionism is nothing more sinister than the process of examining a complex phenomena at a more basic level, and in doing so adding to its "beauteous form" in ways previously unimagined.

Most sociologists, however, still apparently think we murder to dissect. The typical sociological objection to reductionism is exemplified by James Coleman's assertion that when two or more individuals interact, "the essential requirement is that the explanatory focus be on the system as a unit, not on the individuals or other components which make it up" (in Wilson 1998:187). While it is true that the interaction of elements (whether they be chemicals, people, or whatever) often produce effects not predictable a priori from their respective constituent parts, the claim that it is *essential* to focus explanatory efforts only on the whole unit to the exclusion of the parts is unnecessarily constraining. Wilson (1998:187) pointed out in response that biology "would have remained stuck around 1850 with such a flat perspective" if it had taken seriously the claim that "the essential requirement is that the explanatory focus be on the organism as a unit, not on the cell or molecules which make it up." Holistic explanations may be more useful than reductionist explanations in some circumstances, but reductionism has been the "royal road," though not the only road, to progress in science. Thomas Nagel, the doyen of the philosophy of science, claims that non-reductionist accounts simply *describe* phenomena while reductionist accounts *explain* them (in Rose 1999:915).

Social scientists who study broad categories of people do not typically turn their attention toward lower levels of analysis when they have identified categories associated with the problem with which they are concerned. How often have we seen demographic variables such as race, gender, and age implicitly invoked as causes in phrases such as "Gender explains X%, age Y%, and race Z% of the variance"? These variables are descriptors and predictors not explanations, and they beg a multitude of questions. If any attempt at explanation is ventured beyond identifying the associated demographics, it commonly involves invoking higher level constructs such as racism, poverty, or discrimination. Dennett (1995:82) likens this kind of science to a "yearning for skyhooks," by which he means begging for a sort of *deus ex machina* that will lift us miraculously out of scientific difficulty. Invoking higher level variables such as these that are more often assumed than measured is reminiscent of the 19th-century physicists' use of ether for the same purpose. Racism, poverty, and discrimination do exist, of course, but they are too readily invoked as blanket explanations that relieve researchers of their obligation to explore further.

Dennett (1995) contrasts skyhooks with cranes. Unlike skyhooks suspended on nothing, cranes are solidly grounded devices that also serve to lift us out of difficulty with solid, non-question-begging science. There is no excuse for yearning for skyhooks when we have perfectly good cranes available. Invoking higher level categories (or at least not invoking lower level ones) may be true to Durkheim's dictum that only social facts should be used to explain other social facts, but it is poor science. As Lubinski and Humphreys (1997:177) suggest: "Whatever the causes of group differences in social phenomena are, measures of individual differences typically reflect those causes more effectively than does membership in demographic groups." Lubinski and Humphreys (1997) and Walsh (1997) provide several examples of the superiority of lower level measures with reference to demographic variables such as gender, race, age, and SES.

There are certainly times when non-reductionist explanations are more coherent and satisfying than reductionist ones, and we must be careful that we do not lose *meaning* as an essential component to understanding behavior by an overemphasis on mechanistic accounts. Phenomena may be *explained* by lower level mechanisms but they find their *significance* in more holistic regions. Propositions about biological entities such as genes, hormones, and neurons do not contain terms that define the human condition at its most meaningful level. How do we reduce the likes of love, justice, morality, and honor to such terms? We do not, for these things are properties of the whole being, not any one part. We must never confuse a part, however well we understand it, for the whole.

However, understanding the role of neurotransmitters, genes, and hormones certainly can assist us in our attempts to understand the behaviors and traits of interest to us as long as we resist the allure of biological essentialism. I would condemn Dennett's (1995:82) "greedy reductionist" (a person who skips over several layers of higher complexity in a rush to fasten everything securely to a supposedly solid foundation) just as surely as I would a naive antireductionist. Nonetheless, science has made its greatest strides when it has picked apart wholes to examine the parts to gain a better understanding of the wholes they constitute. As Matt Ridley (2003:163), the heavyweight champion of nature *via* nurture, has opined: "Reductionism takes nothing from the whole; it adds new layers of wonder to the experience."

The history of science shows consistently that higher level theories of many phenomena existed long before their underlying mechanisms were discovered. Higher level theories are not necessarily abandoned when lower level theories come along; physicists and engineers still find classical physics quite useful despite relativity theory and quantum mechanics. Unlike social scientists, natural scientists have long recognized the complementarity of reductionist and holistic explanations, and useful observations and hypotheses now go in both reductionist and emergent directions in those sciences. Cell biologists know that at bottom they are dealing with atomic particles and seek to understand their properties. They also know, however, that there are properties of the cell that cannot be deduced from those particles a priori, that they require functional explanations of the whole cell, and how that cell fits into a network of other cells to form a larger whole (the organism). This is why we will always need social science regardless of how sophisticated we become about the genetic and neurohormonal bases of human behavior. Science is eclectic by nature, and can pose questions and offer explanation at several levels of understanding.

Part of the fear of reductionism has always been the intellectual threat the more fundamental sciences are perceived as posing to the autonomy of the more immature sciences. Edward O. Wilson (1990) coined the term *antidiscipline* to describe the relationship between a young science and an adjacent older science. Initially there is tension between the two disciplines, although it is felt most acutely by the younger one, since the "upstart" science poses little threat to the autonomy or the reputation of the established science. As the younger science gains confidence, it feels less threatened, and it begins to experiment with how the ideas and theories of the mature science can be of use to it. After a period of creative interplay, the younger science became fully complementary with its erstwhile antidiscipline. With complementarity accomplished, the younger science begins to realize great gains in theoretical and methodological sophistication. The younger discipline prospers from the thrust provided by the more mature science in ways it could never have done had it not shaken itself free of those more interested in their discipline's autonomy and ideology than in its progress.

CRIMINOLOGY AND BIOLOGY

Criminologists have not been friendly to individual differences as causes of crime. Lawrence Cohen (1987:204) pointed out over 20 years ago that sociology is "the only branch of social science that has . . . failed to recognize openly the possible influence of nature on human behavior, and nowhere is it more evident than in our studies of crime." Ellis and Hoffman (1990:57) contended that the reason "that most criminologists continue to resist the incorporation of biological factors into their understanding of criminal behavior is ideological. As part of their liberal academic tradition, criminologists tend not to blame individuals for their ill behavior, preferring to blame society and its institutions." A survey of favored causes among criminologists supported this contention in that it found the two most agreed upon "causes" of crime were "unfair economic system" and "lack of educational opportunities" (Walsh & Ellis 2004).

To explore the characteristics of individuals engaged in crime would indeed place the blame for their actions on them, which is a really radical idea in a discipline in which many of its adherents seem to claim that everything and everybody are responsible for crime except those who commit it. Hirschi and Hindelang (1977:571) were right when they wrote: "Few groups in American society have been defended more diligently by sociologists against allegations of difference than ordinary delinquents. From the beginning the thrust of sociology has been to deny the relevance of individual differences as an explanation of delinquency, and the thrust of sociological criticism has been to discount research findings apparently to the contrary." It is a strange discipline that seeks to "defend" individuals by denying them the dignity of responsibility for their own actions by painting a picture of them as mere pawns blown helplessly hither and thither by the vagaries of environmental winds.

TYPICAL OBJECTIONS TO BIOLOGICAL THINKING IN CRIMINOLOGY

I will now outline some of the more specific objections criminologists often raise against exploring the biological correlates of individual differences. Many of these objections have been around since the birth of criminology and have been answered many times. The same objections continue to be voiced, however, and thus require further responses. These objections include the following.

Biosocial Theories are Deterministic and Socially Dangerous

We all know that when scientists speak of determinism they simply mean that every event stands in some causal relationship to other events. Surely we are all determinists in this sense. When social scientists use the term *determinism* in the context of criticizing biosocial theorizing, however, they have something more sinister in mind, such as the accusation that it implies that social behavior is a *direct* outcome of genetic programming absent any influence from the environment. Colin Trudge (1999:96) asserts that such accusations represent either mere rhetoric or simple ignorance: "For a start, no evolutionary psychologist [or geneticist or neuroscientist] doubts that a gene is in constant dialogue with its surroundings, which include the other genes in the genome, the rest of the organism, and the world at large." If only those who made accusations of "biological determinism" would take the time to learn something about human biology they would not embarrass themselves with such pronouncements.

As for being socially dangerous, I suggest that far more damage to humanity has been done by those holding "blank slate" views of human nature. The environmental determinism implied by an empty-organism view of human nature is a more sinister form of determinism since the forces supposedly determining behavior are external to the actor. The idea that humans are blank slates is a totalitarian's dream, for if humans have no innate nature they can be knocked into any shape dictated by state ideology (Daly and Wilson 1988a; Fox 1991; Pinker 2002). Stalin, Mao, Pol Pot, and their like murdered in excess of 100 million people in their belief that they could take empty organisms and turn them into the "new Soviet, Chinese, or Cambodian man" (van den Berghe 1990:179). A view of human nature that sees each person as a unique individual born with a suite of biological traits with which to interact with the world is more scientifically defensible and respectful of human dignity than blank slate views that delight political megalomaniacs demanding that everyone be made to their specifications.

Some criminologists may fear that biosocial criminology is resurrected social Darwinism and can likewise be used for anti-progressive purposes. Social justice is a *moral* imperative regardless of what science does or does not have to say about any observed inequalities in society, and human nature will be what it is regardless of whether we acknowledge it. Bigots of all stripes will use any theory—biological, political, religious, or economic— that will give their hatreds whatever support they think they find in them. Nazi beliefs in the inferiority of Jews (more a philosophy than a theory underpinned by any semblance of science) were soundly trounced when exposed to science's self-correcting discourse, but ideologies such as those advocated by Stalin, Mao, and Pol Pot are, almost by definition, impervious to reasoned discourse. Science must be our unfettered guide to understanding human behavior. We cannot let censors or bigots from the left or from the right define scientific agendas, for as Bryan Vila points out, "biological findings can be used for racist or eugenic ends only if we allow perpetuation of the ignorance that underpins these arguments" (1994:329).

Because Crime is Socially Constructed, There Cannot be any Genes for Crime

This is my favorite criticism, since it most clearly illustrates the biological ignorance of those who make it. The argument is captured by Senger (1993:6), who argues: "Those who claim to have found a gene for criminality must explain how any gene knows what is a crime, why, when, and where. They will also have to abolish the concept of guilt, for the born criminal can hardly be more responsible for his supposed criminality than for the color of his eyes." In other words, if crime is a social invention, how can a person inherit a disposition for it? Senger is correct; there are no genes "for" crime, but there are genes that lead to particular traits (e.g., low levels of empathy, IQ, self-control, fear, and conscientiousness, and high levels of sensation-seeking, egoism, and negative emotionality) that increase the probability of criminal behavior, more so in some environments than in others.

It is so fashionable these days to think of almost everything as "socially constructed;" i.e., having little or no objective reality outside of our conceptions of them. Everything is socially constructed at one level; nature does not reveal herself to us sorted into ready-labeled packages, so humans must do it for her. Social construction means only that humans have perceived a phenomenon, named and categorized it according to some classificatory rule (also socially constructed). Classificatory schemes take note of the similarities and differences among the things being classified, and are thus not arbitrary. If our classificatory schemes were arbitrary (empty of

empirically meaningful referents) we would never make sense of anything
and we would all be postmodernists wallowing in incomprehensible display
prose and deconstructing everything in sight.

Crime is socially constructed in the sense that only certain acts are criminalized, but these acts most certainly have a reality beyond being named. The
"socially constructed" argument conflates the distinction between *mala in
se* and *mala prohibita* crimes. *Mala in se* crimes are universally condemned,
and the litmus test for determining what is *mala in se* is that no one (except
in the most bizarre of circumstances) wants to be the victim of one. *Mala in
se* crimes violate the core of human nature, and being victimized by such acts
evokes physiological reactions (anger, helplessness, depression, a desire for
revenge) in all cultures, and would do so even if the acts were not punishable
by law or custom. *Mala in se* crimes engage these emotions not because some
legislative body has defined them as wrong, but because they hammer at
our deepest primordial instincts. Evolutionary biologists propose that these
built-in emotional mechanisms exist because *mala in se* crimes threatened
the survival and reproductive success of our distant ancestors, and that they
function to strongly motivate people to try to prevent such acts from occurring and punishing them if the do (O'Manique 2003; Walsh 2000a).

Mala prohibita crimes are not universally condemned, and the act of
criminalizing them is often arbitrary. However, the legal status of the act
is less important to the understanding of criminal behavior than the willingness of some individuals to violate any standard of fairness or decency
to satisfy their urges and to acquire resources, and to largely discount the
negative consequences to self and others of doing so.

If a Problem is Considered Biological,
Therapeutic Nihilism will Ensue

This argument carries with it the assumption that any behavior said to have
a biological basis is impervious to treatment, and if accepted as such will
lead to the cessation of treatment efforts via environmental improvement.
For instance, in arguing against genetic explanations of problem behaviors,
Diana Baumrind (1993:1313, emphasis added) states: "the purpose of identifying undesirable predispositions of individuals should be to devise more
health-promoting interventions, not to discourage such attempts on the sup-
position that these predispositions are genetically based *and therefore intractable.*" Sandra Scarr (1993:1351) replies that only ignorance of genetics could
motivate such a statement, and that the fear of genetics among many social
scientists would be allayed if they would only learn something about it.

The assumption that it is easier to treat the kind of problems that are correlated with criminal behavior environmentally rather than biologically is
often quite wrong. A number of problems of concern to criminologists have
shown significant improvement after their biological correlates had been
identified and pharmacological treatments were developed for them. With

the advent of antipsychotic medications, many people who would have been incarcerated in mental institutions in previous years now lead reasonably normal lives (Buckley 2004); the treatment of sex offenders with anti-androgen medication such as Depo-Provera and Andocur dramatically reduces recidivism (Maletzky and Field 2003); alcohol and drug antagonists such as Naltrexone and Disipramine have proven to be of great help in the treatment of alcoholics and drug abusers (Kleber 2003); extreme symptoms of PMS and depression have been greatly alleviated by administering serotonin reuptake inhibitors such as Prozac, which has also shown great promise in the treatment of many impulse disorders, including impulsive violence (Bond 2005).

None of these pharmacological interventions are panaceas. They all must be supplemented by psychosocial treatment, but they represent a dramatic improvement over the days when treatment modalities for these problems were exclusively psychosocial (or simply imprisonment, which is hardly the most liberal of solutions to the crime problem). These treatments are aimed at syndromes associated with *criminality*, not *crime*. Problems associated with crime (i.e., the incidence and prevalence of criminal acts) can only be addressed socially, economically, and politically. The most beneficial anti-crime strategies are environmental since they seek to reduce the proportion of a population that becomes criminal.

Conservatives as well as liberals are sometimes upset about alleged biological causes of criminal behavior, feeling that to accept such allegations removes moral responsibility. It does no such thing. Explanations can never become excuses for criminality. As we shall see time and again, certain biosocial variables constitute risk factors for criminal behavior. Whatever those factors are, they are not deterministic causes, and they never relieve the individual possessing them of the obligation to behave decently and obey the law.

Crime Cannot have a Biological Basis Because Crime Rates Change Rapidly while Changes in Genes Require Many Generations

This argument conflates the distinction between *crime* and *criminality*; between the prevalence of criminal behavior in a population and the differential propensity of individuals to engage in it. Beirne and Messerschmidt (2000:221) do this in their criticism of self-control theory when they inform us that the United States has seven to 10 times the homicide rate of most European countries and ask: "Is it because the U.S. population somehow has less self-control, or is it something to do with the nature of social organization in the United States?" Such a criticism is a red herring, and Beirne and Messerschmidt must know that they are confusing the causes of individual variance in criminality with the causes of change in crime rates. It is indisputable that fluctuating crime rates reflect fluctuating social, economic, and political phenomena operating on relatively constant genotypes. Since causal status can only be conferred on factors that vary, fluctuating environments have to be the only causes of fluctuating crime rates.

However, environmental factors lower or raise individual thresholds for engaging in crime, and this is where individual factors come into play. Some individuals would engage in criminal behavior in the most benign of environments while others would remain non-criminal in the most criminogenic of environments. In between these two extremes are individuals whose criminal propensities lie dormant until triggered by appropriate environmental events. It is in explaining why individuals differ in their propensity to commit criminal acts in similar environments, and in explaining person X environment interactions, that biosocial criminology makes its contribution.

Figure 1.1 illustrates criminal propensities interacting with the environments they find themselves in. The horizontal line represents individual propensity for criminal behavior ranging from low to high and the vertical line represents environmental instigation to crime ranging from low to high. Person *A* has high criminal propensity and will cross the criminal threshold line at any level of environmental instigation (he will take advantage of any opportunity and will create his own). Person *B* has low criminal propensities and will only cross the criminal threshold under strong environmental instigation. Person *B* may be a business executive contemplating the opportunity to make millions through stock manipulations with little probability of being caught and punished. A "criminogenic environment" is thus not limited to the squalor of the ghetto, but is rather any environment in which the ratio of potential gain to potential pain favors the former. This cost/

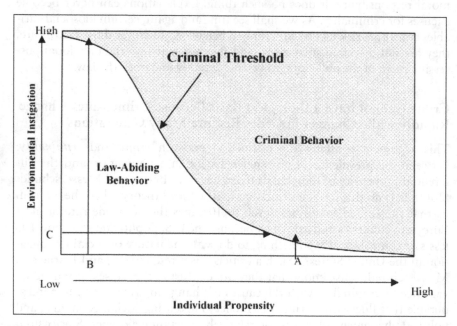

Figure 1.1 Person/environment interaction and criminal behavior threshold. Source: Adapted from Walsh (2009). *Criminal behavior from heritability to epigenetics: How genetics clarifies the role of the environment.*

benefit calculation is conducted within the bounds of each person's unique biopsychosocial makeup (e.g., person *A* is impulsive, dull, and desperate; person *B* is reflective, bright, and greedy). Most individuals will be somewhere between the extremes of *A* and *B*, and so will most environments.

If we begin with the nature of the environment rather than the person, we see that person *C* is in a low criminogenic environment. Since he has crossed the criminal threshold in a crime-constraining environment, we can assume that this person has a high individual propensity for crime similar to person *B*. This is why we will see in Chapter 2 that genes explain more variance in antisocial behavior in advantaged environments than they do in disadvantaged environments.

Biological Theories tend to be Insensitive to People's Feelings

Even if critics do not go so far as to accuse biosocial theorists of having a racist, classist, or sexist agenda, they may feel that the cold stare of science can be discomforting for some, and thus should not be brought to bear on "sensitive" topics. For example, J. Philippe Rushton (1997) has proposed an evolutionary theory explaining the high rates of violent offending found among blacks worldwide. Many social scientists view Rushton's work to be at least insensitive, or at worst, racist. For example, after admitting that the assemblage of data on race collected by Rushton is "not readily dismissible, although many have tried," Daniel Freedman states that his problem with Rushton's work is not with his data, "but with the emotionally distant nature of [Rushton's] scientific presentations." He goes on to opine that in the realm of human studies, "cold science will not do, *for only with love and warmth will the proper things be looked at, the proper things said, and a sympathetic picture of the study participants emerges*" (1997:61, emphasis in original).

This is a strange statement from a respected empirical psychologist who seems to be saying that ignorance is sometimes bliss. What are the "proper things" to be looked at? What are the "proper things" to be said? Do we fudge the data lest what they tell us offend someone? Isn't science supposed to be "emotionally distant," i.e., detached? There is no doubt that many would like to see scientists such as Rushton silenced, and some have tried to do so by treats, intimidation, and even legal action (Rushton, 1994). Such a response to unpopular research is an unconscionable affront to the spirit of science and makes one wonder if we have advanced at all since the Catholic church's treatment of Galileo Galilei because of his championing of Copernicanism 400 years ago. In his defense of science against ignorance, Carl Sagan (1996:430) quoted English philosopher John Stuart Mill's declaration that silencing an unpopular opinion is

"a peculiar evil." If the opinion is right, we are robbed of the "opportunity of exchanging error for truth"; and if it is wrong, we are deprived of a deeper understanding of the truth in its "collision with error." If we know only our own side of the argument, we hardly know even that; it becomes stale, soon learned only by rote, untested, a pallid and lifeless truth.

Unfortunately, this describes most of social science's politically correct "truths."

PARADIGMS, IDEOLOGIES, AND VISIONS

Criminologists could justifiably ignore biological factors in the past because the biological sciences lacked the technology to peer into the "black box" of human nature. This is no longer the case. There are a number of exceptionally ambitious longitudinal studies being carried out today by medical and biological scientists as well as social scientists that gather a wealth of genetic data using DNA from cheek swabs. These include the Dunedin Multidisciplinary Health and Development Study (Moffitt 1993), the National Longitudinal Study of Adolescent Health Study (Udry 2003), and the National Youth Survey (Menard and Mihalic 2001). Neuroscientists now have at their disposal an alphabet soup of brain-imaging technologies (e.g., PET, MRI, fMRI, SPECT) with which they can see *in vivo* the structure and functioning of the brain, and evolutionary scientists are increasingly encroaching on turf formerly occupied by social scientists. We can give these advances a flag-waving welcome as scientific liberators of our discipline or we can ignore them and slink off into postmodernist oblivion.

These advances make criminology ripe for a paradigm shift. In *The Structure of Scientific Revolutions*, arguably the most influential work on the philosophy of science published in the 20th century, Thomas Kuhn (1970) laid out his conception of how science is conducted and how paradigm shifts occur. A paradigm is a set of fundamental scientific assumptions, concepts, values, and practices shared by a scientific discipline that guide its view of discipline reality. It is also a set of problems that have been solved to that discipline's satisfaction by means of these assumptions (1970:10). A paradigm is more than a theory; it is a *tradition,* a macro frame of reference that reigns over a scientific enterprise such as Newtonian physics or Darwinian biology. A number of theories pertaining to specific phenomena that belong in the paradigm's domain and drawn from its assumptions sit comfortably embedded within it.

It is within paradigms that what Kuhn called "normal science" (the ordinary day-to-day activities of scientists) is conducted. Normal science tests hypotheses derived from theories shaped by the contents of the paradigm in which they exist. Thus the idea is to extend the knowledge that the existing paradigm permits, not to test it or to look for novelties within it. Those whose work extends beyond what the paradigm permits are rarely tolerated by the guardians of the paradigm, and what will not fit in the paradigmatic box is "often not seen at all" (Kuhn 1970:24).

In the late 1860s most chemists doing normal science were intolerant of the intrusion of physics, chemistry's "anti-discipline" (Wilson 1990) and its atomic theory into chemistry. Benjamin Brodie (1880) set himself as a guardian of the paradigm, arguing that an "ideal" chemistry requires only explanations involving qualitative (the transformation of substances such as a solid to a gas) and quantitative changes (weight changes due to transformation). Brodie insisted that "hypothetical" constructs called atoms were unnecessary to chemistry because chemical compounds possessed emergent properties not predictable a priori from their constituent parts. But atomism explained so many things formerly left dangling that it caused a paradigm shift, an early triumph of which was the winning of the 1903 Nobel Prize in chemistry by Svante Arrhenius for elucidating the nature of ionic bonding (Knight 1992).

Chemistry soon became integrated with physics and turned its attention to biology to which it brought a molecular approach. Joseph Woodger (1948) became biology's paradigm guardian, arguing that biology has a distinctive way of thinking that is not reducible to that of chemistry. He claimed that biological facts should be explained only by other biological facts (note Woodger's affinity with Durkheim here). Woodger had a stronger point than Brodie, for surely the step from the inorganic to the organic is the sharpest discontinuity in all science. Yet it was only five years after the publication of Woodger's *Biological Principles* (1948) that James Watson and Francis Crick (along with the often forgotten Maurice Wilkins) showed how lifeless atoms combined in extraordinarily complicated ways to form continuity between the inorganic and the organic. Their work decoding the "language of life" contained in the chemical structure of the DNA molecule won them the Nobel Prize for physiology/medicine in 1962.

Arrhenius, Watson, Crick, and Wilkins won their Nobel Prizes by ignoring paradigmatic orthodoxy and by exploring unknown territory at the interstices of different disciplines. As a result, two concepts of modern chemistry and biology once considered to be outside their paradigmatic boundaries (atoms and the chemical structure of genes, respectively) are now central to them. To understand the foundations of their disciplines, modern chemists learn physics and biologists learn chemistry; any suggestion that they should not would be met with puzzlement. I find it no less puzzling that modern criminologists are not required to learn biology. Biologists do not advance hypotheses that contradict principles of chemistry, nor do chemists formulate propositions at odds with the elegant laws of physics, but criminologists loose no sleep over whether their theories and hypotheses cohere with or contradict the established principles of biology.

DOES CRIMINOLOGY HAVE A PARADIGM?

Examinations of criminological textbooks, journals, and course syllabi make it clear that criminology has no paradigm in the sense of an agreed upon set of fundamental assumptions (Blankenship and Brown 1993; Sullivan and Maxfield 2003). We are at best in what Kuhn called the "re-paradigmatic" or "pre-scientific" stage, a stage that is characterized by the collection of masses of facts that are not fully digested. This fact-finding is guided by theory, but we have such an embarrassment of riches in that regard that it is "No wonder different men confronting the same range of phenomenon describe and interpret them in different ways" (Kuhn, 1970:17). We have a Baskin-Robbins discipline with 39 flavors plus, and are both spoiled and confused by choice. For instance, a survey of American criminologists asked them which theory they considered to be "most viable with respect to explaining variations in serious and persistent criminal behavior" (Walsh and Ellis 2004). Twenty-three different theories were represented, a fact that would horrify natural scientists who tend to see crisis when they are unable to reconcile just two or three competing theories. Because all 23 theories (and there are a lot more that failed to get mentioned) cannot be the "most viable," criminologists are obviously seeing different things when looking at the same data.

The situation within criminology conjures up the familiar Indian parable of the blind men feeling different parts of an elephant in order to describe it to the king. Each man described the elephant according to the part of its anatomy he had felt (a wall, and giant snake, a rope, a tree trunk), but each failed to appreciate the descriptions of the others who felt different parts. The men fell into dispute and departed in anger, each convinced of the stupidity, and perhaps the malevolence, of the others. The point is that there are factors independent of the subject matter of the discipline that lead criminologists to feel only part of the criminological elephant and then to confuse the parts with the whole. As with the blind men, criminologists sometimes question the intelligence and motives (e.g., having some kind of political agenda) of other criminologists who have examined different parts of the criminological elephant. Needless to say, such ad hominem attacks have no place in any science.

To the extent that criminology or any other social science has a paradigm (let us call them quasi-paradigms), it is radical environmentalism. This quasi-paradigm is a tradition that for most of the 20th century assiduously denied biology any role in illuminating human behavior. If criminology's paradigm shifts, what will make it happen? This is difficult to answer because there are no constructs within the field in which some definitive experiment can falsify. In other words, there is nothing such as phlogiston (something contained in all flammable materials and which supposedly explained such things as combustion and rusting), luminiferous ether (a medium supposedly required for the propagation of light and electromagnetic waves just like air is needed for the propagation of sound), or a static universe (the idea that the universe is not expanding but rather dynamically stable). These physical science examples were demonstrably wrong in their entirety, not 50% or 90%, but 100%. The environmentalism of mainstream criminology is not wrong, not even half wrong; it is simply incomplete. As we will see throughout this book, the more we learn about the human genome, the human brain, and human evolution, the more we realize how important the environment is. What is wrong with criminology's current quasi-paradigm is the exclusive reliance on environmental variables to explain behavior and the implicit assumption of human equipotentiality, which are ideological rather than scientific assumptions.

THE ROLE OF IDEOLOGY IN CRIMINOLOGICAL THEORY

Arguments about theory between devotees of different paradigms are seldom settled with data. As with the blind men of India, both sides are looking at the same data but see different things, and each side pushes the other to express extreme positions. This is especially true in the social sciences because, in addition to commitment to a particular paradigm and theory, social scientists work more intimately with matters of social and moral concern than do natural scientists, and these matters bring a lot of emotional baggage with them. As well as a paradigmatic and theoretical commitment, social scientists almost invariably bring with them a commitment to a particular sociopolitical ideology to further muddy the picture.

Some social scientists are honest enough to admit that ideology trumps science in their thinking. Charles Leslie is one such person. Upon resigning an editorial position at the journal *Social Science and Medicine* because it published an article on AIDS that he considered racist, Leslie (1990:896) wrote: "Non-social scientists generally recognize the fact that the social sciences are mostly ideological. . . . Our claim to be scientific is one of the main academic scandals." He then goes on to opine: "By and large, we believe in, and our social science was meant to promote, pluralism and democracy." It is indeed an academic scandal to see ideology masquerading as science, but the cure for that is more honest science, by which I mean following the data where it leads us rather than where we would like to go. While I assume that just about everyone likes Leslie's "pluralism and democracy," he wants to support only his vision of what these things mean, even if it means condemning to oblivion work on something as important as AIDS research because it violates his way of looking at the world.

Ideology is a way of looking at the world; a general emotional picture of "how things should be." In a way, ideology is a paradigm, but a more visceral one that goes beyond conformity to a particular theory. Ideological commitment almost guarantees a selective interpretation and understanding of evidence that come to our senses rather than an objective and rational evaluation of the evidence (Barak 1998). Ideology forms, shapes, and colors our concepts of crime and its causes in ways that lead us to accept or reject (or completely ignore) new evidence according to how well or poorly it fits our ideology.

Take, for instance, the fact that a highly disproportionate number of prison inmates come from and return to the same neighborhoods. Some theorists (typically liberals) believe that such a strategy provides only a temporary reprieve and that it will eventually lead to increased crime rates by weakening families and communities and reduce the supervision of children (DeFina and Arvanites 2002). These people will thus decry America's "imprisonment binge" as ultimately counterproductive Others (typically conservatives) will look at the same data and say that there were few families in the traditional sense in such neighborhoods to begin with (Mumola 2000). They may also point out that longitudinal studies show that the presence of criminal fathers in the home increases the risk of antisocial behavior of their children, and that the harmful affects increase the longer they spend with the family (Moffitt 2005), and therefore the net effect of removing such people from the community is positive, not negative. These people will see the "binge" as partly responsible for the large decrease in crime witnessed in the end of the 20th and beginning of the 21st century and thus wise policy.

Ideology is closely related to what Thomas Sowell (1987) calls *visions*. Sowell avers that two contrasting visions in constant conflict with each other have shaped our thoughts about human nature throughout history: the constrained and unconstrained visions. Sowell's visions are ideal types rather than true dichotomies, with many hybrid "visions" such as Marxism in between these extremes. The constrained vision views human activities as constrained by an innate human nature that is largely unalterable. The unconstrained vision denies an innate human nature and believes that it is perfectible. The many differences between the two visions are summed up by the constrained vision's assertion that "this is how the world *is*," and the unconstrained vision's assertion that "this is how the world *should be*."

People holding either vision may believe in the desirability of such things as freedom, equality, and justice, but they argue about what these things mean. Holding different visions leads to asking very different questions about the same issues that both sides see as problematic and requiring attention: "While believers in the unconstrained vision seek the special causes of war, poverty, and crime, believers in the constrained vision seek the special causes of peace, wealth, or a law-abiding society" (Sowell 1987:31). Unconstrained visionaries (mostly liberals) see war, poverty, and crime as aberrations to be explained, while constrained visionaries (mostly conservatives) see these things as historically normal and inevitable, although regrettable, and believe that what has to be understood is the causes of peace, wealth, and a well-ordered society.

Given this, it should be no surprise to discover that criminological theories differ on how they approach the "crime problem." A theory of criminal behavior is at least partly shaped by the ideological vision of the person who formulated it. Sowell avers that a vision "is what we sense or feel *before* we have constructed any systematic reasoning that could be called a theory, much less deduced any specific consequences as hypotheses to be tested against evidence" (1987:14, emphasis in original). Those who feel drawn to a particular theory owe a great deal of their attraction to it to the fact that they share the same vision as the person who formulated it more than to its intellectual content or its empirical support. "Visions" more so than hard evidence all too often lead criminologists to favor one theory over another more strongly than most of them care to acknowledge (Cullen 2005:57).

A number of genetic studies corroborate Sowell's contention that something at the "gut level" underlies our visions of the world by demonstrating that liberalism-conservatism is about 50% heritable (Bouchard et al. 2003; Oniszczenko and Jakubowska 2005). Of course, there are no genes "for" liberalism or conservatism. Rather, attitudes ("visions") are synthesized genetically via our temperaments which serve as substrates guiding and shaping our environmental experiences in ways that increase the likelihood of developing traits (e.g., softhearted versus hardheaded) that help to form our attitudes and color our sociopolitical beliefs (Carmen 2007; Olson, Vernon, and Harris 2001).

Ideology, which Walsh and Ellis (2004) called "criminology's Achilles' heel," was the major factor in explaining which of the smorgasbord of theories criminologists favored in the previously mentioned study. Respondents in the study were categorized by their self-identification as conservative, moderate, liberal, and radical, and these categories were then cross-tabulated with the theory each respondent believed had the most empirical support. The results were highly significant (χ^2 = 177.23, p < .0001, Cramer's V = .65). All those claiming that feminist theory is the most empirically supported were liberal and female, and all who favored Marxist and critical theories identified as radicals. Conservatives and moderates tended to favor either social- or self-control theories, which are squarely in the constrained vision camp in that they do not ask what causes crime, but rather what prevents most of us from committing it.

SIGNS OF CRACKS IN THE QUASI-PARADIGM

At one time even the thought of criminologists embracing biology bordered on the perverse and the unthinkable, and any sort of positive coverage of biology in criminology textbooks published in the 1960s through late 1980s was virtually taboo. For instance, Wright and Miller (1998:9) quote Don Gibbons' 1968 edition of *Society, Crime, and Criminal Behavior*, in which he characterized biological research on crime as "unfruitful," and that "whatever the explanation of lawbreaking, it is not to be found in defective heredity [or] biological taint." Attitudes softened from 1987 through 1997 as knowledge in the natural sciences increased by leaps and bounds and as criminologists became more sophisticated in their thinking about it. In the 1992 edition of Gibbons' text, his assessment of biology in criminology changed to: "There is empirical evidence that lawbreaking is often the product of biological, psychological, and sociological factors operating in complex ways" (in Wright and Miller 1998:9).

Wright and Miller report significantly greater proportions of textbooks in this latter period devoted to biosocial topics relative to the earlier period (and has increased much faster since 1998). They went on to say: "Sadly, twenty recent books link biological explanations of crime to sexism, racism, and fascism, a common tactic used by some criminologists (especially those embracing critical perspectives) to discredit these arguments" (1998:14). The authors of such books were rightly taken to task for resorting to sophomoric name-calling to refute research they do not like and almost certainly do not understand. Today we have sociologically oriented criminologists who, while recognizing that resistance to biology still exists, welcome biosocial theorizing and make supportive statement in textbooks like: "It is clear that the time has arrived for criminologists to abandon their ideological distaste for biological theorizing" (Lilly, Cullen, and Ball 2007:304).

There are a number of other indicators that the quasi-paradigm is shifting. Mark Warr (2002:139), for example, writes that the biosocial sciences are breathing down the neck of sociological criminology, which he says is "disorienting and even threatening" to the majority of the current crop of criminologists. He does not try to dismiss the challenges to criminological orthodoxy made by these sciences as others have done, but rather views the challenge as scientifically healthy, which it assuredly is. The more robust the challenge mounted against biosocial criminology the more muscular it will become. Without an enemy at the gates, sentries grow sleepy and the troops get fat and complacent. Resistance to a new paradigm is a positive thing up to a point, for as Kuhn (1970:65) notes: "By ensuring that the paradigm is not too easily surrendered, resistance guarantees that scientists will not be lightly distracted and that the anomalies that lead to paradigm change will penetrate existing knowledge to the core."

ANOMALIES

Kuhn makes much of the role of anomalies in paradigm shifts. To simplify, an anomaly is some unexpected finding found by those embedded in the paradigm, or more often by those on its periphery, that does not fit into the paradigm's expectations. Although the initial tendency is to suppress any such findings, those that prevail are assimilated into the paradigm and become the expected. When previous anomalies are no longer considered anomalies and become the expected, a paradigm shift has occurred. Most of those engaged in the disciplinary enterprise at the time the shift occurs will experience it as an evolutionary process, but if criminologists of yesteryear could observe the change they would see it as revolutionary.

One particular anomaly is the shock to socialization/developmental research administered by the discovery of the relative unimportance of the shared environment in producing similarities among siblings. Standard socialization theory has long assumed that shared experiences within the family made siblings alike in their psychological development and that the most important of these experiences was parental treatment. Researchers found that children who are treated affectionately are less antisocial than those who are abused and neglected, and that parents who are confident, well-liked, and sociable have children who manage their lives well and get along with others—the paradigm works! Apparently the chemists' phlogiston theory "worked" also, because materials do lose weight (becomes "dephlogisticated") when burned, as the theory predicted it would.

Just as chemists came to realize there were better explanations for this weight loss, many socialization researchers who were formerly strict environmentalists (Harris 1998; Rowe 1994) realized that there were other explanations more attuned to reality for what they had observed. These researchers pointed out that socialization studies typically studied only one child per family (more than 99% of them, according to one estimate [Plomin, Asbury, and Dunn 2001]), ignored the role of genes, and the role of child effects on parents. To examine the role of genes and child effects, research must go beyond sampling a single child in each family as is typically done. Twin and adoption studies are needed to tease apart the genetic and environmental sources of variation in the phenotypical traits being examined, and looking at multiple children in the same family is also needed to alert social scientists to what every parent who has more than one child knows—there are different parenting styles for different children.

David Rowe (1994:2) provides an example that highlights the problem with old-style socialization research. He informs us how medical researchers would design a study to determine the combined effects of psychosocial counseling and a new drug on depression. A well-designed study would require four comparison groups: (1) drug plus therapy, (2) drug alone, (3) therapy alone, and (4) a control group receiving no treatment (see Figure 1.2). Suppose the researchers use only groups 1 and 4, find that members of group 1 (the experimental group) have fewer and shorter depressive episodes than group 4 (the control group). Finding the difference to be statistically significant, they announce that the drug works (ignoring the fact that the group received both the drug *and* counseling). Such a conclusion would not see the light of day in a peer-reviewed journal in the medical field because the findings could be the result of the therapy rather than the drug, or it might only be evident with the drug and therapy administered together. In other words, the only way to tell which of the methods accounted for our observations is to disentangle their combined effects by examining results from groups 2 and 3 as well as groups 1 and 4.

Rowe then recast this in the form of socialization research on the affect of parenting style on children's antisocial behavior. In Figure 1.2, group 1 stands for the joint effects of child's genes and parenting style, group 2 stands for genes only, and group 3 for parenting style only. A control group is not required because obviously no individual can be deprived of genes and a socialization agent in the sense that individuals in the control group in the medical study were denied the drug and the counseling. Social science researchers

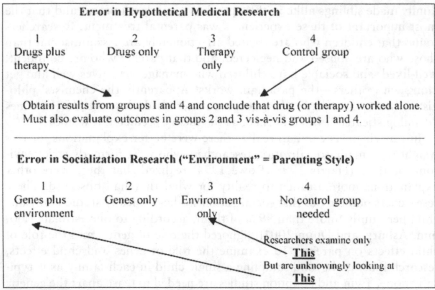

Figure 1.2   Examples of how not to conduct science.

almost always only look at parenting style (group 3) and announce that parenting style affects children's antisocial behavior, thus ignoring genes and the affects of children's evocative behavior on parenting style. Parents influence their children's development in so many ways, but we cannot ignore children's influence on parental behavior and the fact that this influence depends to a large extent on children's genes. Of course, parental responses to children are far from entirely a function of children's evocative behavior, but also of "the parent's genetics, learned modes of behavior, perceptions of the child's needs and characteristics, and socialization objectives" (Maccoby 2000:18). Contemporary developmental scientists are now in agreement that genes influence children's behavior, and thus parents' behavior towards them (Vandell 2000), but it took them a long time after the first cracks appeared to come to this agreement. Many other anomalies will be addressed in later chapters.

Reluctant Apostates

Uncovered anomalies in disciplines formerly smothered by strict environmentalism have led many of their most productive scholars out of the wilderness. Most exciting are the calls for more biologically informed theories made in the context of presidential addresses in the social sciences. Among these we have the addresses to the American Sociological Association by Alice Rossi (1984) and Douglas Massey (2002), Sandra Scarr's (1992) address to the Society for Research in Child Development, Richard Udry's (1994) address to the Population Association of America, and the addresses to the American Society of Criminology by Charles Wellford (1997) and Margaret Zahn (1999) and that of Diana Fishbein to the Academy of Criminal Justice Sciences (1998).

To my knowledge, all of these individuals began their careers as traditional social/behavioral scientists soaked in the strict environmentalism that was in vogue throughout at least the first three quarters of the 20th century. Most of them probably moved toward the biosocial perspective slowly and with some pain. As Kuhn (1970:151) put it: "The transfer of allegiance from paradigm to paradigm is a conversion experience that cannot be forced" (Kuhn, 1970:151). Among the most illustrious of these "converts" who took the painful steps are:

- Alice Rossi, a liberal feminist and past president of the American Sociological Association (ASA), defended the thesis of the interchangeability of the sexes in *Daedalus* (1964). Thirteen years later in the same journal she renounced her former position (1977), and in her ASA Presidential Address (1984) she forcefully impressed upon her colleagues that they will become irrelevant as respectable scientists if they continue to ignore the biological underpinning of sex roles.
- Developmental psychologist Jerome Kagan reports often "wincing at my early credulity." He continued: "For the first twenty years of my career, I wrote essays critical of the role of biology and celebrating the role of the environment. I am now working in the opposite camp because I was dragged there by my data" (quoted in Wright, 1999:93)
- Self-confessed cultural determinist Medford Spiro was also dragged from his biophobia by his data on Israeli kibbutzim. Expecting to find "natural androgyny" there (a major goal of the movement), after three generations of strong indoctrination toward that end, what he did find forced him into "a kind of Copernican revolution in my own thinking" (1980:106). He had expected to find "how a new culture produces a new human nature [but] I found (against my own intentions), that I was observing the influence of human nature on culture" (1980:106).
- Anthropologist Donald Brown (1991:vii) was another dyed-in-the-wool cultural relativist until he wagered fellow anthropologist Donald Symons that he could find exceptions to the sex differences claimed by Symons to be universal. Brown lost the bet and "began to think more carefully about the role that human biology plays in human affairs"
- Sociologist/criminologist Francis Cullen confesses to being disturbed for a long time by "uncomfortable ideas" about biology and behavior and states: "as a sociologist, I have never been fond of research on 'individual differences,' but now their role in human behavior, including crime, is indisputable" (2003:xi). Also see his most recent statement at the beginning of this chapter.
- Psychologist Marvin Zuckerman (2007:xix) dedicated one of his books to the great British psychologist Hans Eysenck, one of the pioneers of biosocial science. Zuckerman credited Eysenck as the scientists who "lured me away from purely social familial explanations of behavior to look more closely at genetic and psychophysiological ones."

This by no means exhausts the list of apostates; there are many more. But on the whole, paradigm changers are apt to be very young, new to the field, or only weakly committed to the reigning paradigm in the first place. While it is true almost by definition that it is the "young Turks" (some of the most brilliant of the current crop include Kevin Beaver, Matt DeLisi, Guang Gao, Satoshi Kanazawa, Matt Robinson, John Shutt, Michael Vaughn, and John Wright,) who will lead the revolution and establish the new paradigm, none of the scientists in my list of converts were very young at the time of their conversion, and most were once strongly committed to the reigning paradigm. I doubt if any of them had an epiphanal experience like Saul on the road to Damascus; all were slowly "dragged by their data" to their conversion experience. This speaks volumes about the honest commitment to science of these scholars rather than to paradigms and ideologies.

Just as each generation sings the songs of their youth, most scientists cling to the paradigmatic melodies drummed into them in graduate school, and tend to experience radical departures from them as so much cacophony from which they must shield their ears. Individuals with a lifetime commitment to a paradigm find themselves psychologically unable to follow the data into alien territory (Kuhn, 1970:151). It is often just too painful to relinquish that which we have held so dear for so long and around which we have developed productive and rewarding careers.

Then there is the practical necessity of learning some genetics, neuroscience, and evolutionary biology: "What—me retool at my age!" Criminologists do not have to become (nor can they expect to become) experts in the deep arcana of these disciplines. All they have to do is learn the rudiments of genetics, neurobiology, and evolutionary biology to the extent that they can read, appreciate, and apply the relevant literature to criminological issues. This is no different from having to learn the rudiments of statistics well enough to conduct credible research. I teach statistics and have written two textbooks on the subject, but no one, least of all myself, expects me to know the subject at the level of a mathematical statistician. Although learning difficult new material is intellectually challenging and exciting, the reluctance of some to take it on is understandable. But as Kuhn intimates, those who fail to do so will find themselves irrelevant: "retooling is an extravagance reserved for the occasion that demands it," and the wise scientist knows when "the occasion for retooling has arrived" (1970:76). What will the wise scientist find? I will let Kuhn have the last word:

> Led by a new paradigm, scientists adopt new instruments and look in new places. Even more important, during revolutions scientists see new and different things when looking with familiar instruments in places they have looked before. It is rather as if the professional community has been suddenly transported to another planet where familiar objects are seen in a different light and are joined by unfamiliar ones as well (1970:111).

2 Genetics and Criminality

The seeds of genetic science were sown in the 19th century by Charles Darwin's theory of evolution, Gregor Mendel's experiments with garden peas, and Francis Galton's statistical work on hereditary genius. Darwin's theory of evolution by natural selection required that there be variation among organism for a variety of traits and a selection process by which the most useful variants were retained and harmful ones eliminated. The theory also required a hereditary mechanism by which these advantageous traits were passed down from generation to generation, but Darwin was unaware of the nature of this mechanism.

Gregor Mendel's experiments provided preliminary details of what this mechanism of selective retention and elimination was. Mendel noticed that the traits of his pea plants (size, color, shape) remained fairly constant over the generations, but a trait occasionally disappeared and reemerged several generations later. He also noted that certain trait attributes were dominant over others, such as tallness over shortness, because the offspring of crossbred tall and short plants were always tall. When his tall–short hybrids were interbred, however, he found that about one-fourth of the offspring were short. Mendel deduced from this that traits such as shortness are not lost when combined with tallness, but are "hidden" and can only be expressed in the absence of the dominant trait. He also concluded from all this that there are two "factors" for every trait, one of which is dominant and the other recessive. We now call Mendel's "factors" *genes*, and their alternate forms (one each from maternal and paternal sources) *alleles*.

Like his cousin Charles Darwin, Francis Galton was ignorant of Mendel's work. Galton's interests lay in the area of hereditary differences in human traits, particularly in cognitive and temperamental traits. Galton's contributions to genetics were theoretical, methodological, and statistical. His *Hereditary Genius*, published in 1869, provided early evidence of the intergenerational transmission of cognitive and personality traits. As far as is known, Galton was the first to stress the importance of using twins to study human traits in order to disentangle the effects of hereditary and environment. Galton and his students and followers such as Karl Pearson, Ronald Fisher, and Sewall Wright invented a variety of statistical methods (correlation, regression, chi-square, analysis of variance, path analysis) for the express purpose of estimating the transmission of genetic variance (Rushton, 1990).

WHAT ARE GENES?

Biologist Richard Dawkins (1982:13) tells us that genes have a "sinister, juggernaut-like reputation" among social scientists. A juggernaut is an inexorable force that crushes everything in its path. If social scientists really do tend to think of genes in this way, then it is reasonable for them to believe that if any problematic behavior is said to be influenced by genes, it is beyond the influence of anything in the environment to prevent, ameliorate, or change. This kind of thinking may go back to Biology 101 in which we learned that genes had alternate forms called alleles that occupy the same locus on a chromosome. We placed these alleles in Punnett squares (big T for tall, little t for short, big S for smooth, little s for wrinkled, and so forth), and learned that if a plant had TtSs alleles, its immutable destiny was to be tall and smooth, no matter what. If we think that genes work on human personality and behavior in the same way, then it is understandable that we should fear the "sinister" notion that genes have anything to do with those things.

The less timid might go to the opposite extreme and dismiss them as irrelevant to human behavior, reasoning that since "genetic" means "inevitable," and since they know that nothing about human social behavior is inevitable, genes cannot be involved in it. Both of these positions arise from dichotomous nature-versus-nurture thinking and both are absolutely wrong. Most human traits do not follow Mendel's binary dominant/recessive rules but are rather distributed continuously in ways represented by a bell-shaped curve. The offspring of tall and short human parents are usually of intermediate height, and the offspring of dark- and light-skinned parents have skin color of intermediate hue. Human height and skin color are examples of *incomplete dominance*, which means that both alleles at a gene locus are partially expressed and produce an intermediate phenotype rather than segregate into recessive and dominant types.

Social scientists can be assured that genes do not code for any kind of behavior, feeling, or emotion; there is no neat cryptography by which certain kinds of genes build certain kinds of brains, which in turn produce certain kinds of behavior. A gene is simply a segment of DNA that codes for the amino acid sequence of a protein (not all geneticists would agree with this definition, but we will not to get into the subtle arguments of specialists). These protein products (enzymes, hormones, or cell-structure proteins) have a lot to do with how we behave or feel, but they do not *cause* us to behave or feel one way or another; they *facilitate* our behavior and our feelings. These substances produce tendencies or dispositions to respond to the environments in one way rather than in another. Even this might be too deterministic a slant; it might be better yet to think of genes as modulators of how we respond to the environment, since the gene products that facilitate behavior and emotions are produced in response to environmental stimuli.

Figure 2.1 is a model of the structure of DNA. DNA is wrapped around a protein core (histones) and contained on our chromosomes packed into the nucleus of each of our trillions of cells. DNA consists of two strings of polynucleotides covalently bonded in a chain and twisted around each other to form the familiar double helix "ladder." Nucleotide bases are aligned as the rungs of the ladder and come in four different variants: adenine (A), thymine (T), cytosine (C), and guanine (G). The two polynucleotides are joined together by the nucleotides bonding with each other in specific ways: C can only pair with G, and vice versa, and A can only pair with T and vice versa. A gene is a group of adjacent base pairs operating in coordination to perform specialized functions. For example, the following sequence of bold-type base pairs make up a section of DNA we call a gene: CTTAGC-CTACGGAAATACGAT. Only about 1 to 5% of DNA codes for proteins; we call these coding regions *exons.* Dispersed between the coding regions are non-coding regions called *introns,* which are removed by splicing and retained in the nucleus of the cell (Altukhov and Salmenkova 2002)

To see how genes respond to our needs as determined by the environment, we will liken the genome to a thermostat. A thermostat senses when the surrounding temperature is above or below its setting. When this information arrives at its sensory mechanism, it activates the furnace or air conditioner to restore the house temperature to the comfort level desired by its inhabitants. The operation of the heating/cooling system thus depends on environmental information (the temperature of the room and the perceptions of what is a comfortable temperature of the person who sets the thermostat) to restore the equilibrium between being too cold and too warm. Genes work like this for us also.

Think of the nucleus of our cells sitting and waiting to instruct its DNA to unwind and transcribe itself into a slice of messenger RNA (mRNA) as the furnace. RNA (ribonucleic acid) is structurally different from DNA. It contains an oxygen atom not present in DNA, and urasil takes the place of thiamine. The afferent nerves sense and transmit information about the state of the organism's internal or external environment to the spinal cord (if a required response is reflexive) or to the brain (if the information needs to be processed before responding) and may be thought of as a set of physiological thermostats. Upon receiving this information, the "furnace" in the nucleus of the call kicks on and sends instructions via its mRNA into the protein building factory in the cellular cytoplasm outside the nucleus (bottom section of Figure 2.1). These instructions are in the form of triplets of bases called codons. Transfer RNA (tRNA) "reads" the coded message and picks up and transports the appropriate sets of bases (anticodons) that

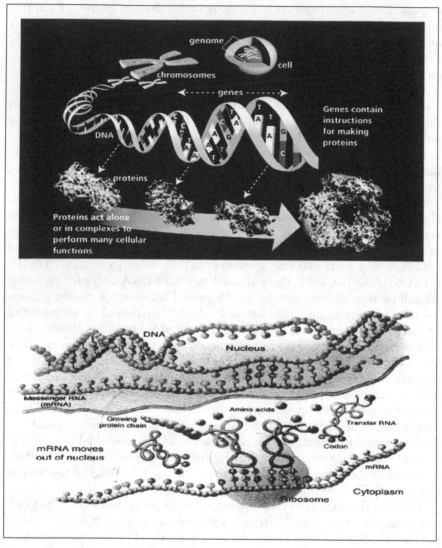

Figure 2.1 The gene from cell to protein. Source: U.S. Department of Energy Human Genome Project, 2001. http//: www.ornl.gov/hgmis.

complement the codons on the mRNA to the mRNA strand, where they are slotted into place by ribosomal RNA (rRNA). When this is complete we have a protein which will help the organism to respond effectively to the environmental challenge.

The protein manufactured depends on the nature of the challenge. If the challenge is a cut finger, genes will automatically manufacture fibrin, which crisscrosses the cut to entangle blood cells, thus preventing excess bleeding.

If it is something requiring conscious decisions, such as whether to fight or to flee, or whether to scratch a sexual itch, several different proteins may be manufactured to facilitate the person's decision. Genes are not little automatons pulling strings in our heads and determining the directions our lives will go in; rather, they help us to get there once the direction has been decided. This is not to say that variability in gene products does not bias us in certain directions, doing so weakly or strongly at different stages of development and within different environmental contexts.

THE CONCEPT OF HERITABILITY

In order to disentangle genetic from environmental effects, behavior genetic models randomize genes to determine the effects of environments, and randomize environments to determine the effects of genes. My discussion of these methods will center on IQ because this is the trait that has been studied more than any other by geneticists and psychologists. Literally thousands of studies of IQ have been conducted around the world with remarkably consistent results. I make no attempt here to defend against accusations that IQ tests are biased or that genes are not substantially involved in intelligence. These allegations have been examined and rejected by the National Academy of Sciences (Seligman 1992), the overwhelming majority of 1,020 Ph.D.-level experts surveyed by Snyderman and Rothman (1988), and the American Psychological Association's Task Force on Intelligence (Neisser et al. 1995). Neuroimaging and molecular genetic studies have increased confidence in the genetic basis of intelligence (Deary, Spinath, and Bates 2006; Posthuma, de Geus, and Boomsma 2003).

A concept of central importance to behavior genetics is *heritability* (h^2). Heritability is a quantification ranging between 0.0 and 1.0 indicating the proportion of variance in a phenotypic trait in a population that is attributable to genes. It is important to note that it varies among populations and within the same population as it experiences different environments. Much of behavior genetics involves partitioning phenotypic variance (the observed, measured variance in a trait) into genetic and environmental components, and is thus important for evaluating environmental as well as genetic effects on the trait in question. All cognitive, behavioral, and personality traits have been shown to be heritable to some extent (Rutter 2007). Turkheimer and Gottesman (1991) go so far as to suggest that $h^2 \neq 0$ be enshrined as the first law of behavior genetics. Heritability *does not*, however, tell us how much of the trait *itself* is genetic.

Heritability does not place constraints on environmental affects; that is, knowing what percentage of the variance in a trait in a population is attributable to genes does not set limits on creating new environments that could influence the trait. Nor does the decomposition of variance into genetic and environmental effects tell us how much of the trait itself is due to genes; it only tells us how much of the *variance* in the trait is accounted for by genes at this time and in this population. Heritability thus estimates the proportion of variance in a trait attributable to *actualized* genetic potential, and whatever the unactualized potential may be, it cannot be inferred from h^2 (Bronfenbrenner and Ceci 1994). Seeds from a prize-winning rose are very likely to realize their maximum genetic potential in a Virginia garden, less so in a Michigan forest, and not at all if planted in a Nevada desert.

Because heritability estimates fluctuate among different populations and in different environments, behavior genetic studies allow us to assess the relative importance of the mix of genetic and environmental factors over time and place. This fact should excite social scientists, but because many still think in nature-versus-nurture terms, it probably does not. Heritability estimates unfortunately do help to perpetuate this false nature/nurture dichotomy among the uninitiated when they see it dividing trait variance into genetic and environmental components. Unlike those who speak in terms of nature versus nurture, however, geneticists are aware that genes and their environment are not separable ingredients in the phenotype, and that heritability is simply a formula that *statistically* transforms them into such. Genes and environments have the same relationship to phenotype as hydrogen and oxygen have to water, and length and width have to area, each meaningless without their complement in terms of the wholes they describe. All phenotypic traits are the result of a complex interaction of genes and environments.

The determination of heritability requires that every genotype be exposed to identical environments because variance can only be attributed to factors that vary. A random sample of genetically heterogeneous corn seeds planted in identically nourishing environments would yield trait variance (sweetness, color, size, etc.) that is *entirely* attributable to genes (i.e., $h^2 = 1.0$). Because the environment was held constant across variable genotypes, it cannot be considered a source of variance. Likewise, a random sample of genetically homogeneous corn seeds planted in diverse environments would yield trait variance 100% attributable to the environment ($h^2 = 0$) because genetic material was held constant across variable environments.

It is necessary to point out that even if h^2 is 1.0 for a trait, it does *not* mean that it is beyond the power of the environment to change. It would be absurd to read "$h^2 = 1.0$" as meaning that the environment has no effect. After all, genes cannot be expressed at all without an environment. As Lykken (1995:85) colorfully reminds us, without the environment one's genome would create "nothing more than a damp spot on the carpet." High heritability tells us that the *present environment at the present time* accounts for very little variance in the trait; it does not tell what other environments *may* affect variance in the trait. Likewise, $h^2 = 0.0$ does not mean that genes have nothing to do with the phenotype because there would be no phenotype without the genes that gave it life and form; $h^2 = 0.0$ only means that genes did not contribute to phenotypic variation.

Let us perform another corn seed thought experiment and select two random samples of heterogeneous seeds and plant them in environments that are each homogeneous within themselves but heterogeneous with respect to each other. That is, we make environment *A* more conducive to the realization of the potential inherent in the genes than environment *B*. Because environments are constant and the genetic material is variable in each sample, phenotypic variance in the cobs from each sample will be entirely attributable to genes ($h^2 = 1.0$). However, although the *within-sample* variance for both samples is entirely a function of the genetic differences among the seeds, the *between-sample* variance is entirely a function of the different environments in which the two samples were grown (example adapted from Lewontin, 1982:132). In other words, *mean* phenotypic differences between the samples were entirely a function of the between-sample environments at the same time as within-group differences were entirely a function of genes. Bringing environment *B* up to the standards of environment *A* will improve phenotype quality commensurately, despite h^2 being 1.0 for *B*'s phenotype.

Because variation can only be attributed to sources that vary, it follows that the more environments are equalized (less variability within them) the more variability will be a function of genes and thus the larger h^2 will be. Conversely, the more variable the environment the lower h^2 will be. A high heritability coefficient for a given trait such as IQ indicates that a society is doing a good job with respect to equalizing the environment relevant to the expression of that trait. Human beings, of course, are genetically heterogeneous and live in diverse environments, making it much more difficult to neatly decompose variance in human traits into genetic and environmental effects. We cannot control environmental effects very well, but we can control genetic variability by the twin-study and adoption methods, as will be demonstrated following.

SHARED AND NONSHARED ENVIRONMENT

In addition to sorting out genetic from environmental effects, behavior genetic studies also enable us to further decompose variance into shared and non-shared environmental effects. Just as shared genes serve to make those who share them more alike, and nonshared genes serve to make them different, shared environments serve to make people similar, and nonshared environments serve to make them different. Shared (or common) environment refers to the environment experienced by children reared in the same family. Shared environmental variables include parental SES, religion, values and attitudes, parenting style, family size, intactness of home, and neighborhood.

Nonshared environment can be familial or extrafamilial. Familial non-shared variables include gender, birth order, perinatal trauma, illness, and parental favoritism. Extrafamilial nonshared factors include having different peer groups and teachers, experiencing different time-dependent cultures, or any other experience unique to the individual. Behavior genetic studies report the percentage of variance in traits attributable to environmental factors, but they do not typically endeavor to determine what those factors are; they leave that to the social scientists.

Environmental features may sometimes be considered either shared or nonshared. For instance, parenting style may not be uniform for all siblings, and may be more a function of the evocative style of each child than anything else. This is supported by studies showing that identical (monozygotic [MZ]) twins *reared apart* assess their affectual experiences with their different adoptive parents significantly more similarly than do fraternal (dizygotic [DZ]) twins *reared together* (Plomin and Bergeman 1991). These findings indicate that just as there is environmental mediation of genetic effects, there is genetic mediation of environmental effects.

In fact, one of the most interesting findings of behavior genetics is that shared environmental effects on cognitive and personality traits, although moderate during childhood, disappear almost completely in adulthood as we become freer to determine our own environments. I am reminded here of the old saying: "The older I get, the more I become myself."

THE TWIN METHOD

Behavior geneticists determine the genetic and environmental sources of variance by comparing interclass correlations for a trait between MZ twin pairs and same-sex DZ twin pairs. The twin-study method takes advantage of the fact that MZ twins are genetically identical, that is, the coefficient of their genetic relationship is 1.0. DZ twins, on average, share half of their genes, so their coefficient of genetic relationship is 0.5. If genes are a source of variation for a trait, then the more individuals are genetically similar the more alike they should be on any given phenotypic trait. If this were not so, it would be logically impossible to calculate meaningful heritability coefficients. Under the assumption of a purely polygenetic model (no environmental effects), the theoretical correlation between a trait and genes between MZ twins should be 1.0, it should be 0.5 between DZ twins and full-siblings, 0.25 for half siblings, and zero for unrelated individuals. Departures from these theoretically expected correlations reflect environmental effects (everything not transmitted by DNA is considered environmental by biologists) plus measurement error.

Table 2.1 shows the average weighted correlations between IQ scores of pairs of individuals with various degrees of genetic relatedness obtained from 111 different studies (Bouchard and McGue 1981). The predicted correlations are those we would expect (ignoring measurement error) if: (1) IQ depended entirely on genes and the environment was irrelevant, and (2) if IQ depended entirely on the environment and genes were irrelevant.

Table 2.1 Average Weighted Correlations for Various Kinship Pairings on IQ and Predicted Correlations Based on Degree of Genetic and Environmental Relatedness*

Relatedness	Actual	Genetically Predicted	Environmentally Predicted
Monozygotic twins reared together	.86	1.00	1.00
Monozygotic twins reared apart	.72	1.00	.00
Dizygotic twins reared together	.60	.50	1.00
Biological Siblings reared together	.47	.50	1.00
Half siblings reared together	.36	.25	1.00
Adoptive siblings reared together	.34	.00	1.00
Siblings reared apart	.24	.50	.00

*Actual r's from Bouchard & McGue, *Science*, 1981, p. 1057.

The fact that both genes and the environment are important to IQ is demonstrated by the actual correlations. Note that the greater the degree of genetic relatedness of pairs the higher the correlation. MZ twins reared together have the highest average correlation (.86), followed by MZ twins reared apart (.72). A strict genetic theory would predict a correlation of 1.0 in both cases, since MZ twin are genetically identical. The pattern of the correlations by degree of genetic relationships across 111 studies involving thousands of kinship pairs provides impressive support for the claim that genes are substantially involved in cognitive functioning as measured by IQ tests. Although we expect minor variation in the magnitude of the correlations from study to study, the same robust pattern is invariably found (Carey 2003; Rutter 2007).

Box 2.1 offers a brief and basic tutorial on the computation of heritability coefficients. The great majority of the 111 studies on which these calculations are based were studies of children and adolescents. As indicated earlier, shared environmental effects (which appear substantial from the calculations in Box 2.1) essentially fall to zero in adulthood. Nonshared environmental effects remain stable, albeit modest, into adulthood, and genetic effects increase, with estimates of adult h^2 for IQ as high as 0.80 (Carey 2003; Plomin and Petrill 1997).

I have presented the most basic method of computing h^2, but the logic is the same for all methods and the results do not vary significantly from more sophisticated methods. Some geneticists argue that the best estimate of h^2 is simply the correlation between MZ twins reared apart (Lykken 1995). Note that the correlations are not squared to express percentage of variance "explained." We are not predicting one twin's score from another; we are looking at the covariance shared by them. The appropriate estimate of shared variance is r rather than r^2 anytime a relationship is due

Box 2.1   Computing Heritability

To compute h^2 we take the correlation for a given trait between pairs
of subjects with one degree of genetic relatedness and compare it to the
correlation obtained from pairs of subjects with a different degree of genetic
relatedness. MZ and DZ twins reared together constitute the two most
common genetic relationship pairs used to calculate h^2. The correlations
are taken from Table 1.1. The formula for calculating heritability from
correlations obtained from MZ and DZ twins reared together is:

$$h^2 = 2(rMZ - rDZ) \text{ substituting } h^2 = 2(.86 - .60) = 2(.26) = .52$$

The difference is doubled because the computed correlations are due to
variance in common genes (v_{CG}) and variance in common environments
(v_{CE}) for both sets of twin pairs. If we subtract rDZ from rMZ, we have
taken away all IQ variance due to common environment and half of the
variance due to common genes:

$[(v_{CG} + v_{CE}) - (1/2\ v_{CG} + v_{CE})]$. Thus rMZ − rDZ (.86 − .60 = .26) equals
only one-half of the variance due to genes and thus must be doubled to
arrive at an estimate of the total variance in IQ attributable to genes.

We may compute h^2 from the MZ and AS (adopted siblings) correlations
as follows:

$$h^2 = rMZ - rAS \text{ substituting, } h^2 = .86 - .34 = .52$$

The coefficient of genetic relationship for adopted siblings is zero; they
share no common genes and so the correlation of .34 between their IQ
scores is entirely the result of their common environment and the .52
difference between the MZ and AS correlations is therefore entirely the
result of genes.

The proportion of variance in IQ accounted for by the environment
can be further broken down into common environmental (c^2) and specific
environmental (e^2) sources. Estimating c^2 from the correlations in Table
2.1 with the following formula: $c^2 = 2rDZ - rMZ$, we get $1.20 - .86 = .34$.
The steps involved are as follows: $c^2 = [2(1/2\ v_{CG} + v_{CE})] - (v_{CG} + v_{CE}) = (v_{CG}$
$+ 2v_{CE}) - (v_{CG} + v_{CE}) = v_{CE}$
The DZ correlation is doubled in this case for the same reason that the
difference between MZ and DZ twins is doubled in calculating h^2 ; i.e., to
equalize the proportion of variance accounted for by shared genes among
DZ twins with the proportion accounted for by shared genes among MZ
twins. Once genetic variance is eliminated by subtraction, we have only
that portion of the variance accounted for by common environment.
Genetics and common environment together account for 86% of the
variance in IQ. The variance accounted for by specific environment is
simply $1 - (h^2 + c^2)$, which is the same as $1 - rMZ$. In the present case, e^2
$= 1 - (.52 + .34) = .14$.

an underlying latent variable (Beatty 2002). In the present case, common genetic structure is the latent variable accounting for the covariance of twins' IQ scores.

HERITABILITY IN DIFFERENT ENVIRONMENTS

Genetic effects are expressed or repressed to varying degrees in different environments. Heritability coefficients are almost always higher in advantaged than in disadvantaged environments. For instance, Rowe, Jacobson, and Van den Oord (1999) and Turkheimer and colleagues (2003) found coefficients for IQ of 0.74 and 0.72, respectively, in advantaged environments, and 0.26 and 0.10, respectively, in disadvantaged environments. Similarly, Rowe, Almeida, and Jacobson (1999) found a heritability coefficient of 0.65 for aggression among adolescents who were one standard deviation above the mean on a measure of family and school "warmth," and one of 0.13 for adolescents scoring one standard deviation below the mean. Another study of adolescent aggressive behavior found h^2 to be 0.70 for youths with low levels of familial conflict, 0.53 for youths with moderate familial conflict, and 0.28 for youths with high levels of familial conflict (Bartels and Hudziak 2007).

Disadvantaged environments suppress the expression of genes associated with prosocial traits and permit the expression of genes associated with antisocial traits, and advantaged environments operate in the opposite direction. Intelligence is like a flower; it needs cultivating to grow; aggression is like a weed, a default option in which doing nothing allows it to grow. In disadvantaged families, parents are typically unwilling or unable to cultivate the intellectual potential of their children or provide them with the tools (proper nutrition, books, personal mentoring) and thus it does not flower. On the other hand, they do little to monitor their children's behavior, so antisocial behavior grows between the cracks. In advantaged environments parents typically endeavor to cultivate the flower of intelligence and to uproot any weeds of antisocial aggression.

Do not make the mistake of assuming that because higher heritabilities for antisocial behavior are found in advantaged environments than in disadvantaged environments that individuals living in advantaged environments possess stronger genetic dispositions toward antisocial behavior. Crime is clearly far more prevalent in disadvantaged environments. Different crime rates in different neighborhoods reflect mean effects of a wide variety of risk factors in those neighborhoods but say nothing about population variance in these factors. It cannot be said often enough: heritability coefficients quantify trait *variation* in a population that is attributed to genetics, not how much of the trait per se is attributed to genes. Although the mean level for genetically influenced risk factors for criminal behavior will be generally low in advantaged environments, individuals in those environments who do cross the criminal threshold need a greater genetic liability to do so precisely because the environment is crime-constraining. The relationship between genetics and antisocial behavior will be less in disadvantaged environments because strong environmental instigation in those areas will mask genetic contributions, even though there are doubtless more individuals with a stronger genetic load for antisocial behavior in those environments than in advantaged environments.

ADOPTION METHOD

The second method used in behavior genetics to disentangle genetic and environmental effects is the adoption method. The adoption method allows us to randomize genes in order to investigate the effect of common environments, and to randomize environments to investigate the effect of common genes. In the first instance, phenotypic trait similarities between genetically unrelated individuals reared in the same home must be entirely a function of their common environment. In the second instance, any similarities between genetically related individuals reared in different homes must be a function of their shared genes. This method also allows us to compare trait similarities between adoptees and their biological and adoptive parents.

Adoption studies tend to show that being raised in an environment more advantageous than that which could be provided by biological parents of adoptees can have beneficial effects on personality and cognitive traits. This is to be expected since most adoptions involve higher- than-average SES couples adopting children from lower-than-average backgrounds. However, adoptees more closely resemble their biological parents than their adopted parents on almost all measures of personality and cognitive functioning. For instance, the Bouchard and McGue (1981) data cited earlier show that the averaged correlation between adopted mother's IQ and adopted children's IQ is 0.19, while the correlation between adopted children and their biological mothers is 0.22. While these correlations are not significantly different from one another, both are significantly different from the 0.42 between biological mothers and their children reared by them. This difference underscores the synergistic effect of genes and environment.

GENE/ENVIRONMENT INTERACTION AND CORRELATION

All living things are designed to be responsive to their environments, and gene/environment interactions (GxE) and correlations (rGE) are important ways by which this occurs. Genes, organisms, and environments form a complex and interacting whole, and in the course of these three-way interactions, individuals create micro-environments attuned (correlated with) to their genetic proclivities. In addition to furthering our understanding of the role of genes in understanding behavior, the concepts of GxE and rGE have yielded enormous benefits to our understanding of the environment's role in shaping behavior; as Baker, Bezdjian, and Raine (2006:44) put it: "the more we know about genetics of behavior, the more important the environment appears to be."

GxE interaction involves the reasonable assumption that different genotypes will interact with and respond to their environments in different ways; i.e., people are differentially sensitive to identical environmental influences. The concept is captured by the saying "the heat that melts the butter hardens the egg." A relatively fearless and impulsive child is genetically more vulnerable to opportunities for antisocial behavior in its environment than will a more fearful and less impulsive child. GxE interaction, however, does not mean that environmental affects are directly mediated by a person's genotype. The influence of an environmental event on the person depends on the person's phenotype *at the time*, and the person's current phenotype is the cumulative result of numerous previous interactions of his or her phenotype with the environment (Turkheimer and Waldron 2000). Certainly, the genotype underlies and guides phenotypic development, but once on a particular developmental trajectory, it is genes and experience (i.e., the phenotype) that interact with the environment.

The concept of rGE correlation avers that genotypes and the environments they encounter are not random with respect to one another and that genetic factors influence complex psychosocial traits by influencing the range of individuals' effective experiences (Moffitt 2005). The concept thus enables us to conceptualize the indirect way that genes help to determine what aspects of the environment will be salient and rewarding to us, but we must keep in mind that rGE is shorthand for what is actually *genes + accumulated experience* (phenotypic)/*environment correlation*. There are three types of rGE correlation: passive, evocative, and active.

Passive rGE refers to the association between genotype and environment provided to individuals. Infants are provided by their biological parents with genes that facilitate certain traits and an environment favorable for their expression—genotype and its environment are positively correlated. A child born to intellectually gifted parents, for instance, is likely to receive genes conducive to above-average intelligence and an environment in which intellectual behavior is modeled and reinforced. The child is thus set on a trajectory independent (passively) of his or her actions. This does not mean that the child does not actively engage its environment; it only means that he or she has merely been *exposed* to it, and has not been instrumental in forming it. The pertinent environment for passive G/E correlation is limited to that shared with parents and siblings during infancy and toddlerhood.

Evocative rGE picks up the trajectory from passive rGE as the phenotype develops, and refers to the way parents, siblings, teachers, peers, and all others in the social environment react to the individual on the basis of his or her evocative behavior. As we have seen, the treatment of children by others is as much a function of children's evocative behavior as it is of the interaction style of those who respond to them. Children bring traits with them that increase or decrease the probability of evoking certain kinds of responses when they interact with others. A pleasant, well-behaved, and compliant child generates different reactions from others than will a moody, bad-tempered, and mischievous child. Socialization is not something parents do to the child; it is a reciprocal process that parents and children do together. Some children may be so resistant to parental control that parents either resort to extreme forms of punishment or give up on the child altogether. Either punitive or permissive responses will serve to exacerbate the child's already antisocial personality and drive him or her to seek social environments in which such behavior is accepted. Individuals similarly disposed usually populate such environments. Evocative rGE thus serves to magnify differences among phenotypes. In other words, differential environmental exposure results in a multiplier effect on the phenotype (Dickens and Flynn 2000), and can be captured by the old saying that "miseries multiply and advantages aggregate."

Active rGE refers to actively seeking of environments compatible with our genetic dispositions. Within the range of cultural possibilities and constraints, our genes help to determine what features of the environment will and will not be salient and rewarding to us. Of course, the most important aspect of the environment is other people, so according to the reactions we evoke from others, they will help us to determine the environments we seek. Thus evocative and active rGE feed back on one another. Active rGE gains momentum as individuals mature and acquire the ability to take greater control of their lives. The affects of genes on forming these environments can be gauged by studies showing that the intelligence, personalities, and attitudes of MZ twins are essentially unaffected by whether or not they were reared together. That is, MZ twins reared apart construct their environments about as similarly as they would have had they been reared together and considerably more similarly than DZ twins reared together (Plomin 2005).

We should note that heritability coefficients are calculated primarily for quantitative traits because quantitative traits always require the shared operation of more genes than qualitative traits. This assures us that there is no direct genetic route from any gene or set of genes to any quantitative trait. Criminality, for instance, is a quantitative variable that is itself an amalgam of other quantitative variables such as negative emotionality, impulsiveness, egoism, low empathy, sensation-seeking, and many others traits that make a person less than desirable as a friend, mate, or employee. Thus heritability coefficients computed for criminality or for any other quantitative trait are actually capturing a wide variety of correlated subtraits.

Criminality is thus a very complex phenotype consisting of a number of building blocks or component parts. Geneticists call these building blocks *endophenotypes* (endo = "within" [the phenotype]). Endophenotypes are considered essential parts of the complex chain leading from genotype to phenotype, but each may have only minor effects on the phenotype, and then only if combined with other relevant endophenotypes and the right environmental variables. The point is, when we get down to the nitty-gritty of molecular genetics, reducing the criminality phenotype to the constituent and less complex endophenotypic parts such as impulsiveness or sensation-seeking makes it easier to identify and study their genetic basis than if we tried to map the more elusive and amorphous phenotype itself (Glahn, Thompson, and Blangero 2002). This is so simply because intermediate traits (the endophenotypes) "sit closer to the genotype in the development scheme" (Gottesman and Hanson 2005:268).

BEHAVIOR GENETICS AND CRIMINAL BEHAVIOR

Studies of rGE and antisocial behavior have focused on parenting practices evoked by the behavior of adopted children (evocative rGE). Antisocial behavior of birth parents serves as the genetic independent variable, and parenting serves as one dependent variable and children's aggressive and conduct disordered behavior as the other. In all studies, adopted children at genetic risk for antisocial behavior consistently received more negative (harsh punishment) parenting from their adoptive parents than did children not at genetic risk. In each case, negative parenting was seen as parental reaction (evocative rGE) to the behavior of their adopted children (Ge et al. 1996; Moffitt 2005; O'Connor et al. 1998; Riggins-Caspers et al. 2003).

GxE studies typically examine the effects of aversive home environments (marital discord, divorce/separation, substance abuse, neglect/abuse) on adoptees who are and who are not at genetic risk for antisocial behavior, again indexed by antisocial behavior of birth parent or parents (Cadoret et al. 1995; Riggins-Caspers et al. 2003). Adverse home environments lead to significant increases in antisocial behavior for adoptees at genetic risk, but not for adoptees not at risk. Genes and environments operating in tandem (interacting) were required to produce significant antisocial behavior; neither was powerful enough to produce it independently. That is, children genetically at risk for antisocial behavior reared in positive family environments did not display antisocial behavior, and children not at genetic risk did not become antisocial in adverse family environments.

Antisocial behavior, especially among adolescents, is an interesting exception to the weak shared environmental influences typically discovered for most human traits. The shifting pattern of genetic and environmental affects at different developmental junctures is seen in studies of juvenile and adult offending. DiLalla and Gottesman's (1989) review of delinquency studies reported weighted average concordance rates of .87 and .72 for MZ and DZ twins, respectively. While this shows a modest genetic effect, the DZ concordance rate suggests a much higher environmental effect. Further, a large study of 3,226 twin pairs found that genes accounted for only 7% of the variance in antisocial behavior among juvenile offenders, but 43% among adult offenders (Lyons et al. 1995). Shared environment accounted for 31% of the variance among the juveniles, and only 5% among the adults (which supports other work indicating the limited *lasting* effects of shared environment on personality and behavior). On the other hand, nonshared environment accounted for more variance than genes for both juveniles and adults (62 and 52%, respectively).

The Danish adoption study (Mednick, Gabrielli, and Hutchings 1984) found weak genetic effects for delinquency across all adoptees, but the 37 adoptees whose biological fathers had three or more criminal convictions (1% of the cohort) accounted for 30% of all convictions among the adoptees. These studies suggest that genetic influences on antisocial behavior among juveniles beneath some unknown threshold may be weak to nonexistent, while above that threshold they may be very strong. It also suggests that pooling subjects with minimal genetic risk for antisocial behavior (the majority of delinquents) with the small minority with a large genetic vulnerability will have the effect of elevating the estimate of the effect of genes overall while simultaneously minimizing it for those most seriously involved.

Figure 2.2 illustrates rGE and GxE in terms of the development of criminal behavior. Because genes affect differential *exposure* to environmental risks via active rGE and differential *susceptibility* to environmental risks via GxE, both processes are always operating and difficult to untangle. In other words, people self-select themselves into different environments on the basis of their genetic preferences when they are able, and because

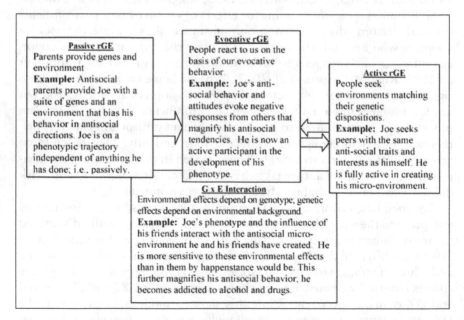

Figure 2.2   Illustrating passive, evocative, and active rGE and GxE interaction. Source: Adapted from Walsh (2009). *Criminal behavior from heritability to epigenetics: How genetics clarifies the role of the environment.*

they self-select a particular environment (like Joe in Figure 2.2) they will be more susceptible to its influence (GxE) than will those there by happenstance. Once in contact with an environment inhabited by antisocial others, the environment may have unique causal effects on those in them.

Meta-analyses and reviews of the heritability of antisocial behavior studies provide a broad estimate of 50% heritability, 20% shared environment, and 30% nonshared environment (Guo, Roettger, and Shih 2007; Miles and Carey 1997; Moffitt 2005; Rhee and Waldman 2002). Lykken (1995) ponders why the heritability of antisocial behavior is so small when it is much larger for the endophenotypes (e.g., fearlessness, aggressiveness, sensation-seeking, impulsiveness, and intelligence) associated with it. He asks us to imagine two hypothetical environments in which: (1) all parents are equally feckless and negligent in their parental duties, and (2) all parents are equally diligent and skilled in their parental duties. To the extent that the only environmental factor that mattered was parenting quality, h^2 would be 1.0 in both instances because parenting under these conditions is a constant across all families. However, there would be much more antisocial behavior in the first case because feckless and negligent parents have created a situation in which only the most fearful, nonaggressive, and conscientious would refrain from antisocial behavior. In the second case, only the most fearless, aggressive, and impulsive children would become antisocial. In the real world, there is a wide variation in the quality of parenting, and parents have greater control over their offspring's behavior than they do over their endophenotypic traits. That, asserts Lykken (1995:109), is why "the heritability of criminality is less than the more basic psychological traits [that are its constituent parts]."

Another reason that studies of juvenile delinquency find little genetic influence is the high base rate for delinquency. It is often claimed that the adolescent male who *is not* delinquent in some way is statistically abnormal (Moffitt 1993). Since the great majority of adolescents engage in behavior that could lead them to become involved with criminal justice authorities, differences in concordance rates between MZ and DZ twins are minimized, thus yielding only small heritability coefficients (DiLalla and Gottesman 1989).

TYPICAL OBJECTIONS TO BEHAVIOR GENETICS' ASSUMPTIONS AND RESEARCH DESIGNS

A number of criticisms of behavior genetic assumptions and research designs have been offered to discount the role of genetics in human behavior. According to Sandra Scarr (1981: 523–24) "Standards of evidence are elevated when one is defying the zeitgeist" . . . our work is subjected to the scrutiny of an electron microscope when the rest of psychology is examined through the wrong end of a telescope." Behavior genetics has become very muscular from standing up to this scrutiny, and has become as meticulous as any branch of science, and certainly more so than any other branch of the social/behavioral sciences.

A number of studies have been conducted to empirically address the alleged problems of behavior genetics and were summarized by Kenneth Kendler (1983). The primary criticism is that MZ twins share more of their social environment than do same-sex DZ twins, and it is this similarity of environment and the similarity of the treatment they receive within them, rather than similarity in genes, that make MZ twins more similar than DZ twins. This criticism assumes that only twins reared together would develop similar behavioral, cognitive, and personality phenotypes. The corollary assumption is that MZ twins reared apart would not develop similar phenotypes. Kendler examines studies that specifically address this criticism by comparing twin pairs whose zygosity was misidentified. Some DZ twins look so much alike that they are often misidentified as MZ twins, and some MZ twins look so unalike that they are often misidentified as DZ twins. If the "similar environment" argument were correct, MZ twins falsely identified as DZ twins would achieve test results showing a similarity level more like DZ twins than MZ twins, and that DZ twins wrongly identified as MZ twins would test more like MZ than DZ twins. This was not the case. Twins behaved consistent with their genetic similarity (their true zygosity) rather than on the basis of social expectations derived from their falsely perceived zygosity. Many studies conducted after Kendler's review have confirmed that the phenotypic similarity of MZ twins is largely unaffected by whether or not they grew up together or apart, and that MZ twins reared apart are considerably more alike than same-sex DZ twins reared together (Cary 2003).

Other reviewed studies attempted to determine if twins with a large degree of physical similarity were behaviorally more similar than twins less physically similar, the assumption being that the more physically alike twins were the more they would be treated alike. Although physical similarity did tend to lead to similarity of treatment, none of the reported studies found any relationship between treatment similarity and similarity of scores on the various behavioral, personality, or cognitive measures utilized. It was also found that MZ twins whose parents made every effort to *emphasize* the twins' sameness did not score any closer to each other on various behavioral, personality, or cognitive measures than twins whose parents made every effort to *minimize* the twins' sameness. Treatment similarity or dissimilarity apparently has no effect on making MZ twins any more or any less similar on a variety of measures.

MZ twins do tend to be treated more alike by parents than DZ twins, but studies show that when parental behavior is divided into parent-initiated and twin-initiated, parent-initiated behavior is similar for MZ and DZ twins, but significantly different when responding to actions initiated by MZ and DZ twins. Parental responses to MZ twin-initiated behavior were significantly more similar than parental responses to DZ twin-initiated behavior, leading researchers to conclude that parents *treat* MZ twins more similarly than DZ twins because MZ twins *behave* more similarly than DZ twins. This example of evocative rGE shows that parents respond to differences in their offspring rather than create them.

Another similar treatment-type argument is the selective placement argument for MZ twins reared apart. This argument states that the similarity between MZ twins reared apart is explained environmentally by the tendency of agencies to place MZ twins put up for adoption in similar homes. There is indeed a small correlation (about .30) between SES of the adoptive homes of MZ twins adopted out separately, but as Lykken (1995) points out, the SES correlation between DZ twins and other siblings reared together is 1.0, and MZ twins reared apart score significantly more similarly than DZ twins reared together on almost every behavioral, personality, and cognitive measure studied.

Kendler also looked at studies assessing the possibility of "reverse bias," i.e., factors that may lead to an *underestimation* of genetic effects. There are three such factors, the first being assortative mating. Assortative mating refers to the tendency of individuals with similar phenotypes to mate with each other at a rate significantly greater than would be expected by chance. Such a mating pattern has the effect of producing DZ twins more genetically similar than would be the case with random mating. Assortative mating has no effect on MZ similarity, however, since they share 100% of their genes regardless of whether they are the products of random or assortative mating. According to Kendler, assortative mating results in about a 11% underestimation of the importance of genetic factors in intelligence because DZ twin correlations are inflated by assortative mating while MZ correlations are unaffected, thus leading to lower heritability estimates than would be obtained under conditions of random mating, because h^2 is obtained by 2(rM2-rD2).

The second factor that may lead to underestimating genetic effects is the "twin transfusion" phenomenon. In this phenomenon, blood transfusion from the donor twin to the recipient twin occurs through placental anastomoses, and results in the recipient twin being born physically healthier and larger than the donor twin. The syndrome may have the effect of underestimating genetic effects because it occurs in the two-thirds of MZ twins who shared the same chorionic membrane (the amniotic sac). All DZ twins are dichorionic, and are thus not susceptible to the transfusion syndrome. Two-thirds of MZ twin are thus exposed to an environmental factor that may make them less similar than their shared genes would have otherwise made them, but DZ twins are never exposed to this factor.

The final factor turns the "equal environments" criticism on its head. Kendler cites two studies in which MZ twins reared apart were more similar in personality than MZ twin reared together. The implication is that MZ twins reared together often attempt more assiduously than DZ twins to be different from one another in order to assert their individuality.

Behavior genetic assumptions and research designs are thus very robust. According to Turkheimer (2000:162), for environmentalist researchers to achieve the same level of precision, they would have to "have 'identical environmental twins' whose experiences were exactly the same, moment by moment, and another variety [DZ twins] who shared exactly (but randomly [as they share randomly 50% of their genes]) 50% of their experiences." There are no "mono-envirogotic" or "di-envirogotic" twins, but we don't need them because behavior genetics informs us of the environmental contribution to human traits as surely as if there were.

HOW NOT TO READ BEHAVIOR GENETICS

Although behavior genetics studies show that the moderate shared environmental effects on cognitive and personality traits in childhood disappear almost completely in adulthood, we should not read this to say that parents have no effect on children apart from the genes they provide them. A number of journalistic accounts of Judith Harris's popular book *The Nurture Assumption* (1998) took Harris's message (that children are resilient and that parents have little influence on their children's personalities and intelligence) to its absurd "parents don't matter" extreme. Behavior genetic studies only show that parental effects on personality and cognitive traits that made siblings somewhat similar while they shared a home fail to survive the period of common rearing, and this is decidedly not the same as saying parents don't matter. Parental affects on their adult children's attitudes, values, behavior, and choice of leisure activities and professions do not disappear, although such affects are surely confounded with genetic affects. The expression of many heritable traits often depends to various degrees on parental factors (Collins et al. 2000). The weak to zero effects of shared environments means only that shared environments have little *similar* effects on siblings for a number of personality and cognitive traits; it does not mean they have had *no* effect.

Carey (2003:404) offers an example of how family affects siblings' traits without making them similar on those traits. He writes of "Sam" and "Steve," the non-twin sons of an assertive and dominant father. Sam is like his father in personality, which results in frequent father/son confrontation. Because of this, Sam becomes more confrontational than he might otherwise have been. Steve has a more submissive personality and accedes quickly to his father's demands, and as a result becomes more submissive than he would otherwise have been. This example of GxE interaction within the family (intrafamilial nonshared environment) magnifies the difference between Sam and Steve with respect to dominance/submissiveness than their genetic differences perhaps predict that they should have been. Even though this effect is called "nonshared" because only effects that make siblings similar are called shared, it is nevertheless a family (nonshared parental) effect. Families matter and they probably matter more than anything else in a child's life. Growing up in a loving and caring home with adequate resources is hugely better than growing up in an uncaring and conflict-ridden home mired in poverty.

The nonshared features of the environment are more salient with respect to the development of an individual's personality and cognitive traits, but parents have a huge say in what those nonshared environments are. Parental social class, for instance, strongly determines the neighborhoods their children grow up in, and thus the kinds of peers they will be exposed to, which is the major source of nonshared environmental effects. Parental monitoring of their children's friendship patterns, and patient and caring supervision of their activities, can go a long way in assuring they take the straight and narrow, as the Lykken (1995) thought experiment discussed earlier points out. The lower heritability coefficients found in disadvantaged environments reflect family effects, and those effects are large.

Finally, Eleanor Maccoby (2000) points out that heritability can remain stable even while the mean level of a variable is rising or falling. In other words, if some environmental factor raises or lowers all (or almost all) scores on a trait to a similar degree thus altering the trait *mean*, it does not necessarily effect the trait *variance*, the basis for computing heritability. As we have seen in our corn, environmental factors can affect a population's trait mean at the same time that genes affect its variance. She points to the monotonic rise in mean IQ over the past several decades in Western societies (the "Flynn effect") that has to be attributable to environmental effects, and that similar changes in the family (single parenting, rise or fall in family income, etc.) can have similar effects on children's traits without effecting heritability estimates.

MOLECULAR GENETICS: BEYOND HERITABILITY

Heritability studies show only that "something genetic" is involved; they do not inform us what genes are involved or how they operate. Another difficulty is that special samples (pairs of individuals with known genetic relatedness or adoptees) are needed to calculate heritability coefficients, and these are not easy to obtain. But we may now go beyond heritability coefficients and into the causal world of molecular genetics. Advances in technology have made it possible for researchers to go straight to the DNA, which can be collected through cheek saliva swabs and analyzed for about $10 per person (Butcher et al. 2004). Molecular genetics is a laboratory science aimed at identifying the molecular structure and function of genes, but it may rely on heritability studies as the first indication that genes for proteins underlying a trait of interest exist (i.e., h^2 significantly greater than zero repeatedly detected).

A basic understanding of molecular genetics may help to clarify the confusion about the terms *inherited* and *heritability*. It is often said that that human beings have too few genes and are too genetically similar for genes to be a major source of variation in personality and behavior. This is something of a red herring. If we were to lump all genes from all primate species from squirrel monkeys to human beings together such that the pool contained 100% of the genes contained in the primate order, we would find that all primate species shared an average of about 98% of their genes with all other primate species (Yoav et al. 2003). However, even the gulf between our closest genetic relative, the chimpanzees, is enormously greater than the 2% (or less) genetic difference might imply. The genetic differences between humans and chimps are quantitative (the amount of gene products) rather than qualitative (different gene products). All humans and chimps, for example, have hair and brains, but chimps have more hair than humans and humans have more brains than chimps.

Quantitative differences in gene products arise mainly due to genetic polymorphisms. *Polymorphism* refers to the differences in allelic combinations at the same chromosomal loci that make us different from one another even though we share the same genes. All humans have all the genes for species-defining physical and mental structures. Thus, all human features are *inherited* because they are coded for in the species-wide pool of DNA bequeathed to us by our ancestors, but they are not all *heritable*. Intelligence is 100% transmitted by genes (inherited), but its expression depends on both genes and environment, so heritability can never be 100% in the same sense that inherited is. Inheritance is therefore about *how* a trait is genetically transmitted while heritability is about *how much* is genetically transmitted. There are no heritability estimates for the presence or absence of human intelligence, aggression, altruism, noses, sex organs, bipedalism, and so on. These are universal human traits, and they function the same way for us all. It is the *variance* in the quantitative properties of phenotypic traits that makes us all different in many ways and is the basis for calculating heritability coefficients.

Polymorphisms are thus variations on a common theme, and may be in the protein molecule itself or in the receptor or transporter molecules for that protein. There are three different types of polymorphisms with the most frequently occurring (about 90%) being *single nucleotide polymorphisms* (SNPs) (Altukhov and Salmenkova 2002). SNPs occur about once per 100 to 300 base pairs. While most are apparently nonfunctional (found in introns), it has been estimated that about 85% of the genetic causes of most behavioral and psychological disorders are attributable to SNPs (Plomin et al. 2001). As the term implies, a difference in just one nucleotide is all that differentiates one allele from another allele. Take the following sequence of nucleotide bases as part of a hypothetical gene in Person 1:

TCACCTTGGA<u>A</u>TGGGCTA

Compare the above sequence of bases with the below sequence from Person 2:

TCACCTTGGA<u>G</u>TGGGCTA

The bolded and underlined nucleotide is the only one of the 18 that is not identical in the two individuals. The difference of one nucleotide may seem trivial, but Beaver (2009:58) explains that in the example given: "the amino acid produced by the tri-nucleotide sequence ATG in Person 1 would be methionine, while the amino acid produced by the codon GTG in Person 2 would be valine." This SNP is the Val158Met polymorphism of the enzyme catechol-O-methyltransferase (COMT) that degrades a variety of neurotransmitters. The one nucleotide difference is not trivial. Valine and methionine influence the function of brain-derived neurotropic factor (a protein that plays an important role in neuronal survival, support, proliferation, and plasticity) and the substitution of one for another has been implicated in a number of neurological problems such as memory deficits, schizophrenia, and attention deficit with hyperactivity disorder (ADHD) (Egan et al. 2003). This may also be so because the valine variant degrades dopamine at about four times the rate of the methionine variant (Beaver, 2009).

The second type of polymorphisms is microsatellites, also known as short tandem repeats. Microsatellites differ from one another in the length of contiguous nucleotide bases on an allele that are repeated a different number of times. The more times the sequence of nucleotides is repeated, the longer the allele. Compare the following repeat sequences for TTA in the following two alleles of a hypothetical gene in which TTA is repeated five times in the top allele and twice in the bottom allele.

TGGATATTATTATTATTATTATTATTA ATTATGTA

TGGATATTATTATTA

The third type of genetic polymorphisms is *minisatellites*. Minisatellites are similar to microsatellites in that they have sections of DNA that are repeated a different number of times. The difference between them is that while microsatellite repeat units consist of only 1 to 6 base repeats, minisatellite repeat ranges are from 9 to 100 (Altukhov and Salmenkova 2002). The dopamine receptor gene (DRD4), for example, has a 48 base pair sequence that can be repeated from 2 to 11 times (Ding et al. 2002). The shorter the repeat, the more responsive the brain is to the neurotransmitter dopamine; the longer the repeat, the less responsive it is. Individuals with 2 or 3 repeats tend to be overstimulated by events that most people (the 4-repeat individuals) find optimally stimulating and thus they seek to withdraw from them or tone them down. Individuals with 5 or more repeat alleles (especially those with 7 or more repeats) are suboptimally aroused by those same events. Suboptimal arousal is subjectively experienced as boredom, and bored people seek to raise the level of stimuli to alleviate it, which sometimes results in criminal or other forms of antisocial behavior. The criminogenic consequences of suboptimal arousal are well known (reviewed in Raine 1997), and the 7-repeat alleles has been associated with many phenotypic traits linked to antisocial behavior such as ADHD, sensation-seeking, and impulsivity (Canli 2006; Congdon and Canli 2005).

QUANTITATIVE TRAIT LOCI

Geneticists often search for multiple candidate genes that may be associated with a quantitative trait via quantitative trait loci (QTL) mapping. QTLs are stretches of DNA that are closely linked to variance in targeted quantitative traits on one or more chromosomes (Plomin 2005). QTLs are detected using either linkage analysis or association analysis. Linkage analysis begins with a behavior, trait, or disease of interest and tries to establish a link between it and specific polymorphisms. It is a "top-down" approach which begins with identifying groups with and without the behavior, trait or disease of interest and then seeks the approximate chromosomal location of marker allels thought to underlie that behavior, trait, or disease.

Association studies use a "bottom-up" approach, beginning with a candidate gene for which there is evidence of a link to a trait or behavior of interest (say from animal studies or from linkage analysis) and tries to establish a firm association between the trait or behavior and the candidate gene. Each discovered QTL may have small effect sizes, but multiple QTLs may be identified and combined into a QTL "set" as genetic risk factors much like various environmental factors are aggregated into environmental risk factors (Plomin and Asbury 2006). QLT risk sets in combination with environmental risk sets determine the level at which the trait(s) in question (say impulsiveness and anger) is/are situationally expressed.

The level of the challenge involved in QLT hunting depends on the number of genes contained on a particular locus as well as the complexity of the traits being examined. Some QTLs for complex traits have effect sizes of only 1% or less with a probability of .01 (Butcher et al. 2004), making the idea of QTL sets a very good one. For instance, a genome-wide search found 29 microsatellite markers out of 374 examined for the comorbidity of conduct disorder and vulnerability to substance dependence (Stallings et al. 2005).

It is important to understand that having discovered QTLs related to a phenotypic trait of interest we have found one possible causal route from a complex system of possible routes. We have not fortuitously found an expressway that takes us unerringly straight from genes to behavior. There is no direct route from genes to nontrivial behavior, only winding detour-ridden back roads riddled with potholes and downed signposts.

In past GxE studies, both G and E were usually latent (at least G was). In this postgenomic era we find an ever-increasing number of studies conducted in which both G and E were specifically identified. Perhaps the most cited of these studies in the criminological literature is the longitudinal cohort study of Caspi and his colleagues (2002). In this study, the measured environmental risk was verified child maltreatment and the identified genetic risk was the monoamine oxidase A (MAOA) promoter polymorphism. The dependent variables were a variety of antisocial outcomes such as a childhood diagnosis of conduct disorder; the most serious being a conviction (verified via police records) for a violent crime by the age of 26.

MAOA is an enzyme that metabolizes a variety of neurotransmitters and comes in long (high transcriptional activity) and short (low transcriptional activity) alleles. The researchers found no main effects for MAOA, slight but significant main effects for maltreatment, and very strong GxE effects. The low MAOA/maltreatment males were only 12% of the cohort but were responsible for 44% of its violent convictions, and the odds of such males having a violent crime conviction being 9.8 times greater than the odds of the high MAOA/nonmaltreated males having one. It should be noted that the maltreatment/nonmaltreatment groups did not differ significantly on MAOA activity, which rules out low MAOA activity contributing to child maltreatment via children's evocative (rGE) behavior.

A study by Foley et al. (2004) with conduct disorder diagnosis being the dependent variable mirrored the Caspi et al. study almost exactly, but other studies have produced mixed or null results (Haberstick et al. 2005). However, the overall conclusion arrived at by a meta-analysis of this body of research was that the interaction between MAOA and maltreatment was a statistically significant predictor of antisocial behaviors across studies (Kim-Cohen et al. 2006). An added twist came from a study by Widom and Brzustowicz (2006), who found that while the high activity MAOA allele buffered whites from the effects of childhood abuse and neglect as it relates to antisocial behavior later in life, it did not protect nonwhites. The authors suggest that other environmental stressors, such as the high density of antisocial others in the neighborhood, may have negated the protective power of the genotype among nonwhites.

Neuroscientists have studied how the MAOA variant works at the level of brain circuitry. Meyer-Lindenberg and his colleagues (2006) examined neural structure and function via fMRI of healthy (noncriminal, nonsubstance-abusing) volunteers identified as possessing the low (n = 57) or high (n = 85) variant of the MAOA polymorphism. Subjects, particularly males, with the low variant showed hyperactivity in the amygdala (fear) and hippocampus (memory) circuits and diminished activity in the prefrontal and cingulate (regulatory) regions during induced emotional arousal (see Chapter 4 for discussion of these brain structures). Not only was the function of these areas different across the different MAOA polymorphisms, but so was their structure (the level of enzymatic activity is thought to influence the wiring of the brain during development). Although these subjects were noncriminal and nonviolent, the structural and functional profiles of the low MAOA individuals reflect a vulnerability to impulsive violence (quick arousal/poor regulatory inhibition) under the right circumstances such as perceived threats.

Researchers are also beginning to examine gene by gene interactions (GxG) in the etiology of different forms of antisocial behavior. Carrasco et al. (2006) examined the effects of the DRD4 and DAT1 (dopamine transporter gene) polymorphism on ADHD and found no independent effects. However, individuals who possessed both the 7-repeat allele of the DRD4 and the 10-repeat allele of the DAT1 were significantly more likely to be diagnosed with ADHD (odds ratio = 12.7) than subjects possessing neither or only one of these alleles. Similarly, Beaver et al. (2007) found no significant main effects for either the DRD2 or DRD4 polymorphisms on conduct disorder or antisocial behavior, but the GxG interaction produced significant effects on both outcome variables.

In another study examining DAT1 polymorphisms, Guo and his colleagues (2007) found that subjects homozygous for the 10-repeat or heterozygous for the 10/9 repeat of DAT1 had trajectories of serious delinquency from 12 to 23 years of age twice as high as those homozygous for the 9 repeat. Because the DAT1 is the major mechanism by which dopamine (DA) is cleared from the synapse by transporting it back into the releasing neuron, it terminates the signals of this important neurotransmitter. The 10-repeat is abnormally efficient in the reuptake process, leaving less DA available for activation (Miller-Butterworth et al. 2008).

EPIGENETICS: THE THIRD WAVE

Additional nuances on the interplay of nature and nurture are provided by the relatively new science of epigenetics. Epigenetics means "in addition" to the genes, and conceptually it is "any process that alters gene activity without changing the DNA sequence" (Weinhold 2006:163). Epigenetic modifications of DNA affect its ability to be read and translated into proteins by making the code more readily accessible or inaccessible, or they may increase or decrease the level of protein products and thus the reaction range of a gene (Gottlieb 2007). DNA itself only specifies for transcription into mRNA, and mRNA itself has to be translated by tRNA and assembled by rRNA. Epigenetic modifications mean that genes are switched on and off by signals from the internal chemical environment and/or by its external physical and social environment according to the challenges it faces. There are some genes that may be shaped by chronic internal or external environmental events so that they are turned on or off by less than the normal environmental instigation, or even in the absence of such instigation. The epigenetics of gene expression may have as much or more influence on developmental individual differences than individual differences in DNA polymorphisms (Kramer 2005).

To analogize the relationship between the genome, polymorphisms, and epigenetics; if the genome is an orchestra and polymorphisms represent the musical variety it can produce, epigenetics represents the conductor governing the dynamics of the performance. The epigenetic "conductor" controls when and what "instruments" (genes) are to be activated and when they are to be silenced, and when they are activated, the gusto with which they will be played and what other instruments will accompany, augment, and modify the "music" (proteins) they make.

Epigenetic regulation is accomplished by two main processes: DNA methylation, which prevents the translation of DNA into mRNA, and hence the protein the gene codes for is not manufactured, and histone acetylation, which loosens the DNA wrapped around the histones, which increases the likelihood of genetic expression (Corwin 2004; Lopez-Rangel and Lewis 2006). Both processes may occur spontaneously, but mostly in response to various internal and external signals, and the resulting regulatory alterations are heritable (thus changes occurring in one generation are passed on to the next without altering DNA sequences) but reversible (Lopez-Rangel and Lewis 2006). Epigenetics suggests the idea of genomic plasticity analogous to neural plasticity which the brain is physically calibrated to environmental events. Although the genome does not possess the level of plasticity that the brain does, epigeneticists aver that it is likewise calibrated by providing the software by which organisms assimilate environmental events without changing the hardware (Pigliucci, Murren, and Schlichting 2006).

Epigenetics contains some suggestive lines of evidence that may open up whole new vistas for criminologists. Mental health researchers, for instance, have been looking into the epigenetic regulation of neurotransmitter receptors in the etiology of schizophrenia and bipolar disorder (Petronis et al. 2000). An MZ twin with a co-twin with schizophrenia will have about a 50% probability of developing schizophrenia compared with a 1% probability in the general population. While this indicates a large genetic effect, the concordance rate is lower than expected given that MZ twins share 100% of their genes. The search for specific genes that predispose individuals to psychosis has not been productive, and the search for environmental effects even less so. The search has now shifted to looking for epigenetics as a possible answer to both the genetic and environmental etiology of psychopathology (Crow 2007).

One study of healthy MZ twins found that they are virtually epigenetically indistinguishable in early life, but as they got older they diverged considerably, with 50-year-old twin pairs averaging four times the epigenetic differences than 3-year-old twins (Fraga et al. 2005). These twins were reared together, so they shared both 100% of their genes and 100% of their rearing environment; thus epigenetic modification has to be attributed to unique environmental (internal or external) events or to stochastic events. It would be a mistake, however, to attribute phenotypic differences among MZ twins to the usual environmental factors that mainstream social scientists study; as previously mentioned, for the biologist, *environment* means everything not transmitted by the DNA (Wong, Gottesman, and Petronis 2005).

THE EPIGENETICS OF NURTURING

Nurturing has long been viewed as critical for the healthy development of children and the establishment of social bonds. The highly dependent human infant is adapted to crave contact stimuli from loving and supportive caretakers as the expected evolutionary environment of the species (Glaser 2000). The epigenetics of nurturing is perhaps the line of research that will prove most germane to criminologists (Rutter 2007).

Working with nonhuman animals, epigeneticists are able to manipulate both genes and environments at will to arrive at cause/effect conclusions with a great deal of confidence. Working with rats, Weaver and his colleagues (2004) zeroed in on the molecular bases and behavioral consequences of different levels of maternal care. Maternal care varies among rat mothers just as it does among human mothers, with the level of rat nurturing indexed by the level of pup licking and grooming (LG) and arch-back nursing (ABN). Examining high- and low-level nurtured pups as adults, the offspring of high LG/ABN mothers had lower hypothalamic-pituitary-adrenocortical (HPA) axis responses to stress (as well as a number of other behaviors such as better memory and learning abilities) and were generally more socially adept than offspring of low LG/ABN mothers (see Chapter 4 for discussion of stress and the HPA axis among humans).

A portion of the pups from each inbred stain was then cross-fostered (high LG/ABN mothers fostering pups born to low LG/ABN mothers, and vice versa) to determine how much of this mother/offspring correlation is attributable to shared genes and how much to the nurturing experience. It was found that in adulthood cross-fostered pups exhibited temperaments and behaviors resembling their adopted mother more than their biological mother, indicating that early nurturing experiences have a profound impact on adult patterns of rat behavior.

Researchers found a number of significant differences in the epigenetic profiles of the high and low LG/ABN pups. High LG/ABN reduces methylation of glucocorticoid receptor (GRs) genes, the genes that determine the number of hippocampal GRs an animal will have. GRs modulate the expression of a variety of neuronal genes and are vital to neuronal homeostasis, and thus to mental health. For instance, animals with high levels of GRs will have greater control of HPA stress responses. In addition to reduced methylation of GRs, pups nurtured by high LG/ABN mothers (regardless of biological relationship) showed significantly greater acetylation of a nerve-growth factor in the hippocampus. The behavior of rat mothers thus led to similar epigenetic modifications in pups, regardless of whether pups were the mother's offspring or cross-fostered, and that these alterations resulted in stable phenotypic differences in adulthood.

Social scientists are somewhat suspicious of applying animal research to humans. Let us be clear: researchers do not study fruit flies, rats, or monkeys because they have a consuming interest in bettering the lot of such species, but because they hope to learn from them something useful about humans. Animal models have often proved pivotal to our understanding of all sorts of human physical and psychological problems. Once a biological mechanism has been demonstrated in one species it is almost always found to be applicable to others (Ridley 2003). Nature is parsimonious; it does not create an entire new genome every time species branch off from the ancestral line. Of course, this does not mean that every gene will have the same effect on humans as it does on rodents; as we have seen, having shared DNA does not mean having identical genetic functioning. Rat pups develop far more rapidly than human babies, and their "critical periods" for incorporating experience-expected events into their neuro-genomic machinery are far shorter (Hensch 2004). For instance, the epigenetic differences occurring as a function of maternal behavior in Weaver et al.'s (2004) study occurred only in the first week of life, after which maternal behavior had no discernible effects.

We do not know the range of human cognitive, personality, and behavioral traits that may be subjected to epigenetic alteration, much less epigenetic inheritance, but some epigeneticists are making statements suggesting that the field may have profound meaning for human development. For instance, Watters (2006:75) quoted epigeneticist Michael Meaney as saying: "We're beginning to draw cause-and-effect arrows between social and economic macrovariables down to the level of the child's brain"; and quoted Lawrence Harper as saying: "If you have a generation of poor people who suffer from bad nutrition, it may take two or three generations for that population to recover from that hardship and reach its full potential."

To the extent that epigenetic effects in humans operate across generations, they may well be the answer to the seemingly intractable 15-point IQ mean difference between white and black populations, as well as any number other problems caused by various types of adversity such as inadequate nurturing. It is too early in the epigenetic game to go much beyond speculation, but the possibilities are as exciting and intriguing as anything that has come along in the behavioral sciences in the past 50 years. But to share this view one has to shed the typical social science view of genetics. As Matt Ridley (2003:6) wrote about unwarranted fear of genes:

> Genes are not puppet masters, nor blueprints. They may direct the construction of the body and brain in the womb, but they set about dismantling and rebuilding what they have made almost at once in response to experience. They are both the cause and consequence of our actions. Somehow the adherents of the "nurture" side of the argument have scared themselves silly at the power and inevitability of genes, and missed then greatest lesson of all: the genes are on our side.

There are far too few genes for them to exercise anything like primary control over human behavior. Complex organisms in complex environments are always better off if left to their own devices rather than being hardwired for every possible situation they could encounter. Think of the almost limitless kinds of situations humans could possibly face, and then think of how absurd it is to think that we could possibly be preprogrammed to respond to them in fixed, undeviating ways. To program such an organism would require a gazillion genes carrying a series of "if this/then do that" algorithms rather than approximately 30,000 that we actually have. What genes actually do regarding our behavior is to provide us with some very general rules, such as "avoid snarling creatures," "be kind to close kin," "fear heights," or even more generally, "seek pleasure and avoid pain," and then leave the specifics to the judgment of the organism. Genes are at our beck and call, constantly responding to our needs as we meet the many challenges of our environments.

Colin Badcock (2000:71) goes so far as to assert that our genes "positively guarantee" human freedom and agency. After all, our genes are *our* genes. They are constantly extracting information from the environment and manufacturing the proteins we need to navigate it, enabling rather than constraining us. They are what make us uniquely ourselves and thus resistant to environmental influences which grate against our natures. This view of humanity is far more respectful of human dignity than the view that we are putty in the hands of the prevailing environmental winds. Rather than viewing biology as a threat to our discipline we should welcome it as an opportunity to collaborate with a very robust ally. The social and biological sciences need each other and belong together. As the history of the physical and natural sciences demonstrate, cross-fertilization of concepts, methods, and theories breeds hybrid vigor in their resulting offspring,

3 Evolutionary Psychology and Criminality

Because all human traits associated with criminal behavior are substantially heritable, we have to ask why genes promoting such traits exist. The genome is the chemical archive of accumulated wisdom that has survived millions of years of the ruthless process of natural selection. Evolutionary biology tells us that genes that currently exist in the gene pool of any species are there because they somehow conferred an advantage on ancestral organisms. Could the genes underlying the traits associated with criminal behavior have survived across the generations because they served a useful evolutionary purpose?

Evolutionary psychology is an approach to psychology that utilizes a Darwinian framework to animate its research agenda. Although the vocabulary of evolutionary psychologists is replete with biological terms and phrases such as natural selection, inclusive fitness, and adaptation, it is even more "environmentally friendly" than behavior genetics. Behavior genetics focuses on differences among people; evolutionary psychology focuses on human similarities. Evolutionary psychology is interested in ultimate-level "why" explanations rather than proximate-level "how" explanations. For instance, endocrinologists might explain male/female differences in dominance seeking and aggression in terms of different testosterone levels whereas evolutionary biologists would explore the adaptive rational for why sex differences in testosterone exist in the first place. Thus, evolutionary criminologists explore how behavior we now define as criminal may have been adaptive in ancestral environments. Because evolved adaptations apply to all members of a species, unlike behavior geneticists, evolutionary psychologists look exclusively for *environmental* sources of variation in criminal behavior. This obviously does not mean that evolutionary psychologists deny human genetic differences. They acknowledge large variation in adaptive traits distributed around adaptive means; it is simply that their concern is central tendency rather than variation.

THE RELEVANCE OF EVOLUTIONARY THEORY TO SOCIAL SCIENCE

Evolution may be very simply defined as changes in a population's gene pool over time by selective retention and elimination of genes. Evolutionary

56 Biology and Criminology

theory lends itself to a great deal of misunderstanding when applied to human behavior, but there is such abundance and variety of evidence supporting it that no scientist seriously disputes it on scientific grounds. Indeed, it is so indispensable to biology that Theodosius Dobzhansky, one of the true giants of 20th-century biology, wrote: "Nothing in biology makes sense except in the light of evolution" (1973:125). For those who object to evolution on religious grounds, it is worth noting that Dobzhansky was both a creationist and an evolutionist who believed that science does not preclude the process of evolution having either an author or an ultimate goal: an Alpha and an Omega.

Although biologists may carry out the bulk of their work without making reference to evolutionary theory, it is available to them as a theoretical umbrella enabling them to link their work to other subfields of biology, and as a guide enabling them to understand their work in ultimate-level terms. Because human beings are as much products of evolution as plants and other animals, a growing number of leading social and behavioral scientists are "gaining enthusiasm for a Darwinian framework, which has the potential to tie together the forest of hypotheses about human behavior now out there" (de Waal 2002:187).

Yet many social scientists have an aversion to evolutionary thinking as it applies to human behavior, apparently believing that humans are "above all that." Such an attitude is scientifically indefensible. Humans are certainly unique in many ways, but so is every other animal species in one way or another. The attitude that humans are so special that they are set above nature and thus require a different set of ontological principles to understand them delays our understanding of human nature. Basic human nature is the sum of human evolutionary adaptations, and an understanding of it is fundamental to any branch of the human sciences.

When we observe some aspect of human anatomy and physiology we correctly infer that it was selected over alternate designs because it best served some particular function that proved useful in assisting the proliferation of its owner's genes. Unless we are believers in divine creation (Dobzhansky's creationism implied a universe in which all the laws of nature were "front-loaded" at the beginning of the universe by a creator and then left alone to do what they were designed to do), we have to make that inference because there is no other scientifically viable explanation for morphological design. Likewise, there is no other scientific explanation for the origin of basic behavioral design, although most social scientists would probably dismiss the idea as "genetic determinism" (Rose and Rose 2000).

Commenting on such attitudes, Kenrick and Simpson (1997:1) state: "To study any animal species while refusing to consider the evolved adaptive significance of their behavior would be considered pure folly . . . unless the species in question is *Homo sapiens*." John Alcock (2001:223) makes a similar point: "To say that human behavior and our other attributes cannot be analyzed in evolutionary terms requires the acceptance of a genuinely bizarre position, namely, that we alone among animal species have somehow managed to achieve independence from our evolutionary history." And Plomin and his colleagues (2003:533) assert: "The behavioral genomic level of analysis may be the most appropriate level of understanding for evolution because the functioning of the whole organism drives evolution. That is, behavior is often the cutting edge of natural selection."

If pressed, social scientists might take the position that while the human behavioral repertoire must have been designed by natural selection, evolved behaviors lost their relevance once the species developed culture: "The beginning of mankind's psychosocial development represents the end of biological evolution" (Ruffie 1986:297). Ruffie does not specify what environmental pressures resulting from our "psychosocial development" might have led to the elimination of alleles underlying evolved behavioral traits that have supposedly been rendered irrelevant. Unless evolved traits become detrimental to survival and reproductive success, the genes underlying them will remain in the human gene pool. This does not mean that evolutionary psychologists consider culture unimportant in explaining human behavior, although they do not view it as a realm ontologically distinct from biology. Evolutionary psychologists simply ask us to remember that "psychology underlies culture and society, and biological evolution underlies psychology" (Barkow 1992:635). It is true that the fine nuances of life are lost as we move from proximate- to ultimate-level explanations, but ultimate-level explanations complement, not compete with, proximate explanations.

Natural Selection

Although the idea that life evolved naturally had been around since at least the time of Plato, it was the 19th-century British botanist Charles Darwin who first organized the evidence into a scientific theory. Darwin's theory (first published in 1859 in *On the Origin of Species by Means of Natural Selection*) has stood the test of time, requiring only a few modifications. Darwin's basic point was that populations of plants and animals grow until they strain the ability of the environment to support all members. The production of excess offspring results in a struggle for existence in which only the "fittest" survive.

Darwin noted that individuals within populations exhibit a considerable degree of *variation* with respect to phenotypes (disease resistance, aggressiveness, color, size, speed, cunning, etc.). Variants of a trait sometimes gave their possessors an edge in the struggle for survival in prevailing environmental conditions. The edge, whatever it may be, meant that those possessing it would be more likely than those not possessing it to survive and reproduce, thus passing the edge on to future generations. For instance, if slightly more aggressive males in a breeding population were more likely to leave slightly more offspring than their less aggressive conspecifics, then the average level of aggression in that population would increase from generation to generation. On the other hand, if most males of the population became so aggressive that individuals often killed one another fighting for mating opportunities, or attacked and killed females and their offspring, aggression would be selected against. Hyperaggressive males would leave fewer offspring than males who obtained mating opportunities by other means. The particular trait variant selected is selected because it best "fit" its possessors into the environmental conditions existing *at the time*; at other times the trait may not confer an advantage. Darwin called this process *natural selection*, because it is nature (the environment) that "selects" the favorable variants and preserves them in later generations.

Natural selection, the differential reproductive success of genotypes, is the engine and organizer of evolution because it continuously adjusts populations to their environments. It is important to note that natural selection does not *induce* variation; it is a process that *reacts* to it by preserving favorable variants. That is, genetic polymorphisms underlying certain traits that better adapted their possessors to the environment as it existed at the time afforded (not guaranteed) their possessors greater reproductive success. Thus, natural selection is "a consequence [not a cause] of the 'struggle for life' or 'existence'" (Dobzhansky et al. 1977:19).

Biologists call these structural, functional, and behavioral adjustments adaptations. It is important not to fall into the trap of thinking that traits are selected *in order* to make organisms more adapted to their environments. Such thinking implies purpose, but evolution has no purpose; it cannot look into the future to divine some plan for optimal adaptation of organisms. Natural selection is very much a trial-and-error process. Environmental conditions set evolution on a particular adaptive trajectory, but if environments change drastically, former adaptations may become maladaptive and may even drive a species into extinction.

The Tautology Criticism

The criticism that natural selection is a logical tautology has been the mainstay of antievolutionists almost from the theory's beginning. In a logical tautology such as "all mothers are females," the predicate adds nothing to the subject because the statement is true by definition. Similarly, the charge of tautology in natural selection is made by stating: (1) Natural selection is summed up as "survival of the fittest." (2) "Which ones are the fittest?" (3) "The ones that survive." Because the consequent is contained in the antecedent, the assertion cannot be falsified. The criticism is true on its face, but loses its power when followed by the assertion that the relational claim that the fittest survive is false.

"Survival of the fittest" is a catchy phrase which succinctly sums up the basic idea of evolution by natural selection for laypersons. Evolutionary biologists rarely use it, and when they do it is as a shorthand statement about a statistical aggregate of observations, not an axiom from which they think they can deduce anything. Fitness is correlated with, but not identical to, survival; it is a measure of reproductive success, not a trait or set of traits that inevitably leads to survival. The superiority of one trait variant over another is *expressed* by its fitness, not *defined* by it in the sense that mothers are defined (in part) by their femaleness. Heritable variation leads to differential reproductive success; this is a predictive statement that can be verified or falsified, and therefore not a circular statement.

John Endler's classical work *Natural Selection in the Wild* (1986), which catalogs 314 direct demonstrations of natural selection (in "the wild" as opposed to in the laboratory) among 141 species, effectively dismantles the tautology argument by setting up a syllogism that partitions the process of natural selection into precedents and results. Each one of these steps has been empirically verified time and time again. The precedents (or initial conditions) are that: (1) there must be phenotypic trait variation in a breeding population with (2) consistent fitness differences between the phenotypes, and (3) heritability of the phenotypic trait(s). The result of this process is a change in the trait frequency distributions across generations. The process is illustrated in Figure 3.1.

The theory of natural selection has enormous explanatory power that has guided research by suggesting testable hypotheses for generations of biologists. This research paradigm has been so successful that it is one of the mysteries of the philosophy of science that the tautology argument keeps popping up. After all, if natural selection was an empirical tautology it would not have had the enormous success that it has. As Stephen Jay Gould so well put it: "We are always ready to watch a theory fall under the impact of new data, but we do not expect a great and influential theory to collapse from a logical error in its formulation" (1976:1). Physical and biological scientists do not have the aversion to logical tautologies that social scientists do. As Matthen and Ariew (2005:356) put it: "We think that it is no more a problem for the theory of natural selection . . . that it has at its heart a 'tautology'—more accurately a (non-empirical truth) of mathematics—than it is for physics that it relies on such mathematical truths as $2 + 2 = 4$."

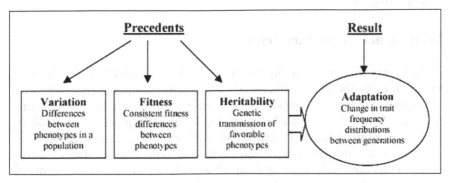

Figure 3.1   Schematic view of John Endler's definition of natural selection.

Neo-Darwinism

Darwin knew all about trait variation, but he had no idea what the source of that variation was, or how it was passed on. He was unaware of the existence of genes and of the work of Gregor Mendel, which would eventually lead to the field of genetics. Today we know that the trait variation among individuals that is the ultimate cause of evolution is caused by genetic mutation, recombination, and flow. The ultimate source of variation is mutation—changes in the DNA itself that produce new alleles. Most mutations are either harmful (in which case they are culled from the gene pool) or neutral (neither harmful nor beneficial). Mutations are changes to DNA or RNA sequences caused by copying errors during cell division or exposure to external mutagens. Occasionally a beneficial mutation occurs that gives its carrier an evolutionary edge in the environment in which it arose. Recombination (reshuffling) occurs during the division of sex cells during meiosis which results in new allelic combinations (not new alleles—only mutation does that) caused by the processes of independent assortment and crossing over. Genetic flow is the introduction of new alleles into a population by an individual moving into it. Today's evolutionary theory is thus gene-based, and is a synthesis of Darwin's theory of natural selection and genetics (the so-called Modern Synthesis).The marriage of genetics to natural selection filled many blanks in Darwin's theory. In addition to understanding the genetic source of trait variation, we now view evolution as changes in the genetic composition of a population from generation to generation, and reserve the term *fittest* to mean the most prolific reproducers. Survival means nothing in evolutionary terms if survivors do not pass on the traits that helped them to survive. The most reproductively successful organisms, regardless of the reasons for their success, leave behind the largest number of offspring, and hence the greatest number of genes, which is, again, what *fitness* means. The genes underlying traits that contributed to reproductive success will therefore be found more frequently in subsequent generations. Evolutionary theory has thus converged on a simple but powerful idea: To the degree a particular type of behavior is prevalent in a population, that behavior is likely to have contributed to the reproductive success of the ancestors of the individuals displaying the behavior.

Thinking in Evolutionary Terms

Evolutionary theory is not shy about revealing the "dark" side of human nature. It talks about reproductive success as the ultimate goal of life, and it lays bare our aggressiveness, deceptiveness, and egoism as evolved strategies that have proved useful in pursuing it. It is not pleasant to think of ourselves in this light, and some would rather burn the message than try to decipher it. But other more positive human characteristics such as altruism, nurturance, and empathy have also evolved because they equipped us with parental and social skills. After all, having offspring does little to perpetuate parental genes if those offspring do not themselves survive to reproductive age in the company of caring and trustworthy others. Indeed, in *The Decent of Man*, Darwin writes about cooperation three times more often than he writes about competition (Levine 2006). But it is the negative traits that most interest criminologists, whose stock in trade is vice, not virtue.

Adaptations, Exaptations, and By-Products

Natural selection produces adaptations, exaptations, and by-products (Hampton, 2004). As was shown in Figure 3.1, an adaptation is a design feature that arose and promoted its increased frequency in the population gene pool through an extended period of natural selection. Evolutionary psychologists are often portrayed as ultra-adaptationists seeking "just so" stories for the existence of every current human trait. It is true that sometimes they go too far (as in trying to explain the "adaptive significance" of male pattern baldness, for example), but this is increasingly the exception. To identify a "true" adaptation we must first identify a mechanism that is clearly currently functional in terms of the evolutionarily relevant goals of survival and reproduction. Secondly, we attempt to determine if the properties of this mechanism solved a recurring adaptive problem present in ancestral environments in a nonrandom way. John Tooby (1999:4) casts the issue of what constitutes an adaptation in hypothesis-testing terms: "To establish something as an adaptation, all one needs to do is to collect evidence that justifies the rejection of the hypothesis that the structure arose by chance (with respect to function)."

Some features, of course, have arisen by chance with respect to function, and it is necessary to differentiate them from adaptive features. Evolutionary psychologists do not assume that everything that is currently useful is an adaptation. All functional features of organisms need not be adaptations per se. Some presently functional features may have had quite different functions at one time but have been co-opted for current use. Features that are useful but which did not arise as adaptations for their present roles have been called *exaptations* or *preadaptations* (Hampton 2004).

To give a readily understandable example of distinction between an adaptation and an exaptation, consider automobile engines and heaters. Automobile engines were designed to move pistons and wheels so that automobiles may convey their occupants from one place to another. A by-product of the activity of engines is heat. Engines were not designed for producing heat, but this by-product was seized upon by engineers to design mechanisms for defrosting the car's windshield and warming the car's interior. The engine's function—that of moving pistons and wheels—is analogous to an adaptation because it solved some environmental problem—that of getting from one place to another. The heat generated by the engine (a by-product of an "adaptation") was seized on and utilized to make moving from place to place both safer and more comfortable. Safety and comfort also solved environmental problems, but they were not the problems engines were designed to solve. The heater mechanism that provides comfort and safety is thus a mechanical analog of a biological exaptation.

Evolutionary by-products are typically incidental and nonfunctional features of adaptations. Noise and exhaust fumes produced by automobiles are only along for the ride because they are the unfortunate by-products of functioning automobile engines. Similarly, belly buttons are simply the attachment point of the umbilical cord which joined fetus and mother via the placenta in utero; it is not designed for anything beyond that—not even for lint collection.

We should not think of adaptations as optimal solutions to all evolutionarily relevant problems. Biologists and engineers could doubtless dream up (but could not create) better solutions to our survival and reproductive concerns than evolution has presented us with. Natural selection, however, does not have the luxury of foresight or access to comparative models; it can only work algorithmically with the genetic variation existing at the time a given environmental problem presents itself. Nor can natural selection anticipate the future. Behavior that was adaptive in the past may not be today, or may even be maladaptive, and behaviors that may be adaptive (fitness promoting) today may not be adaptations in the sense that they have an evolutionary history. "Adaptiveness is always a variable [more or less adaptive] and always relative" (Lopreato and Crippen 1999:119). To claim that something is an adaptation, then, is to make a claim about the past, not the present, and definitely not about the future.

Direct vs. Indirect Motivation

Another common misunderstanding of evolutionary logic is that behaviors alleged to be adaptive are *directly* and consciously motivated by concerns of reproductive success. They are not: "Evolutionary psychology is not a theory of motivation . . . Fitness consequences are invoked not as goals in themselves, but rather to explain why certain goals have come to control behavior at all, and why they are calibrated in one particular way rather than another" (Daly and Wilson1988:7). Our brains were designed to calibrate our behavior toward adaptive ends in the environments in which we find ourselves, not to monitor the fitness consequences of our actions. No behavior can be considered a result of conscious motives to increase fitness, and even vigorous attempts to start a family are hardly driven by the conscious concerns about pushing one's genes into the future, even if the sweating couple are evolutionary biologists. We are adapted to seek the immediate means of achieving specific goals, not ultimate ends (Barkow 2006; Segerstrale 2006). It is the pleasurable means by which reproductive success may be achieved that we are motivated to seek, and there was certainly a tighter fit between conscious means and mindless evolutionary goals in ancestral environments. This is why evolutionary psychologists prefer to use the phrase *adaptation executors* (acting in ways that would have maximized fitness in ancestral environments, but do not necessarily do so today) rather than *fitness maximizers* to refer to the evolved behavior of modern humans.

Similarly, parents do not nurture and love their children because a subconscious "selfish gene" whispers that if they do they will have greater genetic representation in future generations. Parents nurture and love their children because, well, they love them. They do so because ancestral parents who loved and nurtured their children saw more of them grow to reproductive age and pass on the genes underlying the traits we now define as love and nurturance. The neurohormonal properties underlying nurturing behavior are adaptations because they solved a recurring adaptive problem—the survival of offspring (see Chapter 7). Parents who neglected or abused their children compromised their viability, and thus the probability of pushing their own genes into the future. The love and nurturance of offspring greatly increases the probability that parental genes will survive across the generations, but in no sense can this consequence be construed as parents' motivation for lavishing love and care on their children.

The Environment of Evolutionary Adaptation

Evolutionary psychologists often refer to conditions as they existed in the environment of evolutionary adaptation (EEA). There is an unfortunate singularity to *the* EEA that implies to critics that evolutionary psychologists believe that there was only one EEA, and that evolution no longer takes place. There was no one EEA, either in terms of time or geography. Most adaptations that are uniquely human (e.g., language, culture) evolved in Pleistocene epoch (1.8 million to 10,000 years before the present) during which we observed dramatic increases in hominid brain size (Bromage 1987). For the adaptations that we share wholly or partially with other primates (e.g., bipedalism, nurturing behavior), the relevant EEA is the approximately 5- to 6-million-year period before the present. Other adaptations (e.g., aggression, egoism) go back to the very beginning of animal life on the planet. Thus, although it is conventional to speak of "the" EEA, evolutionary psychologists are aware that it should be EEAs—one for each of our adaptations (Crawford 1998:281).

Natural is Good?

The term *natural* is often used synonymously with the terms *good* and *desirable* ("Try the *natural* way to good health." "She has a *natural* beauty."). Scientists should not conflate these two terms, for what is natural is not always good. To conflate the terms is to commit what philosophers of science call the *naturalistic fallacy*. The naturalist fallacy is the fallacy of confusing *is* with *ought*. Nature simply *is*; what *ought* to be is a moral judgment. For example, if an evolutionary biologist claims that forced copulation is a natural phenomenon (i.e., a product of natural selection), they are accused of dignifying it, justifying it, implying that it is inevitable, or even that it is morally acceptable (Dupre 1992).

These critics have not made the distinction between establishing facts and morally evaluating them. To claim that forced copulation is a natural phenomenon is no more a moral statement than to claim that disease, death, floods, earthquakes, and other natural disasters which are likewise unwelcome facts of life are natural phenomena. A number of morally reprehensible traits have been selected for because of their contribution to reproductive success, but this fact does not constitute a moral argument for or against any of those traits. Evolution is morally blind in that it "allows" for the selection of behavioral traits to the extent that they enhance fitness regardless of how morally repugnant they may be. The powerful logic of evolution helps to illuminate the ultimate reasons why these traits exist, and perhaps even suggest ways to control them. It is incumbent upon us to control immoral behaviors regardless of whether or not scientists show them to be products of natural selection, because just as *natural* does not mean *desirable*, it does not mean *inevitable*.

THE EVOLUTION OF CRIMINAL TRAITS

Crime is Normal

In common with many criminologists, evolutionary psychologists view crime as normal behavior (albeit, morally regrettable) engaged in by normal individuals engaged in normal social processes (Kanazawa 2003). If criminal behavior is normal, it follows that the potential for it must be in us all, and that it must have conferred some evolutionary advantage on our distant ancestors. But how can such heterogeneous acts as murder, theft, rape, burglary, and assault be viewed as adaptations when they are usually maladaptive (fitness reducing) in modern environments? Because a behavior is typically maladaptive today (it can land a person in prison for a long time, which is not very conducive to reproductive success), it does not mean that the mechanisms underlying it are not evolved adaptations. Modern environments are so radically different from the environments *Homo sapiens* evolved in that many traits that had adaptive value at the time may not be adaptive at all today. Further, it is the traits underlying criminal behavior, not the specifics of criminal behavior (or of any other social behavior for that matter), that are the alleged adaptations. As David Rowe (1996: 285) has pointed out: "Genes do not code themselves for jimmying a lock or stealing a car . . . the genome does not waste precious DNA encoding the specifics."

Some of the traits of chronic criminals were probably useful in primitive hunter-gatherer bands. Writing about group leaders in such bands, Judith Harris (1998:299–300) points out: "His lack of fear, desire for excitement, and impulsiveness made him a formidable weapon against rival groups. His aggressiveness, strength, and lack of compassion enable him to dominate his groupmates and give him first shot at hunter-gatherer perks." The most desired of these perks were women. Women are attracted to such men not because they are "nice" but because they have status and resources within the group and made good protectors (Buss 2005). Of course, these traits can overshoot their optimum and become liabilities rather than assets, which is often the case when exercised freely in evolutionarily novel modern urban societies.

Criminal behavior is a way to acquire valued resources at little cost to oneself by exploiting others. Exploitive behavior clearly provides evolutionary advantages to a wide variety of animal species (Alcock 2005; Ellis 1998). The general tendency of organisms to seek their own interests above those of others is the adaptation, not any specific manifestation of that tendency. Evolutionary psychologists refer to an exploitive strategy as a *conditional* strategy because although preparedness to engage in it is universal, it is employed only if environmental conditions are deemed appropriate for its use. There are huge differences in people's perceptions of the conditions in which the emergence of exploitive behavior is deemed "appropriate"—this is the point of Figure 1.1 illustrating individual propensities interacting with environmental instigations. Whether such behavior is manifested depends on evolutionarily relevant environmental triggers interacting with individual differences in relevant traits, and with cultural practices constraining it. These threshold distributions are not fixed; they can shift either way in response to changes in the sociocultural environment.

Altruism

From an evolutionary point of view, in order to understand criminality we have to understand altruism. Altruism is an active regard for the well-being of another and is the epitome of prosocial behavior, and thus in many ways the polar opposite of criminality. Evolutionary science is often accused of emphasizing competition and egoism over cooperation and altruism, and of not being able to explain the latter. Reproductive success is the spoils of an unconscious competition to contribute more of one's genes to subsequent generations than other members of the breeding population. Altruism is extending some relevant benefit to others at a cost to the altruist. Altruism was often viewed in former times, even by some biologists, as a fly in the ointment of evolutionary logic. If altruism reduced fitness, why should any animal engage in altruistic behavior? It simply did not make sense that an organism would risk its own fitness for the benefit of others, especially after group selection thinking all but disappeared.

The apparent paradox was solved by William Hamilton's (1964) theory of inclusive fitness (the direct fitness of the individual and the indirect fitness of its genetically close social partners). Inclusive fitness is epitomized in the behavior of eusocial insects such as wasps, bees, and ants of the order Hymenoptera. Colonies of these creatures consist of reproductive females (queens), sterile females (workers), and male drones. Workers do not mate and can only produce male offspring via unfertilized eggs (haploid eggs containing half the number chromosomes in a sex cell). Queens are diploid; i.e., their sex cells contain the full complement of chromosomes. Because males are haploid, they get 100% of their genes from their mother. When a queen mates all of her daughters will have identical paternal genes and 50% of their mother's, leading to a situation in which they have a coefficient of genetic relatedness of 0.5 with their mother and 0.75 with their sisters (half their genes from their mother and identical genes from their father). Specialists like soldier ants and worker bees cannot directly pass on their genes and work and die selflessly for the colony, which represents their nepotistic patch of DNA. There is obviously no conscious decisions made to be altruistic, but if creatures such as these did not evolve a strategy of self-sacrifice their kind would not be with us today (an empirical fact, not a tautology).

The altruistic behavior of Hymenoptera is the quintessential example of the generally accepted position in biology that the gene is the heritable unit upon which natural selection operates. However, nonhuman animal altruism is "biological" altruism because it is entirely defined by its fitness consequences and is presumably exercised without any sort of conscious intentionality. Human altruism is obviously not like this, but inclusive fitness and its close relative, kin selection theory (the tendency to favor close genetic relatives over others), does help to explain how altruism could have been selected for in any species. Much of the altruism we observe in nature is indeed kin-directed, and the theory has a lot of support in human studies (Bateson 1997; Kruger 2003). These gene-based theories cannot help us at all, however, when the beneficiaries of human altruism have no genetic link to their benefactors, and who may even be total strangers to them. Some other and more complex evolutionary mechanisms must be added to the inclusive fitness/kin selection scaffolding.

Robert Trivers (1971) developed the theory of reciprocal altruism in an attempt to account for intentional altruism among non-kin. Reciprocal altruism occurs when an organism provides some benefit to another without the expectation of any kind of immediate payment, although there is an unconscious expectation of future reciprocation. The classical example is that of blood donation by vampire bats. Bats who fail to find a blood meal on a given night will have regurgitated blood donated to them by other members of the group. The tendency is to share blood with those who have previously shared with them, thus returning the favor. Failure to reciprocate usually results in the original benefactor withdrawing future acts of sharing (Wilkinson 1990). Reciprocal altruism, then, is not selfless behavior; it is ultimately designed to benefit the altruist because of the expectation of future reciprocity (Lehmann and Keller 2006).

This is classic tit-for-tat mutual back-scratching in action. A prereqisite condition for tit-for-tat to be a stable strategy is frequent association and the ability to recognize reciprocators and nonreciprocators. Because of the mutual benefits of reciprocal altruism that accrue to all socially interacting species, *Homo sapiens* is a species with "minds [that] are exquisitely crafted by evolution to form cooperative relationships built on trust and kindness" (Allman 1994:147). Altruism and cooperation are tit-for-tat strategies favored by natural selection because of the benefits they conferred to those who practiced them.

But reciprocal altruism no more explains situations in which individuals confer some benefit on strangers (it could be anything from dropping a dollar in a beggar's hat to risking one's life to save another) anymore than inclusive fitness/kin selection does. To distinguish it from biological altruism, this kind of altruism has been called "psychological altruism" (Kruger 2003). If reciprocal altruism is a thoroughly gene-based adaptation, perhaps psychological altruism is an exaptation, seized upon by natural selection to improve upon the operation of reciprocal altruism by infusing it with neuropsychological mediators.

Psychological altruism is apparently motivated by internal rewards, such as guilt reduction, or the joy experienced when beneficiaries express their gratitude for the benefactor's largesse (Brunero 2002; Sober and Wilson 1998). In other words, we act altruistically because we tend to feel good when we do, and because such behavior confers valued social status on us by identifying us as persons who are kind, reliable, and trustworthy. In the ultimate sense, we do so because our distant ancestors who were altruistic and cooperative enjoyed greater reproductive success than those who were not. Possessing the neural architecture that produces warm, fuzzy feelings when we do good things for others is part and parcel of that adaptation (Barkow 1997). Although individual organisms are adapted to act in ways that tend to maximize their own fitness, not to behave for the good of the group, their fitness goals are best realized by adhering to the rules of cooperation and altruism—by "being nice"—and that is "for the good of the group." Thus, intentional "psychological" altruism is ultimately self-serving even if the altruist vehemently denies that it is. This observation does not diminish the value of altruism to its beneficiaries one bit.

Selfishness as understood in the vernacular means a crabbed egoism shorn of any concern for the well-being of others. Such selfishness is ultimately maladaptive because it gainsays cooperation. Selfishness properly understood is the most adaptive of sentiments, because it is precisely by cooperating and being actively concerned for others that we best serve our own interests. As the great biologist Edward O. Wilson put it: "Human beings appear to be sufficiently selfish and calculating to be capable of infinitely greater harmony and social homeostasis. This statement is not self-contradictory. True selfishness, if obedient to the other constraints of mammalian biology, is the key to a nearly perfect social contract" (1978:157). If one insists that psychological altruism is not "real" altruism because it is not entirely selfless, one is unwittingly asserting that "real" behaviors are ineffable, biology-free, and cannot evolve.

Empathy

By defining intentional directed altruism as motivated by internal rewards and shorn of inclusive fitness concerns, we have identified a proximate-rather than ultimate-level mechanism. Robert Trivers (2002:6) contends that "if you start with a motivation you have given up the evolutionary analysis at the onset." We have abandoned evolutionary analysis because natural selection works on the consequences of behavior, not on the motivation behind it, and that while "proximate and ultimate analyses" inform one another, they "are not to be conflated" (de Waal 2008:280). Frans de Waal (2008) maintains that another evolved trait—empathy—channels altruism in social species without undue reliance being placed on cognitive ruminations about such things as reciprocity concerns.

Empathy is the cognitive and emotional ability to understand the feelings and distress of others as if they were our own. The cognitive component allows us to understand the distress of others and why they are feeling it, and the emotional component allows us to "feel" that distress. To the extent that we feel empathy for others, we have an evolved visceral motivation to take some action to alleviate the distress of others if we are able. Altruism can thus be thought of as the co-evolved action component of empathy. The basis of empathy is the distress we feel personally when witnessing the distress of others, and if we can alleviate the distress of others we thereby alleviate our own. Thus empathy also has a selfish component. It is a very good thing that it does because if we were lacking in emotional connectedness to others, we would be like psychopaths, callously indifferent to their needs and suffering.

Frans de Waal (2008) posits that empathy is a phylogenetically ancient capacity predating the emergence of *Homo sapiens* and evolved rapidly in the context of parental care. Empathy is an integral component of the love and nurturing of offspring. Caregivers must quickly and automatically relate to the distress signals of their offspring. Parents who were not alerted to or who were unaffected by their offspring's distress signals or by their smiles and cooing are surely not among our ancestors. Like the diffusion of adaptive love and nurturance of offspring to the nonadaptive love and nurturance of the children of others and to pets, the capacity for empathetic responses, once locked into the human repertoire, diffused to a wider network of social relationships.

What is it that allows us to link the me/other relationship in empathetic responses? There has to be some neural architecture that gives rise to shared representation of affective states. Neuroscientists have located the ability in so-called mirror neurons. Mirror neurons are brain cells that fire (respond) equally whether an actor performs an action or witnesses someone else performing the action. Thus the neuron "mirrors" the behavior of another as though the observer were acting in the same way. It is not simply a matter of being cognitively aware of the actor ("Francis is crying"); it is the actual firing of *identical* neurons in the observer's brain that are firing in the observed's brain. This neuron mimicry can and does often operate outside of the oberver's conscious awareness. It is assumed that this neuron identicalness reflects a correspondence between self and other that turns an observation into empathy.

These neurons exist and their location has been conclusively identified in the brains of macaques, and there is substantial evidence from neuroimaging and electrophysiological studies that they exist in humans (see Agnew, Bhakoo, and Puri 2007; Decety and Jackson 2006 for reviews). The human mirror neuron system seems to be widely distributed throughout the brain and is considered a foundational part of empathetic responses.

Interestingly, females tend to be better than males in reading or "mirroring" the emotions of others. The neuropeptide oxytocin (see Chapter 7) plays an important role in the birthing process and in lactation. It also plays an important role in forming attachment, which means that oxytocin, via its role in attachment formation, is another potential causal mechanism for empathy. The "male" hormone testosterone is thought to act antagonistically to oxytocin, which both supports the observation that females empathize more strongly than males and the notion that the evolution of empathy was driven by caregiving (Herman, Putnam, and van Honk 2006). It has also been shown that high testosterone levels are associated with psychopathy, and psychopaths are notoriously poor in empathetic responding (Blair 2005a).

Altruistic Cooperation Creates Niches for Cheats

Because cooperation occurs among groups of reciprocal altruists, it creates a niche for cheats to exploit. Indeed, the stronger the selection for altruism in a species, the more vulnerable it becomes to "Machiavellian intelligence" (Runciman 2005:132). Cheats are individuals in a population of cooperators who signal their cooperation but fail to reciprocate after receiving benefits. If there are no deterrents against cheating, it is in an individual's fitness interests to obtain resources from others under the assumption of reciprocity and then to default, thus gaining resources at zero cost. "Social parasitism" of this sort has been observed among a variety of nonhuman animal species (Alcock 2005), and its ubiquity across species implies that it has had positive fitness consequences. Among humans, criminal behavior is an extreme form of defaulting on the rules of reciprocity (Lykken 1995). Cheating comes at a cost, however, so before deciding to default the individual must weigh the costs and benefits of cooperating versus defaulting. This is illustrated in the *Prisoner's Dilemma* of game theory (Axelrod 1984).

Suppose two recently acquainted criminal accomplices—Bill and Frank—are being held in jail for an alleged crime. They have both sworn that each would never "rat" on the other. The evidence against them is weak, prompting the prosecutor to approach each man separately and offer him a deal. If Bill testifies against Frank, Bill will be released and Frank will get 20 years, and vice versa. If both testify, both will be convicted and receive a reduced five-year sentence because of their cooperation with the prosecutor. If neither testifies; i.e., they cooperate with each other as they had sworn to do, both will be convicted of a minor crime carrying a sentence of only one year in prison. The dilemma is that Bill and Frank are being held in separate cells so that they cannot communicate with one another and cement their agreement not to default on their promise. Under these circumstances, Bill's best strategy is to testify regardless of what Frank does because it will either get him released (if Frank does not testify) or five years (if Frank does). Both outcomes are far better than the 20 years he will receive if he remains true to his promise but Frank does not. The same holds true for Frank. Each man, following his own best interests, testifies against the other, and receives a sentence of five years. The paradox is that although the payoff for cheating is high when the other actor does not cheat, if both cheat they are both worse off than if they cooperate.

By cheating, each man is behaving entirely rationally (I define rationality as a positive fit between one's ends and the means used to achieve them). But if cheating behavior is so rational, how did cooperative behavior come to be predominant in social species? The answer is that cheating is only rational in the long term in circumstances of limited interaction and communication. Frank and Bill were recent acquaintances who might never see each other again and thus need not fear any repercussions arising from their cheating. Had they been brothers, good friends, or members of a long-standing gang, they most likely would not have defaulted on their promise, and each would have benefited by receiving one year rather than five. Frequent interaction and communication breeds trust among organisms with sufficient intelligence to recognize one another. Under such circumstances, cheating becomes a far less rational strategy because cooperators remember and retaliate against those who have cheated them. Cheating ruins reputations, costs cheaters future cooperation, and can result in punishment, which is why most career criminals either die early or end up destitute (Shover 1985).

Cheats can only prosper in a population of unconditional altruists (game theorists call them *suckers*). Suckers are individual organisms who continue to cooperate with, and extend benefits to, those who have cheated them. Any sucker genotype would soon be driven to extinction by cheats, leaving only cheats to interact with other cheats. Evolution logic predicts that a population of cheats could not thrive anymore than could a population of suckers, and selection for cooperation would occur rapidly (Machalek 1996). Pure suckers and cheats are thus unlikely to exist in large numbers in any social species. The vast majority of social animals, including human beings, are *grudgers*. Grudgers are susceptible to being suckered because they abide by the norms of mutual trust and cooperation and expect the same from others. But if suckered, they retaliate by not cooperating with the cheat in the future, and perhaps even repaying the cheat in kind (Raine 1993). Cheaters interact with grudgers in a *repeated* game of prisoner's dilemma in which players adjust their strategies according to their experience with other players. Cooperation rather than cheating becomes the rational strategy under such circumstances because each player reaps in the future what he or she has sown in the past (Machalek 1996).

As predicted by evolutionary logic, in computer simulations of interactions between populations of cheats, suckers, and grudgers, cheats are always driven to extinction (Allman, 1994; Raine 1993). Why do we continue to see cheating behavior despite threats of exposure and retaliation then? The problem with such simulations is that players are constrained to operate within the same environment in which their reputations quickly become known. Cheats are not constrained to remain in one environment in the real world; they can move from location to location, meeting and cheating a series of grudgers who are unaware of the cheaters' reputation. This is exactly what many career criminals do. They move from place to place, job to job, relationship to relationship, leaving a trail of misery behind them before their reputation catches up to them (Hare 1993; Lykken 1995; Raine 1993). Cheats are much more likely to prosper in large cities in modern societies than in small traditional communities where the threat of exposure and retaliation is great (Ellis and Walsh 1997; Machalek and Cohen 1991).

Raine (1993) cautions us against forming an overly simplistic view of cooperating and cheating from tit-for-tat computer models. Although life is more like a poker tournament than a single game, computer simulations are invaluable for fleshing out the basic logic of evolutionary processes. Real-life strategies are not automated binary strategies (cheat/don't cheat) based on the value of the preceding binary code. Environmental factors such as the stability of the group and cultural dynamics must be considered. For instance, criminologists know that there are communities in which a "badass" reputation is valued by males more than anything else, but even in these communities there must be a certain level of group loyalty and cooperation. We must also consider the fact that there are individual differences in traits underlying cooperative behavior. Two behavior genetic studies found heritabilities of .56 and .68 for altruism and empathy, respectively (Rushton et al. 1986; Rushton, Littlefield, and Lumsden 1986).

DETECTING AND PUNISHING CHEATS

Detection

In order for altruists not to be exploited by cheats, mechanisms to identify and punish cheats had to evolve. The social emotions, such as empathy, shame, and guilt, are the primary mechanisms of cheater detection (Griffiths 1990;

Wiebe 2009), and the primary emotions of outrage and desire for "righteous vengeance" are probably the emotional underpinnings of punishment (Nowak and Sigmund 2005; Walsh 2000). Social emotions have evolved as integral parts of our social intelligence that serve to provide clues about the kinds of relationships (cooperative vs. uncooperative) that we are likely to have with others, and serve as "commitment devices" and "guarantors of threats and promises" (Mealey 1995:525). Barkow (1989:121) describes them as involuntary and invasive "limbic system overrides" that serve to adjust our behavior in social situations. In other words, emotions animate, focus, and modify neural activity in ways that lead us to choose certain responses over other possible responses from the streams of information we constantly receive. The social emotions move us to behave in ways that enhanced our distant ancestors' reproductive success by overriding neocortical decisions suggesting alternatives to cooperation (i.e., cheating) which may have been more rational in the short term, but which were ultimately fitness reducing.

Emotions and rational cognitions are not antagonists; they are two inextricably linked components of all that we think and do: there can be no Kantian "pure reason" (Wilson 1998). It has been well known in cognitive neuroscience for some time that cognition is always suffused with emotions to various degrees, and that they play "a central role in both associative learning and memory" (Masters 1991:307). It is worth noting that that one of the defining characteristics of psychopaths, the quintessential cheats, is their inability to "tie" the brain's cognitive and emotional networks together (Patrick 1994; Pitchford 2001; Weibe 2004).

The social emotions cause positive and negative feelings when we survey the consequences of our actions. Mutual cooperation evokes a deepened sense of friendship, a sense of pride, and a heightened sense of obligation and gratitude that enhances future cooperation. Mutual cheats feel rejected and angry, and when one party cooperates and the other cheats, the cooperator feels angry and betrayed, and the cheater feels anxiety and guilt (Nowak and Sigmund 2005). Because we find the emotions accompanying mutual cooperation rewarding and those accompanying defection punishing, the more intensely we feel the emotions the less likely we are to cheat. Conversely, the less we feel them the more likely we are to prefer the immediate fruits of cheating over concerns of reputation and its effects on future interactions. Emotions thus function to keep our temptations in check by overriding rational calculations of immediate gain.

The continued presence of chronic cheats among us indicates that we have less than perfect ability to detect and punish them and that cheating must have had positive fitness consequences occasionally. Perhaps under certain evolutionary conditions the strategy proved to be so successful that we have a certain proportion of the human population today whose cheating behavior is obligatory rather than contingent. According to some theorists (Pitchford 2001; Weibe 2004), traits conducive to cheating are normally distributed in the population, but there is a small but stable percentage of individuals at one extreme of the distribution for whom cheating is an obligate strategy. These individuals are the few primary psychopaths (3 to 4% of males and less than 1% of females) who presumably have always existed in every society (Lykken 1995).

Such an obligate cheater strategy is likely to evolve alongside the more typical grudger strategy (in which cheating behavior is environmentally dependent rather than obligatory) when its fitness gains are *frequency-dependent*. Frequency-dependent selection occurs when an alternative mating strategy enjoys high reproductive success when few practice it, but low success when it becomes more common. The strategy eventually results in organisms that are genotypically, not just phenotypically, different. The cheating strategy may be sustained as a low-level frequency-dependent strategy in statistical equilibrium with the grudger strategy and will coexist with it as an *evolutionarily stable strategy* (ESS) (Mealey 1995). An EES is a strategy which, when adopted by a large enough fraction of the population, cannot be invaded and eliminated by another.

The greatly fluctuating levels of reproductive success attending a frequency-dependent strategy, combined with the evolution of counterpressures against cheating in the population, assures that obligate cheaters are rare in any population (Moore and Rose 1995). When there are few cheats in a population, each cheat enjoys numerous opportunities to exploit its unwary members, but when many follow a cheater strategy, not only are there fewer "suckers" per cheat; there is a greater awareness of the strategy in the population and thus a lowered probability of its success (Dugatkin 1992; Kinner 2003). However, as cheats become fewer in a population, more opportunities arise for those who remain to prosper, and the whole process recycles.

This process is a coevolutionary "arms race" similar to the coevolution of predator and prey in which the adaptations of one species are molded by the adaptations of the other. The continual competition between cooperators and cheats has molded the sensibilities of both. Just as cooperators have undergone evolutionary tuning of their senses for detecting cheats, cheats have evolved mechanisms that serve to hide their true intentions (Kinner 2003; Trivers 1991). The probable adaptation aiding cheats is a muting of the neurohormonal mechanisms that regulate the social emotions so that cheats have little real understanding of what it is like to feel guilt, shame, anxiety, and empathy. Selection for self-deception would even better enable the cheater to pursue his interests without detection (Dugatkin 1992; Drake 1995). We all have this capacity for self-deception (think of the defense mechanisms in psychoanalytic theory which serve to protect our egos), but most of us are not all that good at maintaining the fiction (Nesse and Lloyd 1992). Because chronic cheats operate "below the emotional poverty line" (Hare 1993:134), they do not reveal clues that would allow others to judge their intentions. Lacking an emotional basis for self-regulation, chronic cheats tend to make social decisions based only on rational calculations of immediate costs or benefits (Mealey 1995).

There is a wealth of evidence that psychopaths have a greatly diminished capacity to experience the social emotions (Viding et al. 2005;

Wiebe 2009), but the proposition that there is a distinct behavioral type in human populations for which deception and exploitation is an obligate rather than a conditional strategy is probably the most difficult evolutionary proposition for criminologists to accept. Many evolutionary scholars disagree with the proposition, but many others do not. Obligate and conditional cheater reproductive strategies, each with its own distinct genetic basis, do exist collaterally in numerous animal species as ESSs (Alcock 2005). There is also some human evidence that psychopathy constitutes a discrete taxonomic class (a categorical rather than continuous variable) *phenotypically* (Harris, Rice, and Quinsey 1994; Skilling, Quinsey, and Craig 2001), but this evidence does not mark psychopaths as distinct from nonpsychopaths *genotypically*.

On the other hand, psychopathy researchers appear to be in agreement that psychopathy is an "experiment of nature," and that "there are currently no reasons to believe that there might be an environmental cause to psychopathy" (Blair 2005b:882). This echoes what the grand master of psychopathy research, Robert Hare, wrote: "I can find no convincing evidence that psychopathy is the direct result of early social or environmental factors" (1993:170). This does not mean that such factors are irrelevant to understanding psychopathic *behavior*. Writing about the famous 19th-century highly talented British adventurer Sir Richard Burton, who was widely considered psychopathic (Lykken 1995), Walsh and Wu (2008:139) write that if he "had been the son of a London butcher instead of the son of an army colonel his 'monstrous talents' may have been utilized for criminal purposes instead of those to which he actually put them and he may have ended up with a rope around his neck rather than the sash of knighthood around his shoulders." Walsh and Wu were not saying that poorer social circumstances would have *caused* Burton's psychopathy, only that they might have led him to *express* it in less heroic ways.

The vast majority of criminals are not psychopaths, and the cheater strategy they employ is conditional rather than obligate. Conditional strategies are evolutionarily more advantageous because of the flexibility they offer. In most nonhuman animal species, cheater strategists (variably labeled as *sneakers, mimics, floaters*, and *satellites*) are typically males who are reproductively disadvantaged in some way (Alcock 2005). The most disadvantaged are the young who have not yet established themselves in the social hierarchy, have limited resources, and who cannot contend physically with older and more powerful conspecifics. The cheater strategy is marginally adaptive because it does provide more reproductive opportunities than would be the case if disadvantaged males simply waited passively for their situation to improve, which may never happen. Because a conditional cheater strategy has some fitness consequences, genes governing the mechanisms that allow us to respond to environmental conditions by changing our behavior (cooperator to cheater, and vice versa) have survived in us all. In a very real sense, then, the antisocial impulse is universal, but is constrained by rules which most of us profit from following most of the time.

Punishment

If the antisocial impulse is universal, so is the urge to punish it. The punishment of cheating conspecifics is noted among a wide variety of nonhuman animal species (Clutton-Brock and Parker 1995). Experiencing and witnessing the exploitive strategies of others lead us to moral outrage. Without this sense of moral outrage we would not react against, and therefore not deter, the murderer, rapist, thief, cheater, and free rider, and any genes inclining individuals to such behavior would have spread more widely than they have (Walsh 2006; Wiebe 2004).

Moral outrage is probably the basis of the desire for revenge, which may be considered a necessary evil for humans in order to make credible threats. If we threaten to retaliate if someone cheats and do not carry out the threat if he does, the cheat has no rational reason for not cheating again in the future. If our ancestors made nothing but idle threats, the impulse to take what is not ours would be far stronger than it is today. The more the potential for retaliation is a reality the less likely we are to actually harm one another because the would-be miscreant is likely to take that potential seriously. Moral outrage buttressed by retaliatory action is a plausible candidate as the basis of our sense of justice (Sigmund, Fehr, and Nowak 2002; Walsh 2000).

As with any universal desire, the desire to punish has a built-in physiological basis. We recognize justice because we recognize injustice, and we recognize injustice by *felt* outrage at experiencing injustice, not by reference to some legal rule. That is, we *feel* injustice by virtue of evolved physiological responses to it prior to articulating conceptions of it. As Walsh (2000:853) put it: "The positive feelings accompanying the punishment of those who have wronged us, coupled with the reduction of negative feelings, provide powerful reinforcement [for punishing], as suggested by the popular saying 'Vengeance is sweet.'"

There is experimental evidence for this from brain-imagining studies showing increased blood flow to areas of the brain that respond to reward, suggesting that punishing those who have wronged us provides both emotional relief (assuaging outrage) and reward (the feeling of satisfaction) for the punisher (de Quervain et al. 2004). Similar experimental research has provided strong evidence for Durkheim's theoretical insight about the functional role of punishment. This body of research has concluded that punishment has been crucial for the evolution of cooperation among humans (Fehr and Gachter 2002).

VIOLENCE AND STATUS

When evolutionary biologists explore the behavior of any species, their first question is: "What is the adaptive significance of this behavior?" Violence is a major concern of criminologists, and they might want to know how violence was adaptive in evolutionary environments, what its function is, and what environmental circumstances are likely to evoke it. Violence evolved to solve some set of adaptive problems; if it did not it would not be part of our behavioral repertoire. The bloody history of *Homo sapiens* demonstrates how easily and frequently humans are moved to create situations that lead to violence. Perhaps the selective pressures for violence in humans exerted themselves as a consequence of our practice of stockpiling resources (land, food, territory, and weapons) which others who desired them found could be appropriated by violent means (Buss and Duntley 2006).

Calling violence an adaptation does not mean a return to the old psychoanalytic "hydraulic" notion of aggression in which it was assumed that aggression slowly builds up in individuals until it is vented. Under this model, one could safely vent a little steam before the whole pent-up thing violently exploded. This implied that aggression is obligatory and will be expressed one way or another regardless of environmental context. Evolutionary psychologists, on the other hand, view aggression and violence as a context-contingent strategy triggered by evolutionarily relevant contexts; i.e., circumstances akin to the adaptive problems our distant ancestors experienced and which resulted in a desired benefit (Buss and Duntley 2006).

It is a central tenet of evolutionary theory that the human brain evolved in the context of overwhelming concerns for resources and mate acquisition. When food, territory, and mates are plentiful, the use of violent tactics to obtain them is a risky and unnecessary waste of energy, but when resources are scarce, acquiring them any way one can, including the use of violence, may be worth the risk. Given this, credible threats of violence (not necessarily violence itself) are related to reproductive success in almost all species of social animals. Reacting violently when others intrude on one's territory, resources, or mates would be very useful in evolutionary environments when you just couldn't call 911 for the police to settle the problem. Having a reputation for violence would be better because others would be aware of it and avoid your resources. In other words, in environments where people are expected to take care of their own beefs, violence or the threat of violence works to let any potential challenger know that it would be in his best interests to avoid you and your resources and look elsewhere. All this is why a "badass" reputation is so highly valued in certain subcultures, why those with such a reputation are always looking for opportunities to validate it, and why it is craved to such an extent that "Many inner city young men . . . will risk their lives to attain it" (Anderson 1994:89).

Status is not necessarily associated with aggressive and violent tactics (typically, quite the opposite today in most social contexts), but it almost certainly was more so in our ancestral environments (Chagnon 1996; Wrangham and Peterson 1998). Because status brought more copulation opportunities, genes inclining males to aggressively pursue their interests (which sometimes meant becoming violent) enjoyed greater representation in subsequent generations. From the evolutionary point of view, violence is something human males, as well as males in numerous other species, are designed by nature to exhibit under evolutionarily relevant circumstances.

In cultures where polygyny and low paternal investment exist we find homicide rates greatly exceeding those of any modern society, such as the Agta rate of 326 per 100,000 and the Yanomamo rate of 166 per 100,000 (Ellis and Walsh 2000:71). Homicide translates directly into reproductive success among the Yanomamo, with males who have killed the most in intervillage warfare having, on average, three times as many wives and children than those who have killed least or not at all (Buss 2005; Chagnon 1988).

However, natural selection has not strongly favored violent battles over access to females in long-lived species such as ours because males in such species have time to move up status hierarchies and acquire mates via courtship rather than risking their lives in desperate mating battles (Alcock 2005). The male/female body-size ratio in a species is an indicator of its mating history. Selection for size and strength among males results in them becoming much larger (up to 300% in some species) than females. Such a large degree of sexual dimorphism reflects a polygynous mating history in which dominance is established by physical battles among males. The fossil record shows that early hominid males (*Australopithecines*) were 50 to 100% larger than females (Geary 2000). The fairly low degree of sexual dimorphism for body size among modern *Homo sapiens* (males are only about 15% larger than females, on average) indicates an evolutionary shift from violent male competition for mates to a more monogamous mating system and an increase in paternal investment (Plavcan and van Schaik 1997).

Violence imposes a cost on those who use it, such as the possibility of death or severe injury. So perhaps some alternative strategy would be better if others coveted one's resources, such as sharing them. Sharing resources is the hallmark of reciprocal altruism that has shaped us into a cooperative species, but what happens when we meet cheats who want it all and resort to violence to assert their claim? Anyone unprepared to match violence with violence under those circumstances may loose their life as well as their resources. Natural selection has provided us with the ability to switch to a violence mode quickly when we have reason to believe that things we value may be taken from us (Gaulin and McBurney 2001; Kelly 2005). This propensity is most useful today in disorganized neighborhoods where a tradition of settling one's own quarrels without involving the authorities is entrenched; that is, in neighborhoods in which social institutions that control, shape, and sublimate violent tendencies are absent or enfeebled.

Individuals living in "respectable" neighborhoods in which the fairness of the criminal justice system is taken for granted rarely have to resort to violence to gain what they want or to protect what they have. Violence is maladaptive for them; natural selection has favored flexibility over fixity of human behavior, which is why behaving violently is very much contingent on environmental instigation.

As previously noted, natural selection has not strongly favored violent competition in long-lived species because males in long-lived species have plenty of time to move up status hierarchies and acquire resources in more peaceful and cooperative ways. But what if individuals' interpretations of their experiences tell them that they do not have "plenty of time" and that they cannot advance in life through cooperative and peaceful means? Criminologists frequently note that impulsivity and discounting the future are maladaptive. Wilson and Daly (1997:1271), however, suggest that discounting the future "may be a 'rational' response to information that indicates an uncertain or low probability of surviving to reap delayed benefits, for example, and 'reckless' risk taking can be optimal when the expected profits from safer courses of action are negligible." In other words, when people perceive little opportunity for legitimate success and when many people they know die at an early age, then living for the present and engaging in risky violence to obtain resources makes evolutionary sense. Males compete for status by whatever means are available to them in the cultural environments in which they live.

Wilson and Daly (1997) tested their assumption with data from 77 neighborhoods in Chicago for the years 1988 through 1995. They hypothesized that neighborhoods with the lowest income levels and the shortest life expectancies (excluding homicides) would have the highest homicide rates. Life expectancy (effects of homicide mortality removed) ranged from 54.3 years in the poorest neighborhood to 77.4 years in the wealthiest, and the attending homicide rates ranged from 1.3 to 156 per 100,000, a huge 120-fold difference. Wilson and Daly interpreted the data as reflecting escalations of risky tactics that make sense from an evolutionary point of view given the conditions in which people in disorganized neighborhoods live. Natural selection has equipped humans to respond to extreme levels of inequality and short life expectancy by creating risky, high-stakes male–male competitions that too frequently result in homicide. From a moral point of view, this is something to be condemned, but to the extent that such contingent responses are the products of natural selection they are not pathological from the point of view of evolutionary biology. As Bob Dylan sang: "When you got nothing, you got nothing to lose"; and as Robert Wright (1995:69) wrote: it is surprising "how far to the left one can be dragged by a modern Darwinian view of the human mind."

SPECIFIC EVOLUTIONARY THEORIES OF CRIME

We now briefly explore specific evolutionary theories of criminal behavior. The four primary evolutionary theories are *cheater theory, conditional adaptation theory, adaptive strategy theory*, and *differential K theory*. The foundation for all four theories is reproductive strategies (mating effort versus parenting effort) and the tactics that flow from them. Mating effort is the proportion of one's reproductive effort invested in finding sexual partners (the *cad* strategy); parenting effort is the proportion invested in caring for offspring (the *dad* strategy).

Empirical research shows that a heavy emphasis on mating effort is linked to criminality. A review of 51 studies found that 50 of them reported a significant positive relationship between the number of sex partners and criminality, and a review of 31 other studies found that age of onset of sexual behavior was significantly inversely related to criminal behavior in all 31 (Ellis and Walsh 2000). A British cohort study found that the most antisocial 10% of males in the cohort fathered 27% of the children (Jaffee et al. 2003), and anthropologists tell us that there are striking differences in behavior between members of cultures that emphasize either parenting or mating strategies. Behaviors considered antisocial in Western societies such as low-level parental care, hypermasculinity, transient bonding, and violence are normative in cultures emphasizing mating effort the world over (Ember and Ember 1998).

Molecular genetic studies are getting into the act also. Guo and his colleagues (2007) found that males who were homozygous for the 10-repeat dopamine transporter (DAT1) gene had 80 to 100% (depending on age category) more sex partners than males who were homozygous for the 9-repeat version. Another molecular genetic study found that the same DAT1 polymorphism that was significantly related to number of sex partners was also significantly related to antisocial behavior among a sample of 2,574 males between the ages of 18 and 26 (Beaver, Wright, and Walsh 2008).

Cheater Theory

We have already encountered aspects of cheater theory in the discussion of psychopathy. The theory rests ultimately on the broad asymmetry between the reproductive strategies of males and females. Females have a much lower reproductive ceiling than males, although almost all females will probably reproduce. The major factor in female reproductive success has been to secure and hold on to the assistance of a mate to raise her offspring. There is much more variability in male reproductive success, with some males leaving no offspring, and others fathering large numbers. This is particularly so in polygynous species and polygynous human cultures, and probably so in some human EEAs. Given lower variation but greater reproductive certainty, females have evolved a mating strategy inclining them to be far choosier about whom they will mate with than males are (Buss 1994; Geary 2000).

Male reproductive success is potentially greater the more females they can mate with and males have an evolved desire for multiple partners. Males can respond to the more reticent female strategy in one of two ways: they can comply with female preferences and assist a single female to raise their offspring or they can either trick or force a female to have sex and then move on to the next female. Most heterosexual males have probably falsely proclaimed love and fidelity and used some form of coercion to obtain sexual favors at one time or another, but the vast majority will eventually settle down and assist a female in raising their young. This is strategy is facilitated by the social emotions, particularly love (Aron et al. 2005; Fisher 2002). The second strategy is likely to be followed by males who are deficient in the social emotions, such as chronic criminals and psychopaths across the life span (Walsh 2006; Wiebe 2004). The basic point of cheater theory is that criminal activity is facilitated by the same traits that make for the successful pursuit of a cheater sexual strategy. It is important to stress that cheater theory does not, nor does any other theory presented here, postulate that criminal behavior reflects a defective genome; rather, it reflects a normal, albeit morally regrettable, alternative.

Conditional Adaptation Theory (CAT)

CAT proposes that people adopt different reproductive strategies based on early childhood experiences (Belsky and Draper 1991). CAT proposes that individuals will tend to achieve early menarche and adopt an unrestricted (promiscuous) sexual strategy if they learned during their childhood that interpersonal relationships are ephemeral and unreliable (as indexed by such things as parental divorce and witnessing others engaging in short-term relationships). Individuals who learned the opposite will tend to adopt a more restricted strategy. Needless to say, neither strategy is consciously chosen, but rather flows from subconscious expectations based on early experiences of the stability/instability of interpersonal relationships.

Unlike other evolutionary theories of crime, CAT includes features that allow for predictions about the involvement of women as well as men in antisocial behavior. Thus, in addition to the *cad* vs. *dad* dichotomy to describe male reproductive strategies, we have the *whore* vs. *Madonna* dichotomy to define female mating strategies. As we have seen, a number of lines of evidence show that both male and female criminals and serious delinquents have significantly more sex partners and begin sexual activity earlier than do persons in general.

David Rowe (2002a:62–63) provides a thumbnail sketch of the traits useful in supporting the "cad" mating strategy, which are the same traits useful in pursuing criminal activities.

> A strong sexual drive and attraction to novelty of new sexual partners is clearly one component of mating effort. An ability to appear charming and superficially interested in women while courting them would be useful. The emotional attachment, however, must be an insincere one, to prevent emotional bonding to a girlfriend or spouse. The cad may be aggressive, to coerce sex from partly willing partners and to deter rival men. He feels little remorse about lying or cheating. Impulsivity could be advantageous in a cad because mating decisions must be made quickly and without prolonged deliberation; the unconscious aim is many partners, not a high-quality partner.

Because the traits underlying the respective strategies are highly heritable, individuals may vary in their susceptibility to adopt a particular sexual strategy for genetic reasons rather than early childhood experiences (Belsky 1999). In other words, the negative and ephemeral relationships observed among the sexually unrestricted may be a *consequence* of their strategy rather than a cause. Children receive a suite of genes as well as an environment from their parents, and the similarity of parent/offspring sexual strategies may have more to do with shared genes than shared environments (Walsh 2000c). Empirical support for this interpretation is supplied by Rowe (2002b), who found heritability coefficients of 0.50, 0.54, and 0.28 for age at menarche (CAT proposes that females exposed to early relationship transience will achieve early menarche), nonvirgin status, and age at first intercourse, respectively.

Alternative Adaptation Theory (AAT)

AAT proposes that humans are arrayed along a continuum regarding where they have a tendency to focus their reproductive efforts largely for genetic reasons (Rowe 1996). At one extreme is mating effort and at the other is parenting effort. The best demographic predictors of where effort is focused are gender and age, which are also the best demographic predictors of crime and delinquency. Males and the young emphasize mating effort and females, and older persons emphasize parenting effort. The suite of traits useful for focusing on mating effort, such as deceitfulness, impulsiveness, and hedonism, are also useful in pursuing criminal activity. Conversely, traits useful for parenting effort, such as empathy, conscientiousness, and altruism, are also useful for noncriminal, or prosocial, activity.

AAT makes the same predictions CAT does regarding early onset of sexual behavior and number of sexual partners, but would explain the relationship by indicating that both criminal activity and a high level of mating effort are sustained by the same suite of heritable traits. Also, unlike CAT, Rowe places little emphasis on child rearing, pointing out that behavior genetic studies consistently show that common rearing environment (which is stressed by CAT) has little or no lasting influence on an individual's personality or intelligence.

Anthropological evidence supportive of aspects of both CAT and AAT is provided by Harpending and Draper (1988), who contrasted reproductive strategies in two cultures located in drastically different ecological environments. The first group is the !Kung bushmen, who inhabit the inhospitable Kalahari desert in South Africa, and the second is the Mundurucu, who inhabit the resource-rich Amazon basin. Because conditions are harsh in the Kalahari, life is precarious, cooperative behavior is imperative, and parenting effort is favored over mating effort. The Mundurucu's rich ecology frees males for fighting, for raiding other groups, and to engage in competition for females. Mating effort is favored over parenting effort among the

Mundurucu. What is most interesting from a criminological perspective is that cultures emphasizing mating effort exhibit behaviors that would be considered antisocial in Western societies, such as low-level parental care, aggressiveness, "protest masculinity," and transient bonding. (Ember and Ember 1998; Harpending and Draper 1988). These behaviors, however, are adaptive in such cultures, and may well be adaptive in certain subcultures of modern industrial societies also.

Social scientists would undoubtedly tell us that "culture" explains the different behaviors of the !Kung and Mundurucu and leave it at that. However, such an explanation begs the question of what lies behind these two cultures that makes them different. As long as social scientists view culture as an autonomous causal agent, that is, as "a shared set of presuppositions, values, and the like that was a more or less arbitrary selection from the basket of possible human values" (Turner 2007:364), we can never understand much about group differences in behavior. If there was no universal human natures the stories from ancient and distant cultures would mystify us, but as Grosvenor (2002:434) reminds us: "It is the existence of perennial traits that enables us to understand , for example, the motivations of characters in the plays of Shakespeare or Sophocles, even though they were written in times [and in cultures] radically different from our own." A coherent explanation of cultural differences requires an understanding of human nature and the fitness imperatives imposed on it. The individuals of both cultures have similar evolved adaptations, but are constrained to execute those adaptations in greatly different environments. Only an understanding of human nature can help us to appreciate the different psychologies underlying the social behavior of these two groups in ways that would lead to predictions about the behaviors of other groups inhabiting similar niches. We should repeat Barkow's (1992) admonition here not to forget that psychology underlies culture and that evolution underlies psychology.

Differential K Theory (DKT)

DKT is both more complicated and more controversial than the other theories because it makes predictions about race, as well as age and gender. Specifically, it predicts that blacks will commit more crimes than whites, who will, in turn, commit more crimes than Asians. This racial pattern of crime rates is consistently found worldwide (Ellis and Walsh 2000; Eysenk and Gudjonsson 1989; Rushton 1995). DKT is a life-history theory based on the mathematics of population biology embedded in a continuum of reproductive strategies ranging from r to K (Molles 2008). The extreme r strategy (r stands for the intrinsic rate of population growth) is characterized by maximum egg production and no parenting effort. The r strategy is most characteristic of short-lived creatures such as insects and fish, whose extreme fertility ensures genetic continuity even if the vast majority of offspring perish before reaching reproductive age. The K strategy (K stands for the carrying capacity of the environment) emphasizes parental care over offspring numbers to ensure genetic survival, and is characteristic of long-lived species, particularly humans.

The r and K strategies covary with many heritable traits helpful in maintaining them (Rushton 1995). For instance, Ellis (1987) showed that traits identifying r-selection are more typical of criminals, and that traits used to identify K-selection were more typical of noncriminals. His review of numerous studies revealed that persons with serious delinquent and criminal histories appeared to have the following six r-selected traits to a greater extent than persons in general: (1) shorter gestation periods; (2) earlier onset of sexual activity; (3) greater sexual activity outside bonded relationships; (4) less stable bonding; (5) lower parental investment (high rates of abuse, neglect, and abandonment); and (6) shorter life expectancy.

This is where charges of racism appear. If criminals in general are higher on traits used to define r-selection species, and if the black > white > Asian pattern of crime prevalence is consistently found, does it imply a black > white > Asian gradient of r-selection? Answering this question requires the rank-ordering of large samples from each population on as many indicators of r and K selected traits as possible and their mean scores compared. Rushton (1995) did this by factoring 26 traits associated with r or K selection into five higher order concepts—social organization, personality/temperament, maturation rate, intelligence, and reproductive effort. Socialization is heavily involved in some of these, such as achievement and sexuality, but minimally, if at all, in others traits such as morphology, speed of physical maturation, and gamete production. If racial differences are random with respect to r/K indicators, then most research results will be null, and the remainder will be about equally split between negative and positive results. Reviews of the literature have shown this has not been the case across the hundreds of studies that have examined these characteristics and traits in a variety of contexts (Ellis 1987; Rushton 1995, 2004). It is important to note that although population differences on these indicators are demonstrably there, they are not large, and there is considerable overlap across the races. We also need to realize that DKT when applied within-species is not an example of "genetic determinism," as Rushton (1994:42) made clear: "Although genes provide the initial set point, environmental factors move individuals up or down the continuum of reproductive strategies."

Because humans as a species are highly K-selected, subsequent work has tended to eliminate the r end of the continuum and concentrate on K traits that are differentially distributed on a continuum. Aurelio Figueredo and his colleagues have conducted numerous studies on their life-history theory in which DKT is central (reviewed in Figueredo et al. 2006). Using factor analysis, they consistently find that individual scores on K-selected traits form a strong principal component in their analysis that they call the K-Factor. Entering the K-Factor into regression models along with sociological variables such as class, they typically find that the K-factor accounts for around 90% of the reliable variance in a whole host of socially problematic behaviors, including crime. When discussing one of their studies involving MZ and DZ twins, Figueredo et al. (2006:256–257) concluded that "These results point to the existence of a single, highly heritable latent psychometric common factor (the K-Factor) that, as predicted by evolutionary ecological theory, underlies both the phenotypic and genetic covariances among a wide array of behavioral and cognitive life-history traits." Recall that life-history strategies are essentially reproductive strategies ranging from high levels of mating effort (low K) to high levels of parenting effort (high K). This research agenda underlines the usefulness of evolutionary thinking and is exciting stuff which, unfortunately, is taking place outside the confines of mainstream criminology.

BEHAVIOR GENETICS AND EVOLUTIONARY PSYCHOLOGY

As previously noted, evolutionary psychology and behavior genetics are animated by two separate concerns: the former by the central tendency of traits, and the latter by their variation. Evolutionary psychologists explore how behavior we now define as criminal may have been adaptive in EEAs. Because adaptations apply to all members of a species, some evolutionary psychologists may view substantial heritability of important traits linked to criminality as challenging the tenets of adaptationism (Daly 1996) since natural selection drives heritability of the most vitally important traits in a species to zero. Observations such as this lead some evolutionary psychologists to distance their discipline from behavior genetics. According to behavior geneticist David Rowe (in Horgan 1995), some evolutionary psychologists disavow behavior genetics because it makes their work both politically and scientifically easier. If all differences in human behavior can be ascribed to the environment, and all similarities to genes, then model construction is simplified and political correctness is maximized. These are very poor reasons for ignoring such a closely related discipline. Surely evolution is the domain of population genetics more so than of psychology, and certainly we cannot apportion behavioral differences and similarities into separate disciplinary specialties so simply.

What of the claim that substantial heritability of traits associated with criminality challenges the tenets of adaptationism? Most evolutionary psychologists appear to see no contradiction in viewing behaviors designed to solve adaptive problems as also being heritable. As we have seen, it is true that there can be no heritability for the presence or absence of morphological structures or species-defining characteristics, but in terms of cognitive traits, temperament, personality, and behavior, natural selection in fluctuating human environments would have almost certainly favored genetic variation over fixity in these characteristics (Bailey 1997; Segal and Macdonald 1998). Indeed, the Hardy–Weinberg Equilibrium (the foundational law of the genetic basis of evolution) informs us that there is always a reservoir of genetic variation (Beaver 2009). This reservoir is a necessary requirement for evolution to occur if future conditions dictate it. Thus I am in agreement with a number of scholars who have worked in both disciplines who maintain that evolutionary and behavior genetic explanations are complementary explanations that together provide a deeper understanding than either provides separately (Bailey 1997; Crawford 1998; Scarr 1995; Segal and McDonald 1998). After all, the much-heralded *Modern Synthesis* is the synthesis of Darwinian and Mendelian thought.

4   The Neurosciences and Criminality

Whether a behavior-motivating stimulus arises from within the person or from the environment, it is necessarily mediated by the brain, the most immensely complicated structure in the universe. "In the human head there are forces within forces within forces, as in no other cubic half-foot of the universe we know," wrote Nobel Prize–winning neuroscientist Roger Sperry (in Fincher 1982:23). While only about 2% of body mass, this blob of jelly consumes 20% of the body's energy as it perceives, evaluates, and responds to its environment (Shore 1997). Because the brain is where genetic dispositions and environmental experiences are integrated and become one, the basics of neuroscience must be part of the criminologist's repertoire of knowledge. Powerful brain-imaging technologies such as PET, MRI, and fMRI allowing for in vivo assessment of brain structure and function have resulted in an explosion of new information on the brain over the past two decades. Although we are a long way from fully understanding this "enchanted loom," we cannot ignore what is known, particularly aspects of it that are relevant to criminology. Robinson (2004:72) goes as far as to say that any theory of behavior "is logically incomplete if it does not discuss the role of the brain." The insights criminologists can derive from the neurosciences will not only buttress our theories, but may also strengthen our claims for preventative *environmental* intervention.

THE BASIC BRAIN

According to neurobiologist Bruce Perry (2002:81), there are three key brain systems relevant to survival and reproductive success. They are mechanisms that (1) facilitate responses to threats to our well-being, (2) facilitate mate selection and reproduction, and (3) facilitate the protection and nurturing of the young. This chapter focuses on mechanisms 1 and 3; number 2 is addressed in Chapter 7. Frequent references to brain anatomy in this and subsequent chapters make it necessary to briefly describe areas of the brain (in greatly simplified fashion) that are relevant to neurologically based theories of criminality. Discussion of the brain structures and functions mentioned here will be expanded on as we discuss particular theories.

There are a number of ways to divide the anatomy of the brain, but I will use the division favored by evolutionary biologists to reflect the brain's evolutionary stages and developmental sequence: reptilian (brain stem and its immediate projections), paleomammalian (limbic system), and neo-mammalian (cerebrum) (MacLean 1990).

The brain is part of the nervous system, which is divided into the central nervous system (CNS), composed of the brain and the spinal cord, and the peripheral nervous system (PNS), which carries information to and from the CNS via the various sensory organs. The most primitive part of the brain begins at the top of the spinal cord and includes the brainstem and the reticular activating system (RAS). The RAS is a sort of information-filtering system that broadly determines consciousness, arousal, and alertness, and is a little finger-sized bundle of neurons located at the core of the brain stem between the medulla oblongata and the midbrain and feeds arousal stimuli into the thalamus for distribution throughout the brain (see Figure 4.1).

Wrapped around the reptilian system is a set of structures collectively known as the limbic system. Among the many structures of the limbic system are the amygdala, hippocampus, and anterior cingulate gyrus (ACG). Each of these structures is distributed bilaterally (one on each side of the brain). The amygdala's primary function is the storage of memories associated with the full range of emotions, particularly fear. After the amygdala processes an emotion it directs other brain structures to initiate responses geared to specific situations. The hippocampus is specialized for storing and processing visual and spatial memories such as facts and events. Connections between the amygdala and hippocampus help to focus the brain

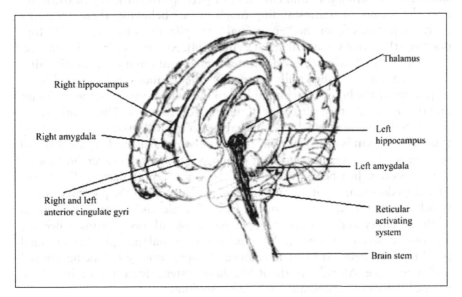

Figure 4.1 Schematic image of the brain showing various parts of interest.

on what the organism has learned about responding to the kind of emotional stimuli that has aroused it, thus regulating it with memories of prior experiences ("Do I run, fight, scream, talk, ignore?").

The curves of the cingulate gyri surround the wishbone-like structure of the limbic system and provide a connection between the limbic system and the cerebral cortex. The ACG has a number of functions, including mediating between possible responses "suggested" by the hippocampus and the amygdala. It plays a role in self-control by aiding the hippocampus to rein in negative emotions; e.g., people prone to aggression and violence show attenuated activation of the ACG (Davidson, Putnam, and Larson 2000). According to Laurence Tancredi (2005:36), the ACG "provides for civilized discourse, conflict resolution, and fundamental human socialization." We are most concerned with its role in emotional learning, forming attachments, and in regulating endocrine functions.

The limbic system evolved in conjunction with the evolutionary switch from a reptilian to a mammalian lifestyle, which includes the addition of nursing and parental care, audiovocal communication, and play, and it is especially concerned with emotion and motivation (Buck 1999). Because emotion is not "rational," it was once thought that limbic system activity was a primitive evolutionary "throwback" that was opposed to culture and required cerebral inhibition (McDermott 2004). The evidence today now points overwhelmingly to the position that the emotions perform many functions vital to social and cultural evolution (Nichols 2002; Phelps 2006). Emotions require rational guidance (not inhibition), just as cognitions require emotional guidance. If we ever doubted the intimate relationship between rationality and emotion, or perhaps the primacy of the latter over the former in understanding much of our behavior, there are many more projections from the "emotional" amygdalae to the "rational" hippocampi than vice versa (Richter-Levin 2004). As a basis for social interaction, emotions clearly preceded rationality in evolutionary time; rationality is a later addition to the vitally important role of emotions in social life.

Just above the brain stem and linked to the limbic system lie the thalamus and the hypothalamus, the oldest of the limbic structures. The thalamus is a dual-lobed structure that receives signals from the RAS and serves as a relay station for all kinds of sensory information except smell, and organizing and sending the incoming messages to the appropriate areas of the cerebral cortex for processing. Just beneath the thalamus is the hypothalamus. The hypothalamus does many things, among which is controlling the pituitary gland, which affects the functioning of the endocrine glands. Through its control of the pituitary and two other sets of physiological mechanisms called the autonomic nervous system (ANS) and the hypothalamic-pituitary-adrenal (HPA) axis, it helps to regulate emotional expression by balancing arousal and quiescence. Along with the RAS, these systems feature prominently in arousal theories of criminality to be discussed later.

The most recent evolutionary addition to the brain is the cerebrum, which forms the bulk of the human brain. The cerebrum is divided into two complementary hemispheres, each with their own specialized functions, and connected at the bottom by the corpus callosum. It is generally accepted that the right hemisphere is specialized for perception and the expression of emotion (particularly negative emotions), and that the left hemisphere is specialized for language and analytical thinking (Parsons and Osherson 2001). The outer layer of the cerebrum is the cerebral cortex, the "thinking" brain that organizes and analyzes information from other structures and formulates and relays back to them the appropriate responses.

From the perspective of the behavioral scientist, the most important part of the cerebral cortex is the prefrontal cortex (PFC), "the most uniquely human of all brain structures" (Goldberg 2001:2). The PFC occupies approximately one-third of the human cerebral cortex, a proportion greater by far than in any other species, and is the last brain area to fully mature (Romine and Reynolds 2005). This vital part of the human cortex has extensive connections with other cortical regions, as well as with deeper structures in the limbic system. Because of its many connections with other brain structures, and because it is involved in so many neuropsychological disorders, it is generally considered to play the major integrative, as well as a major supervisory, role in the brain. The PFC is also vital to the forming of moral judgments, mediating affect, and for social cognition (Romine and Reynolds 2005; Sowell et al. 2004). Given these functions, it is no surprise that biosocial criminologists have given a lot of attention to the PFC.

Making up all these anatomical structures are hundreds of billions of nerve cells called neurons. All our thoughts, feelings, emotions, and behavior are the result of communication networks composed of these neurons. Other more numerous brain cells called glial cells nourish and support the neurons. Each neuron consists of the cell body (soma), an axon, and a number of dendrites. The cell body contains the nucleus that carries out the metabolic functions of the neuron. The axon originates in the cell body, and the dendrites are branch-like extensions of the cell body. Axons serve as transmitters sending signals to other neurons, and dendrites serve as receivers picking up information from neighboring neurons. Each of the over 100 billion neurons "form over 100 trillion connections with each other—more than all of the Internet connections in the world!" (Weinberger, Elvevag, and Giedd 2005:5). Many lower level brain structures such as the brain stem come with neural connections complete at birth (in the parlance of neuroscience, they are "hardwired"), but development of the higher brain areas in the cerebrum, and to a lesser extent in the limbic system, depends on making connections between neural networks after birth.

These communicating structures are not physically connected. The receiving and sending of messages takes place in microscopic fluid-filled gaps between axons and dendrites called synapses. Information from the cell passes along the axon electrically until it reaches the presynaptic knob, at which time it is translated into chemistry as tiny vesicles burst open and spill out one or more of a variety of neurotransmitters. Neurotransmitters cross the synaptic gap to make contact with postsynaptic receptor sites, where the message is translated back into an electrical one either for further transportation or inhibition of the message. Once they have passed on their messages, excess amounts are pumped back up into the presynaptic knob or degraded by enzymes. Neurotransmitters that have been linked to the probability of antisocial behavior include norepinephrine, serotonin, and dopamine, as well as enzymes such as monoamine oxydase and catechol-O-methyltransferase, which degrade them after they have performed their tasks. The anatomical and chemical systems and circuits described earlier are fully interactive entities, each affected by and affecting the other to various extents. Figure 4.2 is a schematic illustration of synaptic transmission.

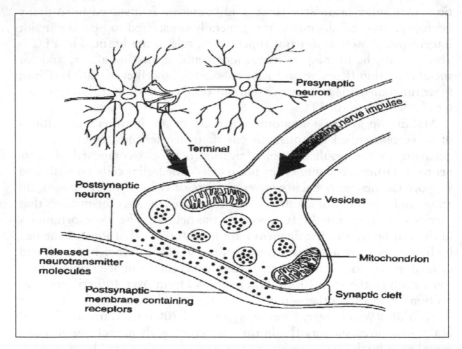

Figure 4.2   The process of synaptic transmission. Sketch of two neurons at the top with an enlargement at the bottom showing the release of an unspecified transmitter into the synaptic cleft. Source: Ellis & Walsh, 2000, p. 288. Reprinted with permission of Ellis and Walsh.

NEURAL SELECTIONISM AND CONSTRUCTIVISM

There are nature/nurture arguments in the neurosciences surrounding the issue of brain development (albeit, not about naive dichotomies). The debates are about contributions to brain development of processes intrinsic to the brain (in the genome) relative to extrinsic (environmental) contributions. Neuroscientists who come down most strongly in favor of innateness are generally called *selectionists* or *nativists*, and those more in favor of the power of extrinsic factors to mold the brain are generally called *constructivists* or *connectionists*. Both positions agree that the environment is hugely important in the development of the brain; the argument is not "*whether* the environment thoroughly influences brain development, but *how* it does" (Quartz and Sejnowski 1997:579; emphasis in original).

Selectionism

Selectionists are strongly influenced by evolutionary theory and by artificial intelligence theory. Their argument boils down to the assertion that the brain cannot have evolved as a general all-purpose computing machine because there was never any general problem in environments of evolutionary adaptation that our ancestors had to solve. Selectionists maintain that the human brain was cobbled together piecemeal over evolutionary time in response to a series of problems (all subproblems of the one central problem all living things share—how to survive and propagate their genes) faced by our ancestors (Cummins, Cummins, and Poirier 2003; Geary 2005). The neurophysiology underlying each brain function is thus seen as an adaptation designed to solve some *specific* problem. Nativists point out that damage to circuits serving specific brain functions produce specific impairments, not an overall impairment (Dellarosa-Cummins and Cummins 1999; Gazzaniga 1998). A popular metaphor for this discrete yet functionally integrated collection of specialized modules (sets of neural networks) is that of a Swiss army knife (Cosmides and Tooby 1992). For the evolutionary neuroscientist, the mind is a series of specialized modules with a basic logic specified by DNA.

This does not mean that selectionists are "genetic determinists" claiming that our minds are preprogrammed to respond stereotypically to everything they will encounter, that there is a gene for every trait, or that learning is not important. We have already seen that there are nowhere near enough genes to specify how one should respond to the almost limitless potential situations humans may have to respond to, and selectionists are well aware of that fact. But how can we reconcile the assertion that the mind has a genetically programmed logic with the claim that the environment thoroughly influences brain development? To answer this question we have to turn to the so-called *frame problem* of artificial intelligence theory.

The human mind is not a blank slate that must learn everything through experience; it is fertile with specific built-in assumptions (call them by the old-fashioned term *instincts* if you like) about the nature of the species-relevant environments that it will encounter. We selectively attend to some kinds of information more readily than to others because we possess emotion-driven motivational systems that move us to do so. This was not fully appreciated until researchers in robotics and artificial intelligence tried to program "common sense" into their computers, "the ultimate tabula rasa" (Pinker 1997:15). Without highly detailed and specific "if/then" instructions, robots tend to respond to novel encounters with the environment with bizarre and often self-destructive responses, such as trying to push their hands through their own bodies (Spelke 1999). The frame problem for artificial intelligence researchers is that devising a decision-making system that can reorganize itself automatically in the face of complex novel systems has so far has presented a seemingly impenetrable barrier for them (Hendricks 2006). Natural selection, however, solved the frame problem for living things eons ago by providing them with brains created piece by piece over evolutionary time as winners emerged in the relentless competition for survival and reproductive success.

There are an enormous number of ways that an organism can respond to environmental stimuli, some of which are advantageous and some of which are highly disadvantageous. The problem is recognizing the stimuli that are relevant to survival and taking appropriate action. Think of how members of different species might respond to insects, heights, snakes, water, cats, temperature, dung, dead carcasses, open/closed spaces, or any number of other environmental features. Members of each species respond to environmental stimuli in species-specific ways with built-in assumptions (neural algorithms) that "frame" the response problem for them, thus helping them to attend to tasks of adaptive importance whether they are common or novel. Responses vital to survival in any species' typical environment cannot be left to the vagaries of experiential learning. Responses are certainly strengthened and fine-tuned with experience, but selectionists aver that the general response pattern was framed by natural selection. As Elizabeth Spelke (1999:1) put it: "The frame problem precludes the possibility of bootstrapping oneself up from a position of no assumptions." Unlike robots, animals have epigenetic rules exquisitely crafted by natural selection to perceive and respond to the environment by sorting stimuli into positive and negative categories according to their potential for harming or assisting them in their survival and reproductive goals. We categorize so effortlessly and automatically that the response process is referred to in the vernacular as "common sense." It was only after artificial intelligence types tried to incorporate such common sense into their contraptions that it became obvious how truly remarkable common sense really is.

CONSTRUCTIVISM

Constructivists reject both the image of the brain as a general-purpose computational machine containing only generic learning procedures and the extreme view of domain-specific modules embedded in the brain carrying a priori predispositions. The essence of the constructivist position is that evolution has shaped our minds by providing us with a highly plastic (malleable) brain able to construct itself from the experience rather than providing us with innate modules finely tuned for specialized tasks. Constructivists also claim that the nativist model assumes that we operate with brains forged in the Pleistocene and have not evolved at all in the Holocene, yet there is evidence that the brain continues to evolve as new genetic variations affect its structure and function (Evans et al. 2005; Mekel-Bobrov et al. 2005). They also point to studies showing that the brain can recruit supposedly specialized areas to perform tasks usually assigned to other areas when those areas have been damaged. The most dramatic example of this is the ability of one brain hemisphere to assume the functions of the other in children who have had one hemisphere removed if surgery is performed during the period of the brain's maximal plasticity (Elman et. al. 1996). However, constructivism is not a return to a tabula rasa view of the mind or a total denial of nativist principles. Indeed, major proponents of this view explicitly state that "assumption-free learning is impossible" (Quartz and Sejnowski 1997:584).

The argument between these two schools of thought with respect to this issue is thus one of degree only, and it is not inconceivable that natural selection has provided both selectionist and constructivist solutions to adaptive problems (Collier 2005; Hurford et al. 1997). It is certainly not necessary for criminologists to fully understand and appreciate the arcane subtleties that separate these two positions to understand and appreciate the message that brings them together. The message is that genes have surrendered much of their control over human behavior to a more plastic, complex, and *adaptive* system of control called the human brain, and that the development of the brain depends greatly on environmental input. Put otherwise, we and the experiences we encounter will largely determine the patterns of our neuronal connections, and thus our ability to successfully navigate our lives. Genes play a crucial part in specifying the trillions of connections that exist among the brain cells of human beings and the environment provides the other equally crucial part (Mitchell 2007; Penn 2001).

THE BRAIN AND ITS ENVIRONMENT

About 50 to 60% of all human genes are involved in brain development (Mitchell 2007; Shore 1997). The genes specify basic brain architecture, build the cells, synthesize the various neurotransmitters and enzymes, and

provide the framing assumptions. The process of neuron birth, migration, and differentiation are also genetically determined, but they can be
adversely affected by environmental events such as maternal substance
abuse, malnutrition, or exposure to noxious substances. While genes carry
an immense amount of information, they are too few in number (only 20 to
30 thousand, according to the International Genome Sequencing Consortium [2001]) relative to the billions of neurons and the trillions of connections they will eventually make with one another.

If genes alone had been assigned the task of specifying neuronal connections, we would all be hardwired drones incapable of adapting to novel
situations, just like the robots who tried to push their hands through their
bodies. The more varied and complex the environment, the more organisms
inhabiting them must change and adapt; human environments are much too
varied and much too complex for hardwired brains. Furthermore, because
genes are subject to epigenetic modulation, it is possible that the modularity
of the brain is not entirely a product of natural selection during the Pleistocene. These specialized modules may have emerged epigenetically through
learning and cultural experience (Panksepp et al. 2002).

Neuroscientists distinguish between two brain developmental processes
that *physically* capture environmental events in the organism's lifetime in the
same way that genes captured environmental events in the genome during the
life span of the species. These two processes are called *experience-expected*
and *experience-dependent* (Black and Greenough 1997; Edelman 1992;
Schon and Silven 2007). Experience-expected mechanisms are hardwired
and reflect the phylogenic history of the brain; experience-dependent mechanisms reflect the brain's ontogenic plasticity. Otherwise stated, every member of a species inherits species-typical brain structures and functions that
are produced by a common pool of genetic material, but individuals will
vary in brain functioning as their genes interact with the environments they
encounter to construct those brains (Depue and Collins 1999; Gunnar and
Quevedo 2007). An example that best reflects the distinction between the
two processes is language. The *capacity* for language is entirely genetic (a
hardwired experience-expected capacity), but what language(s) a person
speaks is entirely cultural (soft-wired in experience-dependent fashion).

Language acquisition also helps us to further understand the nativist/
constructivist debate. All humans have a hardwired "language acquisition
device" that is domain-specific—it does not participate in any other task
such as helping us to learn calculus. The possession of such a device is why
we learn our native tongue with almost no conscious effort. We do not have
a domain-specific module devoted to acquiring calculus skills because the
kinds of things for which we find calculus useful today were irrelevant to
the fitness concerns of our hominid ancestors. To learn calculus we have to
recruit something akin to a "general purpose learning device" that involves
numerous brain areas not domain-specific (and perhaps even some that
are), which is why learning calculus requires considerable effort.

Experience-expected processes have evolved as neural readiness during certain "critical" or "sensitive" developmental periods to incorporate environmental information that is vital to an organism and ubiquitous in its environment. That is, natural selection has recognized that certain processes such as sight, speech, depth perception, affectionate bonds, aversion to insects and waste products, mobility, and sexual maturation are vital, and has provided for mechanisms (adaptations) designed to take advantage of experiences occurring naturally within the normal range of human environments. Pre-experiential rudimentary brain organization (built-in assumptions) frames or orients our experiences so that we will respond consistently and stereotypically to vital stimuli (Geary 2005; Black and Greenough 1997). Natural selection has removed heritable variation for these processes (their alternatives failed evolutionary selection tests long ago), making them stable across all members of a species.

Experience-dependent brain development depends on experience acquired in the organism's developmental environment. Much of the variability in the wiring patterns of the brains of different individuals depends on the kinds of physical, social, and cultural environments they will encounter. It is not an exaggeration to say that "experience-dependent processes are central to understanding personality as a dynamic developmental construct that involves the collaboration of genetic and environmental influences across the lifespan" (Depue and Collins 1999:507). These processes best reflect the constructivist argument that the brain literally wires itself (perhaps via epigenetic mechanisms) in ways directly reflecting its experience. Although brain plasticity is greatest in infancy and early childhood, a certain degree is maintained across the life span so that every time we experience or learn something we shape and reshape the nervous system in ways that could never have been preprogrammed.

Synaptogenesis

The process of wiring the brain is known as *synaptogenesis*, a process that occurs both according to a genetic program and the influence of the environment. The neonate's cerebral cortex consists of small and underdeveloped neurons with few dendrites. During the first few months, dendrites proliferate and specialized glial cells wrap around axons to begin the process of myelination. Myelin is a fatty substance that insulates the axons and makes for speedier transmission of electrical impulses—nerve conduction velocity increases from about 6.5 feet per second to about 164 feet per second when an axon becomes fully myelinated (Casear 1993). The process of dendritic growth and axonal myelination continues to some degree throughout life, but proceeds at an explosive rate during infancy and toddlerhood. Since they are the most vital brain regions in terms of sheer survival, the experience-expected "lower" brain regions (reptilian and limbic regions) are the first to be myelinated, and some "higher" brain areas (especially the PFC) are not fully myelinated until well into adult-hood (Paus et al. 1999; Sowell et al. 2004). This observation has impor-tant consequences for criminological theory.

The most important issue relating to synaptogenesis is not the birth of a set of synapses composing a neuronal module, but whether it will survive the competition for synaptic space in the brain. The most active period of synaptogenesis is infancy and early childhood, with the number and density of synaptic connections being higher than they will ever be (Rakic 1996). About half of these connections will eventually be eliminated. Although the brain creates and eliminates synapses throughout life, creation exceeds elimination in the first two or three years. Production and elimination are roughly balanced thereafter up until adolescence, after which elimination exceeds production (Giedd 2004; Sowell et al. 2004).

This process of selective production and elimination has been termed *neural Darwinism* by Nobel Prizewinner Gerald Edelman (1987, 1992). Edelman posits a selection process among competing brain modules (popu-lations of dendrites and synapses). In order for neuronal natural selection to take place, there must be an excess of synapses available, just as natural selection requires an excess of genetic variation. Neuronal connections are selected for retention or elimination according to how functionally viable (adaptive) they prove to be in the organism's environment. Just as environ-mental challenges select from a population's repertoire of genetic variation those that help it and its progeny to meet those challenges, the brain's rep-ertoire of excess connections are selected for retention or pruning accord-ing to the input pattern provided by the environment. The brain's neuronal populations thus evolve in somatic time very much like species evolve in geological time by selective elimination and retention.

Retention of neural networks is very much a use-dependent process (Penn 2001). Each neuron sends out its antennae searching for potential connections. Many connections are made in this promiscuous environment, but most turn out to be "one-night stands." The connections that are sustained are those that exchange information frequently and strongly. In the competition for scarce synaptic space, the game "is biased in favor of the [neuron] populations that receive the greatest amount of stimulation during early development" (Levine 1993:52). Experiences with strong emotional content are accompanied by especially strong impulses, and if these impulses are frequently made, the neu-rons involved become more sensitive and responsive to similar stimuli in the future (Shi et al. 2004). Frequently activated neurons are thus primed to fire at lower stimulus thresholds once voltage-dependent neurological tracks have been laid down. This process is encapsulated in the neuroscientist's version of the evolutionist's "survival of the fittest": "The neurons that fire together, wire together; those that don't, won't" (Penn 2001:339).

Experience-dependent neuronal organization has been experimentally tested in the lab. It has long been known that animals raised in stimulus-enriched environments have significantly greater neurological development than animals reared in stimulus-deprived environments (Lu 2003). Cloned (thus genetically identical) animals subjected to different environmental experiences show more between-pair variability in dendritic density and branching patterns (arborization) than between the left and right sides of the brain of the same individual animal (Molenaar, Boomsma, and Dolan 1993). If genes alone were responsible for synaptogenesis and dendritic arborization, then branching patterns would be identical in cloned animals regardless of the environments to which they had been exposed.

It is safe to say that the evidence is unequivocal on two major points about brain development: (1) the brain is always a "work in progress," and (2) development is "use-dependent." However, although synaptic selection does depend on the strength and frequency of activation, all cortical modules are not *equal* candidates for selection. Recalling the studies from Chapter 2 showing that monozygotic twins reared apart assess their affectual experiences with their adoptive parents significantly more similarly than do dizygotic twins reared together, the concept of gene/environment correlation assures us that some modules have a greater probability of selection than others via channeling affects (Mitchell 2007). Temperamentally pleasant infants evoke responses (cuddling, playing, kissing) from others that will lay down neuronal pathways to the infants' pleasure centers. Temperamentally unpleasant infants are at risk for evoking responses (infrequent holding, neglect, and possibly abuse) that may lay down pathways leading to displeasure centers. Because the temperaments of parents and their biological children are typically positively correlated, the probability of pleasant stimuli evoking pleasant responses and unpleasant stimuli evoking unpleasant responses is increased. Brains are wired in a probabilistic epigenetic way, and this pattern of development has been invoked as an explanation for why idiosyncratic environmental events (nonshared environment) are consistently found to be more powerful in accounting for variation in most human traits than shared environment (Molenaar, Boomsma, and Dolan 1993; Smith 1993).

WHY BONDING AND ATTACHMENT ARE NEUROLOGICALLY IMPORTANT

It has been pointed out that "the affection dimension of child rearing appears to pull in more correlates with child behavior than any other dimension" (Rowe 1992:402). Humans possess powerful neuroendocrinal structures that demand the formation of affectual bonds: "Human newborns, and altricial animals in general, are adapted to receive support for physiological homeostasis and motivation from the contact stimuli of an affectionate caretaker in a protected environment" (Trevarthen 1992:225). These bonds are experience-expected and in their absence the relevant synaptic connections are pruned (Glaser 2000). The literature lists many negative outcomes associated with the failure to form these bonds, including the relative inability to form lasting adult bonds (Lee and Hoaken 2007; Zeifman and Hazan 1997). What evolutionary advantage does the acquisition of affectual bonds confer, and what might be the neuropsychological consequences relevant to criminal behavior of not forming them?

We might begin by noting that hominids experienced a rapid rate of selection for intelligence after separating from the apes. Brains are particularly voracious in their appetites for energy; thus selection for increased brain size would only result from extreme pressures, and the human brain is much larger than should be reasonably predicted for a species of our body size. On the basis of a large number of interspecies studies showing that group size is linked to various measures of brain size, a number of evolutionary biologists and neuroscientists propose that the intellectual demands of living in large complex groups drove the selection for what they call the "social brain" (Dunbar and Shultz 2007; Lindenfors 2005; Zelazniewicz 2007). This social brain was needed to negotiate relationships, to understand what others are thinking and feeling, to cooperate to secure resources and for mutual defense.

From the approximate 1.5 million years that separated *Australopithecus afarenis* and *Homo erectus,* hominid cranial capacity doubled from a mean of 450 cc to a mean of 900 cc, and by another 70% to about 1350 cc, from *Homo erectus* to modern *Homo sapiens* (Bromage 1987). The selection for intelligence, and the cranial capacity to store it, placed tremendous reproductive burdens on females. The human female birth canal could not accommodate the birthing of an infant whose brain was 60% of its adult weight, as it is in newborn macaques, or even 45%, as in newborn chimpanzees (Hublin and Coqueugnoit 2006). The pelvis of *Australopithecine* females was probably shaped by natural selection to satisfy upright posture and bipedalism (which has the effect of narrowing the birth canal) more than for increased fetal brain size, thus precipitating a conflict between our ancestral females' obstetric and postural requirements (Buck 1999; van As, Fieggen, and Tobias 2007). Evolutionary conflicts such as this are not uncommon; natural selection works on trajectories already in motion, and it cannot anticipate future needs. Selection for larger pelvises may have been one strategy tried, but pelvises large enough to allow the passage of human infants as developmentally advanced as other primate infants may have severely hindered locomotion and placed both mother and infant at the mercy of predators.

The evolutionary mechanism that partially solved the obstetrics/posture conflict (human females still have more difficulty giving birth than other species because of this problem) was for human infants to be born at earlier and earlier stages of development as cerebral mass increased. Human infants experience 25% brain growth inside the womb, and 75% growth outside the womb (Perry 2002). Neuroscientists call these two growth periods *uterogestation* (gestation in the womb) and *exterogestation* (gestation outside the womb). Such a high degree of developmental incompleteness of the human brain assures a greater role for the extrauterine environment in its development than is true of any other species.

If a species is burdened with extremely altricial young, then there must have been strong selection pressures for neurohormonal mechanisms designed to assure the young would be nurtured for as long as is necessary. These mechanisms assure that there will be a "continuous symbiotic relationship between mother and child [across the two gestation periods]" (Montagu 1981:93). Ashley Montagu calls these mechanisms by their popular name—*love*. He goes on to add: "It is, in a very real and not in the least paradoxical sense, even more necessary to love than it is to live, for without love there can be no healthy growth or development, no real life. The neotenous principle for human beings—indeed, the evolutionary imperative—*is to live as if to live and love were one*" (1981:93, emphasis in original). Modern neurobiology and clinical medicine provide abundant evidence that love confers enormous benefits in terms of healthy functioning in human beings through its impact on central and peripheral nervous system functioning (Esch and Stefano 2005).

The concept of neoteny further helps us to understand why *Homo sapiens* is a species exquisitely adapted to fine tune itself in response to problems and opportunities present in its environment. *Neoteny* literally means *holding youth*, and conceptually it means "the retention of embryonic or juvenile characteristics by retardation of development" (Bjorklund 1997:155). Human beings are highly neotenous, losing far fewer embryonic and juvenile features (both morphological and behavioral) than other animals, including, as Montagu suggests, the need for love. Neoteny is an adaptation resulting from perhaps only a few minor mutations at genetic loci involved with developmental timing, and may have been all that it took for our ancestors to branch off the evolutionary line we shared with the apes (Feder and Park 1989). The analysis of hominid skeletons from *Australopithecines* to modern humans has led many evolutionary scientists to conclude that neoteny has been a major, perhaps *the* major, determinant of human evolutionary direction (Bromhall 2003; Groves 1989).

Neoteny has many important evolutionary advantages. Because human development is so delayed, we are given more time to become adaptively affected by environments in flux to a far greater degree than any other animal. Developmental delay means that unlike, say, snakes or alligators, we don't need hardwired brains to make us respond appropriately (instinctively) to a limited set of environmental exigencies. What is "appropriate" in human environments varies so much that we have to learn what it is, and that takes time (this in no way gainsays what was said earlier about built-in assumptions that bias learning in certain directions).

While our delayed development provides us with many windows of opportunity, the negative is that it also provides for windows of vulnerability.

There are two major disadvantages from the criminologist's point of view. The first is that the most neotenous feature of *Homo sapiens* is the associative areas of the prefrontal cortex responsible for emotional evaluation, and regulation is the last part of the brain to fully develop (see Chapter 6). The second disadvantage is that we do indeed need Montagu's "continuing symbiotic relationship" in order to take full advantage of the opportunities that neoteny affords us, but some of us fail to get it. Those who fail to form this relationship often fall prey to many developmental and behavioral problems, including criminal behavior. The relationship between criminality and childhood abuse and neglect is one of the best documented in the literature (Lee and Hoaken 2007; Perry 2002). The criminological literature rarely addresses how maltreatment influences brain development; we thus have an inadequate understanding of how deleterious childhood deprivation really is.

THE NEUROBIOLOGY OF CHILD MALTREATMENT, STRESS, AND BRAIN DEVELOPMENT

The literature on the neurobiology of childhood maltreatment is huge and complex. For example, Pollak (2005:741) writes that its effects differ widely according to "intensity/severity, chronicity/duration, and developmental timing of experience may be critically related to the different forms of psychopathology." He also added differences in type of maltreatment such as physical versus emotional or sexual abuse, and abuse versus neglect, as well as gender. Abuse (an event) influences brain development at experience-dependent stages while neglect (an ongoing caregiver/child relationship) deprives young children of affectionate input in times of experience-expected stages of maturation (Glaser 2000). We cannot parse the various types and effects of maltreatment according to which of these components children are exposed to in this limited chapter, but suffice to say that all types of maltreatment are stressful.

Stress is a state of psychophysiological arousal in response to perceived challenges to an organism's physical, emotional, or mental well-being. It is a normal adaptive part of life without which we would be seriously handicapped in our ability to meet these challenges. Individuals who experience average levels of stress during childhood most likely possess brains so calibrated as to better navigate the travails of life as adults than those who have been assiduously protected from almost all stress (Meaney 2001). It is protracted and toxic stress which does the damage to vital memory storage/behavioral regulatory regions such as the amygdala and hippocampus. Efficient stress management is the goal, but people differ in their ability to manage stress partly for genetic reasons and partly for experiential reasons that have been captured in the brain's circuitry. Child maltreatment is a major stressor experienced at a stage of life when the brain is most plastic, and frequent activation of stress response mechanisms may lead to their dysregulation, leading in turn to a number of psychological, emotional, and behavioral problems (Gunnar and Quevedo 2007; Lee and Haken 2007; Perry 2002).

Response to stress is mediated by two separate but interrelated systems each controlled by the hypothalamus: the autonomic nervous system (ANS) and the hypothalamic-pituitary-adrenal (HPA) axis. The ANS has two complementary branches, the sympathetic and parasympathetic systems. When an organism perceives a threat to its well-being the hypothalamus directs the sympathetic system to mobilize the body for vigorous action. Pupils dilate, the heart and lungs accelerate their activity, and digestion stops, among other things, all of which is aided by pumping out the hormone epinephrine (adrenaline). This is an almost instantaneous response. The parasympathetic system restores the body to homeostasis (the return of physiological functions to their "set points" of acceptable range values) after the organism perceives the threat is over. The ANS will be discussed again later in the chapter.

The HPA axis response is much slower than the ANS response and lasts longer because its impact on the brain occurs through changes in gene expression (Gunnar and Quevedo 2007). It is activated in situations that call for a prolonged series of thoughtful responses rather than the visceral immediacy of the fight-or-flight response activated by the ANS. As the name implies, the HPA axis response begins with the hypothalamus feeding various chemical messages to the pituitary gland (master of the endocrine system), which leads to further chemicals that stimulate the adrenal glands to release the hormone cortisol. Unlike epinephrine, which does not cross the blood/brain barrier, the brain is a major target for cortisol (van Voorhees and Scarpa 2004).

Frequent and chronic HPA axis arousal may lead to hypercortisolism (overproduction of cortisol) or hypocortisolism (underproduction). *Hyper*cortisolism leads to anxiety and depressive disorders and is most likely to be found in maltreated females (van Voorhees and Scarpa 2004). Hypercortisolism suggests a failure of the system to adjust to chronic environmental stressors. *Hypo*cortisolism, on the other hand, suggests an adaptive adjustment to chronic aversive circumstances; i.e., frequent stressful encounters habituate the organism such that it does not perceive further encounters as quite so stressful; thus both the HPA axis and ANS responses become blunted. Hypocortisolism has been linked to early onset of aggressive antisocial behavior (McBurnett et al. 2000), to criminal behavior in general (Ellis 2005), and is more likely to be found in maltreated males (van Goozen et al. 2007). Consistent with previous studies, O'Leary, Loney, and Eckel's (2007) study of cortisol and psychopathic personality traits found that males high in psychopathic traits lacked stress-induced increases in cortisol displayed by males low in psychopathic traits. Blunted arousal means a low level of anxiety and fear, something which is very useful for those committing or contemplating committing a crime.

The development of hyper- or hypocortisolism is one example of the process of *allostasis*. Unlike homeostasis, allostasis describes the body's attainment of equilibrium by *altering* the acceptable range of physiological set points (rather than returning them to their previous state) to adapt to extreme acute stress or chronic stress. Thus allostasis literally means to achieve physiological stability through change According to Goldstein and Kopin (2007:111): "Adaptations involving allostasis to cope with real, simulated, or imagined challenges are determined by genetic, developmental and previous experiential factors. While they may be effective for a short interval, over time the alterations may have cumulative adverse effects."

Frequent or prolonged events leading to allostatic responses are termed *allostatic load*. Goldstein and Kopin (2007) liken this to having heating and cooling systems running simultaneously in the same apartment, a situation guaranteed to waste energy and to contribute to the wear-and-tear of both systems. In addition to effecting neurohormonal functioning, allostatic load compromises the cardiovascular and immune systems (Bremner 2000), which may well explain a large chunk of the variance in health outcomes between people raised in middle-class environments and those raised in environments where violence is an everyday occurrence.

Our understanding of the effects of abuse and neglect on the developing human brain has relied heavily on animal models, including Gary Kraemer's (1992) seminal laboratory work with rhesus monkeys. There are those who balk at extrapolating findings derived from monkeys to human beings, but as mentioned in Chapter 2, nature is parsimonious in its preservation of mechanisms that work so that it does not have to start all over again as species branch off from the parental line. Mechanisms might need additional tuning as they cross species lines, but evolution can only work with what is in the gene pool. Kraemer's work deals with how affective and attachment failure influences neurohormonal mechanisms, and how these mechanisms, in turn, affect behavior in one primate species. Such failures can be expected to influence humans even more so because human primates possess brains that are more plastic than monkey brains, "even more primed to respond to the environment" (Shore 1997:25).

Neural pathways forged from the early experiences among experimentally deprived monkeys have been identified via aberrant electrophysiological activity for decades. It has also been known for some time that chronic stress can produce neuron death via the frequent production of cortisol in many species, including humans (Perry 2002; Teicher et al. 1997), and that children with chronic high levels of cortisol experience more cognitive, motor, and social development delays than other children (Gunnar and Quevedo 2007; Perry 2002). Kraemer's work covers decades, and was built on the earlier work done at the Harlow Primate Laboratory at Madison, Wisconsin. Harry Harlow's work (1962) focused on the behavioral aberrations of monkeys raised in isolation. Advances in theory and technology since Harlow's time enabled Kraemer to explore the affects of such rearing (non-attachment) in terms of brain structure and functioning. Among his many findings were the following:

- Reductions in cortical and cerebellar dendritic branching.
- Altered electrophysiology in the cerebellar and limbic regions.
- Dysregulation of biogenic amine systems, particularly a reduction in norepinephrine concentrations in cerebrospinal fluid.
- A neurobiological and behavioral "supersensitivity" when exposed to pharmacological agents or novel stimuli that act on biogenic amine systems.
- Changes in brain cytoarchitecture (the cellular composition of a structure).
- The failure to organize emotional behavior in response to stressors.

What all this boils down to is that the lack of attachment experiences during early development, even in the absence of active maltreatment, may cause permanently altered neurophysiological systems that will adversely affect the organism's ability to interact with its world adaptively. We should not be too alarmist about slightly negative or dysfunctional human environments, however. Studies such as Kraemer's elucidate mechanisms by which normal development may be diverted, but they also show that for diversion to occur among genetically normal animals the deprivation must be quite severe. Among genetically "normal" individuals, adaptive behavior will develop within a wide range of "average expectable environments" (Scarr 1992, 1993). Humans are fairly resilient creatures, and neoteny affords children reared in negative or dysfunctional environments the opportunity to learn positive ways of responding to the world (to rewire their brains) in spite of early experiences (Bjorklund 1997).

This observation should not divert our attention away from rearing environments that are truly horrendous and do leave permanent neuropsychological scars. As Perry and Pollard (1998:36) point out: "Experience in adults *alters* the *organized* brain, but in infants and children it *organizes* the developing brain" (emphasis added). In brains organized by stressful and traumatic events, future events, even neutral or positive ones, will tend to be relayed along the same neural pathways etched out by those events and be interpreted in negative or even hostile ways. For better or for worse, well-grooved synaptic pathways established in early life are more resistant to pruning than pathways laid down later in life. These pathways have been stabilized, and thus they subconsciously intrude into our transactions with others across the life span.

REWARD DOMINANCE THEORY

Reward dominance theory is a neurological theory based on the proposition that behavior is regulated by two opposing but complementary mechanisms, the *behavioral activating system* (BAS) and the *behavioral inhibition system* (BIS). The BAS is sensitive to signals of reward and the BIS is sensitive to threats of punishment (Kruesi et al. 1994). The BAS can be likened to an accelerator motivating a person to seek rewarding stimuli, and the BIS can be likened to a brake, which, in response to punishment cues, inhibits a person from going too far in that pursuit. The BAS motivates us to seek food, drink, physical, emotional, and sexual pleasures, and the BIS tells us when we have had enough for our own good. A normal BAS combined with a faulty BIS, or vice versa, may lead to a "craving brain" that can get a person into all sorts of physical, social, moral, and legal difficulties, such as addiction to gambling, food, sex alcohol, and drugs (Ruden 1997). Because most of these rewards are natural (unconditioned), they evoke natural responses such as salivation, consumption, and sexual arousal (unconditioned responses); they thus constitute classical (Pavlovian) conditioning, although they become embedded in operant conditioning circumstances as they are actively sought out (Day and Carelli 2007).

The anatomical and chemical substrates of the BAS and BIS have been roughly mapped out. The BAS is primarily associated with dopamine (DA) and with mesolimbic system structures such as the nucleus accumbens, a structure particularly rich in neurons that produces and respond to DA (Day and Carelli 2007). The BIS is primarily associated with serotonin (5-HT), and with limbic system structures such as the hippocampus that feed into the prefrontal cortex (Goldsmith and Davidson 2004). DA and 5-HT act both as neurotransmitters and neuromodulators depending on where in the brain they are working. Neurotransmitters influence the intensity of messages from sensory receptors and neuromodulators filter the information to be processed. Regardless of where it operates in the brain, DA facilitates goal-directed behavior, and 5-HT generally facilitates neural processes subserving the inhibition or modulation of that behavior (Depue and Collins 1999; Yacubian et al. 2007). Figure 5.3 shows the location of the nucleus accumbens and the DA-rich ventral tegamental area, as well as the PFC.

Reward dominance theory is reminiscent of Freud's theory of the id, ego, and superego components of personality. The id obeys the pleasure principle and is the biological raw material representing drives and instincts for acquiring life-sustaining necessities and life's pleasures (the BAS). Like a spoiled child, the id craves instant gratification of its desires and cares not whether the means used to satisfy them are appropriate or injurious to self or to others. The superego strives for the ideal and represents all the moral and social "brakes" (the dos and don'ts) internalized by the person during the process of socialization (the BIS). The ego's function is to sort out the conflict between the antisocial demands of the id and the overly conformist demands of the superego. The ego (BAS/BIS balanced) obeys the reality principle; it does not deny the pleasure principle; it simply adjusts it to the demands of reality.

Reward Dominance and Alcoholism

Alcoholism presents a classic example of reward dominance. Alcohol is a brain-numbing depressant, but at low dosages it is a stimulant because it raises DA levels while at the same time it decreases 5-HT, thus reducing impulse control and increasing the likelihood of aggression (Martin 2001; Ruden 1997). Alcohol also reduces inhibition by affecting the neurotransmitter gamma-aminobutyric acid (GABA), which is an inhibitor of internal stimuli such as fear, anxiety, and stress (Buck and Finn 2000), resulting in reduced anxiety about the consequences of our behavior (Martin 2001). Alcohol's direct effects on the brain can thus help us to reinvent ourselves as "superior" beings: the fearful to become more courageous, the self-effacing to become more confident, and the timid to become more assertive. As the wisest of our Founding Fathers, Ben Franklin, is said to have remarked: "Beer is positive proof than God loves us and wants us to be happy." But too much of a good thing can lead to the disease of alcoholism and an awful lot of heartache.

Alcoholism is a chronic disease marked by the incapacity to control alcohol consumption despite psychological, social, or physiological disruptions. Alcoholism is a state of altered cellular physiology caused by the repetitive consumption of alcohol that manifests itself in physical disturbances (withdrawal symptoms) when alcohol use is suspended. Nonalcoholic heavy drinkers are able to stop self-destructive drinking in the face of compelling life circumstances such as marital problems or job losses, but alcoholics are not able to sustain control over time unless treated.

Numerous theorists have hypothesized that alcoholism and criminality, while not synonymous, are linked because they share a common cause, namely, a reward dominant brain (Fishbein 1998; Gove and Wilmoth 2003; Martin 2001). Alcoholics have an old saying that one drink is too

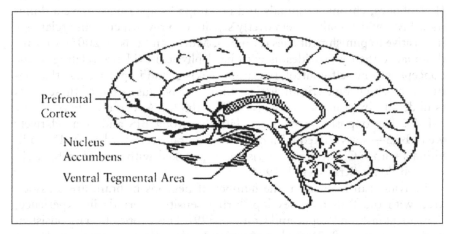

Figure 4.3 Brain areas relevant to reward and executive functions. Source: National Institute of Drug Abuse. The Brain's Drug Reward System, 1996.

many and a hundred drinks are not enough. This seemingly contradictory statement informs us that a single drink activates the pleasure centers in the nucleus accumbens and that this one drink leads to such a craving for more that even a hundred drinks will not satiate.

Alcoholism researchers have shown that there are two different types of alcoholics: Type I and Type II. According to Crabbe (2002:449): "Type I alcoholism is characterized by mild abuse, minimal criminality, and passive-dependent personality variables, whereas Type II alcoholism is characterized by early onset, violence, and criminality, and is largely limited to males." Type II alcoholics start drinking (and using other drugs) at a very early age and rapidly become addicted, and have many character disorders and behavioral problems that *precede* their alcoholism. Type I alcoholics start drinking later in life and progress to alcoholism slowly, and if they have character defects, these are typically induced by alcohol and are not permanent (DuPont 1997). Thus Type II alcoholics may have inherited abnormalities of the 5-HT and DA systems and/or on their enzymatic regulation that may be driving both their drinking and their antisocial behavior (Buck and Finn 2000; Demir et al. 2002). Heritability estimates for Type II alcoholism are about 0.90, and about 0.40 for Type I alcoholism (Crabbe 2002; McGue 1999). Alcoholism provides an excellent example of the difficulty of nailing down specific genes "for" a specific problem behavior/condition. There are a bewildering number of genes that control the neurotransmitters and enzymes implicated in alcoholism. For instance, Li, Mao, and Wei (2008) examined 2,343 lines of evidence from peer-reviewed journals and identified 316 alcohol addiction-related genes and 13 addiction-related pathways (the molecular routes and interactions among neurotransmitters and enzymes) to alcohol addiction.

The normal brain is roughly equally sensitive to both reward and punishment, but for some people one system dominates the other. Like Type II alcoholics, chronic criminals and psychopaths apparently have a dominant BAS, which makes them overly sensitive to reward cues and relatively insensitive to punishment cues (Franken, Muris, and Rassin 2005). Reward dominance theory provides us with hard physical evidence relating to the concepts of low self-control, sensation seeking, and impulsiveness that figure in much current criminological thinking, because each of these traits is underlain by either a sticky accelerator (high DA) or faulty brakes (low 5-HT). The opposite profile—a relatively strong BIS and/or a relatively weak BAS—is associated with obsessive compulsive disorder (OCD). Individuals with OCD are characteristically wracked with unreasonable guilt and self-doubt (Kruesi et al. 1994).

Individual differences in the number of neurons in brain areas associated with the BAS may develop during sensitive periods in experience-dependent fashion (Depue and Collins 1999). This proposition is consistent with Kreamer's (1992) work pertaining to the affects of non-attachment on neurological dysregulation. Serotonergic mechanisms, however, seem to be more affected by immediate rather than distal environmental events. For instance, poor parenting is correlated with low 5-HT levels in children (Pine et al. 1997), although parenting could be a reaction to children's impulsive behavior induced by low serotonin rather than a cause of their children's low 5-HT, or it could be a reciprocal relationship with each feeding on the other.

THE ANS AND CONSCIENCE

Conscience is the result of a complex mix of emotional and cognitive mechanisms coming together that enables individuals to internalize the moral rules the social group. People with strong consciences feel guilt, shame, and anxiety when they violate these rules; those with an impaired conscience will experience none of them. As Kochanska and Aksan (2004:304) put it: "An impaired conscience is the core aspect of conduct disorders, antisocial development, and psychopathy. Conversely, the capacity for remorse and empathy, an appreciation of right and wrong, and engaging in behavior compatible with rules all mark successful adaptation." The relative sensitivity of a person to reward and punishment via BAS/BIS mechanisms should in large part predict the strength of his or her conscience, as should ANS functioning.

Jeffrey Gray (1994), a pioneer of reward dominance theory, added a third system of behavior control, the fight/flight system (FFS), to the BAS/BIS model. The FFS is part of the ANS, but Gray wanted to distinguish its specific fight/flight function from the other more basic "housekeeping" functions of the ANS. Because researchers continue to use the term *ANS* rather than *FFS* when referring to arousal, I will do so also. The fight/flight function of the ANS is of great importance in understanding the emotional half of the acquisition of a conscience.

Differences in the emotional component of conscience are observed as early as 18 months, long before children are able to cognitively reflect on their behavior as morally right or wrong. These differences reflect variation in ANS arousal patterns which allow for the learning via classical conditioning of the prescriptions and proscriptions of moral behavior (Kochanska and Aksan 2004; van Bokhoven et al. 2005). Unlike operant conditioning, which is active (it depends on the actor's behavior) and cognitive in that it forms a mental association between a person's behavior and its consequences, classical conditioning is mostly passive (it depends more on the level of ANS arousal than on anything the actor does) and is visceral in nature; it simply forms a subconscious association between two paired stimuli.

It is through visceral associations between behavior and the response of others that we develop the "gut level" emotions that make up the "feeling" superstructure of our consciences. These social or secondary emotions are retrofitted to the same neurophysiological machinery that drives the hard-wired primary emotions of fear, anger, sadness, and joy. After the primary emotions are elaborated and refined by cognition during socialization, we fear the pain of punishment and the shame of rejection when we transgress and welcome the joy of acceptance and affection when we behave well.

Children must learn which behaviors are acceptable and which are not (the knowledge part of our conscience). When children are reprimanded for misbehaving they experience a variety of unpleasant physiological events. The degree to which these emotions affect them depend on the severity of the reprimand interacting with the responsiveness of the ANS (Dolan 2004). Assuming an ANS adequately responsive to discipline, children will eventually name similar behaviors that generate the same ANS responses as bad or wrong, and will feel guilt if they even contemplate doing them. If they refrain from forbidden behavior in the future, it is not because they have rationally calculated the cost/benefit ratio involved, but rather because their internal response systems (the emotional component of conscience) discourage it by generating unpleasant feelings.

Thresholds for ANS are normally distributed, with most individuals by definition clustered around the mean and a small minority of individuals on each tail. Individuals with a *hyper*arousable ANS are very fearful and condition easily because arousal produces punishing visceral feelings associated with anxiety and fear. Easily aroused persons who even contemplate actions contrary to the expectations of others will quickly engage the sympathetic system. The decision not to act on the impulse engages the parasympathetic system to restore the ANS to homeostasis. Acquiescing to the conformity demands of socializers (both specific others and the "generalized other") prevents sympathetic ANS arousal, which powerfully reinforces conforming behavior via the reduction of fear and anxiety.

A hyperarousable ANS is a potent protective factor against criminal behavior. A review of 40 studies of ANS activity (measured by skin conductivity and/or heart rate) and criminal and antisocial behavior found 38 studies that supported the link, and two that had nonsignificant results (Ellis and Walsh 2000). A number of subsequent studies have shown that youths raised in criminogenic environments who remain free of their criminogenic influences are shy, timid, and evidence hyperactive ANS arousal under conditions of threat (Boyce et al. 2001; Lacourse et al. 2006). Such youths are actually less likely to commit antisocial acts than youths with hyporeactive ANS arousal reared in noncriminogenic environments (Brennan et al. 1997).

In sum, individuals with a readily aroused ANS are easily socialized—they learn their moral lessons well. They do so because sympathetic ANS arousal ("butterflies in the stomach") is subjectively experienced as fear and anxiety. Children learn that when they behave well they do not incur the wrath of their socializers and fear and anxiety do not appear. An ANS that is easily aroused and generates high levels of fear and anxiety is a protective factor against antisocial behavior. Individuals with relatively unresponsive ANSs are difficult to condition because they experience little fear, shame, or guilt when they transgress, even when discovered and punished, and thus have no built-in visceral restraints against further transgressions. They do not receive visceral reinforcement for conforming behavior because they are not aroused to fear in the first place and thus cannot receive positive reinforcement for conforming behavior in the form of a return to ANS homeostasis (Bell and Deater-Deckard 2007). Having knowledge of what is right or wrong without that knowledge being paired with emotional arousal is like knowing the words to a song but not the music.

PREFRONTAL DYSFUNCTION THEORY

Another neurologically specific theory of criminal behavior is prefrontal dysfunction theory. As mentioned earlier, the prefrontal cortex (PFC) is responsible for a number of functions such as making moral judgments, planning, analyzing, synthesizing, and modulating emotions. The PFC provides us with knowledge about how other people see and think about us, thus moving us to adjust our behavior to consider their needs, concerns, and expectations of us. These PFC functions are collectively referred to as *executive functions* (Fishbein 2001), and are clearly involved in prosocial behavior if functioning normally. If the PFC is compromised in some way, however, the result is often antisocial behavior.

The association between the frontal lobes and pro- and antisocial behavior is often illustrated by the dramatic case of Phineas Gage in the 19th century. Gage had a tamping iron blasted through his skull at the PFC, and this terrible accident drastically changed his personality. This case gave early neuroscientists their first clues about the functions of the PFC, although it is important to realize that it was an extreme example and that damage does not have to be massive or even anatomically discernible to negatively impact behavior. PFC damage could be at the cellular level and be the result of genetic factors that affect neuron migration during the earliest stages of frontal lobe development, or maternal substance during pregnancy (Perry 2002).

Brain imaging studies consistently find links between PFC activity and impulsive criminal behavior. A PET study comparing impulsive murderers with murderers whose crimes were planned found that the former showed significantly lower PFC and higher limbic system activity (indicative of emotional arousal) than the latter and other control subjects (Raine et al. 1998). The nonimpulsive murderers had prefrontal functioning similar to the control subjects, but they also showed excessive right-side limbic activity. An MRI study (Raine et al. 2000) found that males diagnosed with antisocial personality disorder (APD) had 11% less prefrontal gray matter volume than subjects in two control groups (a "healthy" group and a substance-dependent group) matched for SES, ethnicity, IQ, and head circumference. The antisocial group also evidenced reduced ANS activity during exposure to stress. The APD group self-reported a significantly greater number of violent crimes than either control group. The observed PFC and ANS deficits predicted APD versus non-APD status independent of 10 demographic and psychosocial risk factors, and were by far the most powerful predictors. Another MRI study found that psychopaths (n = 16) labeled "unsuccessful" (caught) had a 22.3% reduction in prefrontal gray matter compared with 23 control subjects. "Successful" (uncaught) psychopaths (n = 13), however, did not differ in gray matter from controls. Higher psychopathy scores were associated (r = −.39) with low gray matter volume (Yang et al. 2005). It is important to note that these findings relate to acts of impulsive reactive violence, not planned proactive violence.

Cauffman, Steinberg, and Piquero (2005) combined reward dominance and PFC dysfunctions theories in a large-scale study of incarcerated youths in California and found that seriously delinquent offenders have slower resting heart rates (indicative of low fear and low ANS arousal) and performed poorly relative to nondelinquents on various cognitive functions performed by the PFC. In another study combining both theories, Yacubian (2007) and her colleagues monitored activity in the PFC and the ventral striatum areas via fMRI scans among 105 healthy male volunteers. Some subjects showed high levels of activity and some showed blunted activity in anticipation of similar (monetary) rewards. Examining subjects' genotypes, the researchers found that variation in neural activity was mediated by variation in genetic polymorphisms related to DA activity, namely, the dopamine transporter gene and catechol-O-methyltransferase (COMT—an enzyme that degrades DA in the synaptic gap). Studies such as this provide solid neurological and genetic evidence of the bases for the reward dominance theory of addiction to all kinds of things, including addiction to crime (see Chapter 5).

Reward Dominance and PFD Theories and Child Abuse/Neglect

Although the set points of BIS/BAS and PFC functioning are genetically influenced, the plasticity of the human brain suggests that early experience may be more important than genetics in this regard. The hippocampus and amygdala are the most plastic areas of the limbic system, and the PFC is the most plastic of all brain structures (Perry 2002; Teicher et al. 1997). Recall that plasticity essentially refers to the physiological calibration of the brain in response to environmental experience. Traumatic events may alter set points in the many brain areas that respond to threats. The evidence for trauma-induced alteration of brain function is strongest in controlled animal experiments (recall the methylation of glucocorticoid receptor genes that varied according to the quality of the nurturing that rat pups received discussed in Chapter 2). However, given the greater human reliance on learning to navigate life, abuse and neglect may be more strongly associated with dysregulation of various neural structures in abused children (McBurnett et al. 2000).

The possible affects of abuse and neglect on limbic system structures such as the hippocampus and amygdala have particular relevance for the BIS. The hippocampus is a crucial component of the BIS. Protracted abuse and neglect can alter its threshold setting via the effects of excess cortisol secreted in response to stress so that it often fails to perform its appropriate inhibitory role (Teicher et al. 1997). Similarly, frequent "kindling" (arousal, excitation) of the amygdala in response to abuse and neglect may eventually lead to subconvulsive seizures (Mead 2004). Subconvulsive seizures are the result of kindling in the amygdala (the most seizure-prone area of the brain) that often may not have any immediate environmental derivation. The term *episodic dyscontrol* is used to describe behavior resulting from these seizures. Ellis and Walsh (2000) reviewed 111 studies pertaining to the relationship between subconvulsive seizures and criminal behavior and found the relationship to be invariably significant for impulsively violent and acting-out behaviors, but not for any kind of planned criminal behavior.

In terms of the affects of abuse and neglect on PFC functioning, Perry (1997) suggests that children who spend a great deal of time in a low-level state of fear tend to focus consistently on nonverbal cues of imminent danger because the brain has been habituated to do so. This tendency has been referred to as "frozen watchfulness" (DeLozier 1982:98). Both Perry and DeLozier suggest that a cognitive profile in which performance IQ is significantly greater than verbal IQ (P > V) is a marker of this tendency. A significant (12 points at p < .01) P > V discrepancy has been consistently linked with chronic criminal offending (Miller 1987; Moffitt 1996), and a significant V > P profile has often been found to be a marker of prosocial behavior (Walsh 2003). One study found that P > V males were 29 times more likely to be found in prisons than V > P males, although both IQ profiles are found to be about equally represented in the general population (Barnett, Zimmer, and McCormack 1989). A study of 513 juvenile delinquents found that boys who were born illegitimate and raised by a single mother had higher abuse/neglect scores, higher delinquency scores, and a greater P > V discrepancy (mean discrepancy score = 19.22) than boys born and reared under other circumstances (Walsh 1990).

Other lines of evidence support Perry's insight. Teicher and his colleagues (1997:197) write that childhood memories may be preferentially stored in the right hemisphere because abuse and neglect may be "associated with greater left-hemisphere dysfunction, which may lead to greater dependence on the right hemisphere. Increased right frontal function, in turn, may lead to enhanced perception and reaction to negative affect." They further note that that the right hemisphere plays a particularly significant role in the perception, processing, and expression of negative emotions. This is supported by an EEG study that found that right PFC activation is related to BIS strength, and left PFC activation is related to BAS strength on the BIS/BAS scale (Sutton and Davidson 1997). Sutton and Davidson (1997:209) conclude that "individuals with tonically active right prefrontal regions may be predisposed to become vigilant for threat-related stimuli [DeLozier's frozen watchfulness], concurrently inhibiting behavior, organizing resources for behavior withdrawal, and experiencing negative affect (i.e., BIS activity)."

The most difficult data to reconcile are those concerning trauma and changes in arousal patterns and the data linking abuse/neglect to later criminal behavior. There is evidence that abuse may lead to increased ANS and HPA axis responsiveness, and that chronic stress and/or acute trauma can result in post-traumatic stress disorder (PTSD) and hyper ANS responsiveness and hypercortisolism among adults (Pitman and Orr 1993). If abuse/ neglect can lead to a hyperarousable ANS, and if such an ANS response has been linked to a lower probability of antisocial behavior, then abuse and neglect could be naively construed as a protective factor against antisocial behavior. As we have seen, there is contrary evidence indicating that abuse and neglect may result in hyporesponsivity of the ANS and hypocortisolism. Raine (1997) reconciles these two lines of evidence stating that an important factor is when the stress is experienced. If the organism experiences chronic stress during the organizational phase of brain development, as opposed to acute traumatic stress experienced as an adult or adolescent, then that stress should lead to ANS hyporesponsivity and hypocortisolism as an "inoculation" to stress.

We may conclude that there is abundant evidence that a significant association exists between child maltreatment and the kinds of neurophysiological abnormalities that have been associated with criminal behavior, but it is difficult to determine the direction of the causal relationship in any absolute sense. Given the growing awareness of gene/environment correlation processes among researchers, many of them now voice their awareness of the equally plausible hypothesis that congenital neurophysiological abnormalities may lead to acting-out behavior on the part of children. Acting-out behavior, in turn, may increase the likelihood of maltreatment on the part of caregivers (Heck and Walsh 2000; Perry 2002).

Most of us are doubtless reluctant to entertain the painful notion that certain children could be somehow responsible for their own abuse by evoking negative responses from their caregivers. For practical purposes (policy recommendations) it is not necessary to do so, for we know that infants and children are adversely affected by maltreatment and lack of attachment that have nothing at all to do with their behavior. It has been estimated that as many as 375,000 infants born each year have been exposed to cocaine in utero, and that in some inner-city hospitals as many as 50% of pregnant women test positive for cocaine (Mayes et al. 1995). Children in orphanages around the world have long been known to suffer numerous physical and psychological problems (reviewed in Frank et al. 1996). These children had no opportunity to influence their mothers' drug habits or their own orphan status.

For instance, Russian and Romanian orphans who had been in orphanages an average of 16.6 months before being fostered to American families for an average of 34.6 months showed significantly lower base levels of the neuropeptides vasopressin and oxytocin (a class of chemicals that function as neurotransmitters or hormones that produce feelings of peace and calm) compared with a control group of American children reared by their biological parents (Wismer Fries et al. 2005). They also showed significantly lower levels after experimental interaction with their mothers, which normally increases neuropeptide levels. These results reveal mechanisms that may account for the well-documented fact that children reared without frequent tactile comfort become vulnerable to difficulties in forming secure relationships with caregivers and forming social bonds in the wider society. Of course, this outcome is not inevitable since idiosyncratic genetic, epigenetic, and experiential processes will create many individual differences with respect to these outcomes (a small proportion of the adoptees in the aforementioned study actually had oxytocin levels higher than the average of the control group).

Neglected and orphanage children also lack something as basic as tactile stimulation (which releases neuropeptides), the importance of which is made palpable by the fact that the infant can only experience and express love or its absence through its body. Primates are programmed to secrete the opiate-like neuropeptides in situations of social comfort, and a lack of frequent bodily contact between mother (or other caregiver) and child has be interpreted as abandonment, because the infant can only "think" with its skin. The infant's contact comfort experienced from the sensitive responses of others during times of duress tells it that everything's O.K., "She's there for me," I'm safe," All's right in my world." Ashley Montagu summarizes the relationship between tactile stimulation and human development when he writes: "The kind of tactuality experienced during infancy and childhood not only produces the appropriate changes in the brain, but also effects the growth and development of the end-organs in the skin. The tactually deprived individual will suffer from a feedback deficiency between skin and brain that may seriously affect his development as a human being" (1978:208).

The neurological evidence strongly supports liberal calls for nurturant strategies to crime control, such as paid maternal leave, nutrition programs, home visitation programs, and head start–type programs and so forth. Methwin (1997) provides a number of examples of cost-efficient child development programs yielding good results that could be implemented on a national scale. Data like these, plus the relatively hard data from the neurosciences, doubtless have more power to influence tightfisted lawmakers and a skeptical public than heartstring appeals from humanists, and as Vila (1997:18, emphasis in original) points out: "Keeping adequate resources flowing toward child development programs is a *social investment strategy that pays compound interest.*"

5 The Anomie/Strain Tradition and Socioeconomic Status

THE SOCIAL STRUCTURAL TRADITION

Whether explicit or implicit, all criminological theories contain a view of human nature. The early 20th century saw the beginning of sociology's dominance of criminology, which largely took as its own its parent discipline's assumptions about human nature. One of these assumptions is that human nature is socially constructed; i.e., it has no (or very little) essential content beyond providing a blank slate on which "society" scratches is prescriptions and proscriptions (Guo 2006). This is the meaning of the oft heard remark in social science circles that "Man has no nature, he only has a history" (Ruffie 1986:297). Another assumption is that human beings are naturally good until corrupted by society. Given these two assumptions, the task of sociological criminology is to explain why inherently good social animals commit antisocial acts. If crime is alien to good social impulses, its causes must be sought outside of the individual. This search typically involves looking for flaws and defects in society, such as a discriminatory class system, racism, value conflicts, and capitalism, and largely discounts the existence of any flaws or defects in the individual criminal.

Many criminological theories emphasize social structure; that is, how society is organized by social institutions and stratified by social roles and statuses. Social structure is the framework that shapes the patterns of relationships members of society have with one another. Structural theories favor external "out there" reality as being of primary importance in determining human social behavior; i.e., they tend to work "top down" from assumptions made from a general model of society and to deduce the everyday experiences of individuals from them. Because of this, they have a tendency to slip into ways of thinking that render the individual almost irrelevant. An extreme example of this is Rodney Stark's (1996:140) defense of structural explanations over what he calls "kinds of people" explanations when he wrote: "Surely it is more efficient and pertinent to see dilapidation, for instance, as a trait of a building rather than as a trait of those who live in the building." The implication of this statement is that buildings are maintained or trashed by forces acting independently of the values and behavior of those who inhabit them.

We begin with anomie/strain theory for three reasons: (1) It is arguably the most long lived of all existing criminological theories; (2) it began as the most macro of theories emphasizing whole societies and illustrates the epidemiological approach of continually reducing levels of analysis as the lessons learned at one level are exhausted, and (3) it is concerned with the central concept of sociology: socioeconomic status (SES). The theory began with Durkheim's (1895/1982) and Merton's (1938) analysis of whole societies and has gone through subcultural (Cohen 1955) and social psychological (Agnew 1992) phases, and may be on the verge of entering a biosocial phase (Agnew 2005; Walsh 2000b).

Emile Durkheim

The concept of anomie was introduced by Emile Durkheim, one of the most influential sociologists of all time. Despite his reputation among his contemporaries as a metaphysician, and unlike many sociologists after him, Durkheim had a view of human nature, and it was a naturalistic one at that (Lopreato and Crippen 1999; Schmaus 2003). His view was close to the classical view that humans first and foremost seek to maximize their pleasures and minimize their pains. Although Durkheim never used evolutionary terminology, he viewed human nature similarly; i.e., as something that "is substantially the same among all men, in its essential qualities" (1951:247). He also saw humans as clearly adapted to attend to their self-interest, and clearly adapted to live in social groups, which mitigates self-interest. The balance between these two adaptations is a central feature of Durkheim's thought.

Although Durkheim viewed humans as similar in their "essential qualities," he realized that "One sort of heredity will always exist, that of natural talent" (1951:251). Thus, although everyone is more or less equal in their desires, not every one is equally capable of achieving them. This presents a problem for the individual (and by extension for society) because for Dukheim human appetites are insatiable, and "No living thing can be happy or even exist unless his needs are sufficiently proportioned to his means" (1951:246). Happiness can thus be viewed as a ratio of expectations to accomplishments, which must at least equal 1.0; anything less than unity breeds unhappiness (or in later terminology, "strain'). Because all persons do not possess the same means to accomplish the things necessary to satisfy their appetites, "A moral discipline will therefore still be required to make those less favored by nature accept the lesser advantages which they owe to the chance of birth" (1951:151). This moral discipline is provided by society, which "is the only moral power superior to the individual" . . . for "it alone has the power necessary to stipulate law and to set the point beyond which passions must not go" (1951:149).

According to Durkheim, society does not exist by rational agreement, as many so-called contractual theorists such as Hobbes and Rawls assert, but rather it exists by virtue of a "pre-contractual solidarity" based on an emotional sense of belonging to a community and on moral obligation to it. Evolutionary psychology would again find much to agree with in Durkheim's vision, for all indications are that *Homo sapiens* evolved over millions of years living in small hierarchically structured social groups with strong attachments to one another, and have probably always been more emotional than rational (Barkow 2006; de Waal 1996; Krebs 1998). Durkheim referred to this sense of belonging and moral obligation that was the basis for social solidarity as the "collective conscience" or "collective consciousness."

Durkheim was preeminently concerned with the effects of modernization on social solidarity, and with how individuals responded to these effects. In the *Division of Labor* (1951a), he explicitly applies Darwin's account of specialization (division of labor for Durkheim) as a solution to the struggle for survival to human society. Modernization essentially involves the progression from traditional societies characterized by uniformity, primary group interactions, strong normative agreement, and mechanical ("instinctive") solidarity, to modern industrialized societies characterized by mostly secondary group interactions, weak norms, and organic solidarity. Societies characterized by mechanical solidarity possess a strong collective conscience and exert strong pressure on individuals to conform to its norms. Primarily because of an extensive division of labor, people in modern societies have less common "social likeness," the collective conscience is weakened, and it thus they exert less pressure for social conformity. The greater degree of social integration enjoyed by members of mechanical societies cushions the trials and tribulations of life for them. Conversely, the looser bonds experienced in organic societies provide less social support, which may result in greater stress and frustration and in a greater probability of deviant responses to it.

Modern societies are characterized by a great deal of change. Each change tends to detract from shared social likeness and thus weakens the collective conscience. Rapid social change leads to a state of affairs Durkheim referred to as *anomie* (*a* = "without" *nomos* = "norms"), a condition of social deregulation and a weakening of social solidarity. Anomic social conditions result in diminished feelings of participating in a shared community and of a sense of obligation to fellow citizens that tends to release egoistic self-interest. Under such conditions we can expect a rise in all sorts of deviant behavior, including crime, as people seek to satisfy their appetites unrestrained by a sense of shared belonging and shared morality.

Udry (1995) is probably correct in his assessment that Durkheim was not the extreme "social factist" that sociologists have made him out to be. Durkheim was careful to distinguish between social facts and biological and psychological facts, and between crime and criminality: "From the fact that crime is a phenomenon of normal sociology, it does not follow that the criminal is an individual normally constituted from the biological and psychological points of view. The two questions are independent of each other" (1982:106). There is no doubt that crime is sociologically normal in that it occurs in all societies, even in societies characterized by mechanical solidarity, and at all times. There is also no doubt that crime rates fluctuate with social conditions, and are thus firmly in the category of social facts.

Criminality, on the other hand, is a property of individuals signaling willingness to use and abuse others by any means for personal gain. Because criminality is a continuously distributed trait, as social bonds weaken and norms break down in response to social, political, and economic changes, the threshold for this willingness is lowered, and more and more people cross the line from law-abiding to law-breaking behavior, as is illustrated in Figure 1.1. Anomic conditions serve as "releasers" of criminal behavior, which occurs at lower thresholds for some individuals than for others. This is where biosocial theories are needed because just as individual-level theories cannot explain crime rates, macro-level theories cannot explain why some individuals commit crimes while others in the same environments do not. Durkheim himself seemed to make this point when he remarked: "Thus, since there cannot be a society in which the individuals do not diverge to some extent from the collective type, it is also inevitable that among these deviations some assume a *criminal character*. What confers upon them this character is not the intrinsic importance of the act but the importance which the common consciousness ascribes to them" (1982:101, emphasis added).

According to Durkheim, then, there are a number of individuals in any society that possess a character that motivates them to act in ways that society defines as criminal. As Cohen and Machalek interpret Durkheim's position: "Thus, although the definition of a behavior as criminal is strictly a matter of social labeling, the *root causes of the behavior itself* are to be traced to individual character traits" (1994:293, emphasis in original). It is not clear whether it is Durkheim's position that all criminal definitions are arbitrary, or that he recognized that there are clearly crimes that are inherently (mala in se) harmful, such as murder, rape, and the theft of resources, because they militate against the evolutionary imperatives of survival and reproduction. Furthermore, people can and do use and abuse other people for personal gain regardless of whether or not the means used have been defined as criminal, and it is this propensity that defines criminality independent of the social labeling of an act as criminal.

Durkheim sounds very much like an evolutionary psychologist in that he was clear that it is humanity's insatiable appetite for resources (wealth, status, prestige) that underlies criminal activity. Resources have been the coin of reproductive success throughout our evolutionary history, and we should thus expect our appetite for them to be substantial (although surely the satiation threshold varies from person to person). Such appetites may surely lead to widespread cheating were it not for some mechanism to hold it in check. Because we are a social species, we have evolved tit-for-tat strategies (reciprocal altruism) and a set of moral norms with our evolved social emotions to support them and to constrain our self-interest within acceptable boundaries. Whenever events occur that tend to weaken these moral norms, selfish appetites are released to flood society with crime and other forms of deviance. As an evolutionary psychologist might paraphrase Durkheim, in modern evolutionarily novel environments and the enormous degree of social change they generate, a vast number of niches are created in which individuals can pursue conditional cheater strategies.

Durkheim also sounds like an evolutionary psychologist in writing about the coevolution of cheating and cooperative behavior in his belief that crime (cheating) is necessary for the evolution of social solidarity (cooperation). Although crime and deviance beyond a certain level is socially harmful, at "normal levels" Durkheim characterized it as "a factor in public health, an integrative element in any healthy society" (1982:98). Crime and other forms of deviance, particularly those forms that undermine norms of reciprocal altruism, serve to define the boundaries of right and wrong. Punishment of such acts reinforces those norms and strengthens the collective conscience. Although Durkheim saw punishment primarily as a social ritual, he recognized that the urge to punish is inherent in human nature, and that it serves an expiatory role for the individual (Walsh 2000b).

Robert Merton

Robert Merton extended Durkheim's concept of anomie in his famous paper *Social Structure and Anomie* (1938). Because Merton was writing in mid-20th-century America and Durkheim was writing in late-19th-century France, we can expect some discontinuity of thought. However, although some have criticized Merton for his theoretical departures from Durkheim, Passas (1995:93) argues that Merton presents "significant lines of continuity," and is an "appropriate extension of Durkheim's ideas." A major difference between the two theorists is that Merton's theory limits itself to the role of culture and social structure and ignores individual differences altogether. The "root" cause of crime is not to be found within individuals whose insatiable appetites must be kept under control by strong normative controls, but rather in sociocultural contradictions. Merton agreed with Durkheim that the inability to attain resources legitimately generates unhappiness (strain), and sometimes leads to efforts to obtain them illegitimately. Unlike Durkheim's view of society as a positive force restraining individuals from adopting illegitimate options, however, Merton viewed it as a negative force that motivated some to adopt illegitimate options.

Additionally, Merton viewed acquisitiveness as traits generated by a culture driven by an overweening concern with monetary success rather than as an intrinsic property of human nature. He argued that American culture feeds the notion that all citizens should aspire to the "American dream," the attainment of which symbolizes a person's character and self-worth. At the same time these cultural goals are being touted as goals for everyone to strive for, he maintained that social structure restricts access to legitimate means of attaining them to certain segments of the population. Also, unlike Durkheim, for whom anomie was an occasional condition rising and declining according to levels of social stability, Merton viewed anomie as a permanent condition of capitalist American society that is generated by this disjunction between cultural goals and structural impediments to attaining them (Merton 1968).

The meta-theoretical essence of Mertonian anomie/strain theory is that people are social animals who desire to follow social rules and will only resort to breaking them when placed under great pressure or strain. For Merton, crime is woven into the fabric of American society (and presumably into the web of all capitalist societies) because it arises from *conformity* to its values—a way disadvantaged people get what they have been taught to want—not from deviation from its values due to social deregulation. Otherwise stated, while for Durkheim the freeing of natural greed, acquisitiveness, egoism, deviance, and crime are *consequences* of anomie, for Merton they are socially constructed traits that are the *causes* of anomie (Passas 1995).

Merton considered American culture's emphasis on success goals to be applicable across class lines, so all members of society share the supposed strains of attempting to achieve common cultural goals. However, because of limited access to legitimate means of achieving these cultural goals, the lower classes will be more strained than others. Thus, Merton used the cultural argument to explain why crime rates are high in the United States in general, and the structural argument to explain the concentration of crime in the lower classes (Bernard 1987).

Merton provided his well-known typology (conformity, ritualism, retreatism, rebellion, and innovation) of people's adaptations to anomic strain. But what is it that sorts members of society into these different modes of adaptation? People do not consciously decide that they are going to follow one mode rather than another, but tend to subconsciously slip into them over time largely based on their social class position. According to Merton, the upper and middle classes have access to legitimate means of attaining desired success goals, but the lower classes do not. Even people who are successful may feel strain as they compare themselves to others who are more successful, but it is the lower classes that feel the bite most strongly.

Given the emphasis on the attainment of monetary success and middle-class status as the overweening goal stressed by American culture, it is curious that strain theorists have not explored the correlates of occupational success and of the different ways people have of coping with strain (but see Agnew following). Merton's (1938) famous "plus or minus" table of adaptations merely states that people "accept" or "reject" cultural goals and institutionalized means of obtaining them. It is easily deducible from Merton's writings, however, that people sort themselves into one mode or another based on their perceptions of, and attitudes about, their chances of achieving middle-class success goals legitimately. These perceptions and attitudes are assumed to be class-linked, and class is assumed to be both given and relatively static. This essentially means that social class is the cause of social class. Unless they want to adopt the position that SES is an uncaused "first cause," anomie/strain theorists have to come to terms with the notion that SES is as much a dependent variable as it is an independent variable.

Anomie/strain theorists also have to come to terms with the fact that people handle strain differently. By definition, most people adopt the conformist mode, many others adopt the ritualist mode, and only a comparative few adopt the retreatist and innovative modes. Among those who do, most will do so only temporarily during adolescence (Moffitt 1993). Even among the lower classes, where strain is felt most acutely, most will adopt ritualist rather than criminal lifestyles, as Merton readily acknowledged. Thus, some people cope with strain poorly and destructively while others cope with it well and constructively. Those who cope poorly are not likely to achieve middle-class status, regardless of the SES of their parents, while those who cope well will have at least a sporting chance, regardless of their class origins.

Robert Agnew

Early extensions of Merton's theory by Cohen (1955) and Cloward and Ohlin (1960) noted the individual's lack of interest in and/or inability to pursue the legitimate means of attaining middle-class success as causes of crime and delinquency as much as sociocultural barriers. Cloward and Ohlin (1960: 96) wrote of the inability of lower-class youths to defer gratification, their impulsivity, and sensation-seeking, and their preference for "big cars," "flashy clothes," and "swell dames." Writing about the status frustration lower-class youths experience, Cohen (1955: 66; emphasis in original) states that these youths come to define as meritorious "the characteristics they *do* possess, the kinds of conduct of which they *are* capable."

These individual characteristics, not being consistent with the social factist paradigm, were de-emphasized by others with a desire to defend criminals and persistent delinquents from any allegations of being different (Hirschi and Hindelang 1977). Cloward and Ohlin (1960:117) also fell back on the more traditional sociological tactic of denying that delinquents are any different from nondelinquents in the abilities required for occupational success: "There is no evidence . . . that members of delinquent subcultures are objectively less capable of meeting formal standards of eligibility [for middle-class occupations] than [nondelinquents] . . . the available data support the contention that the basic endowments of delinquents, such as intelligence, physical strength, and agility, are the equal of or greater than those of their nondelinquent peers."

The equality of delinquents (especially of persistent delinquents) with their nondelinquent peers on the first of these "endowments" is empirically false. It may or may not be true that delinquents are equal or superior to nondelinquents on strength and agility, but unless one aspires to be an athlete, acrobat, lumberjack, or some such occupation, strength and agility, without the temperament and intelligence to match, have little relevance for occupational success in modern society.

Robert Agnew's (1992) general strain theory (GST) resurrects individual differences, and is thus more faithful to Durkheim than to Merton (Barak 1998). Traditional strain theory was concerned only with the strain that resulted from being prevented from achieving positively valued goals, or more specifically, from achieving monetary success legitimately. Agnew (1992) expands strain beyond that generated by failure to achieve positively valued goals to include strain resulting from the removal of valued stimuli, and from the presentation of negative stimuli. Every imaginable source of strain is included in GST, and its magnitude varies with social location. But for Agnew, the important factor is not strain per se, but rather how one copes with it.

Agnew (1992) recognizes this in his discussion of negative affect (the tendency of persons to subjectively experience strain negatively and to react to it with frustration and anger) as an important intervening variable. Negative affect, or *negative emotionality* as it is referred to in psychology, has been found to have heritability coefficients of between .48 and .60 in various populations (Krueger, Markon, and Bouchard 2003; McGue, Bacon, and Lykken 1993). Thus, Agnew pushes the theory into the realm of person-environment interaction, and thus into the realm of gene-environment interaction. The personal characteristics that Agnew cites as important as insulators against the negative consequences of strain are ". . . temperament, intelligence, creativity, problem-solving skills, self-efficacy, and self-esteem" (1992:71). In addition to being insulators against strain, these variables also have obvious applicability to achieving occupational success, and thus to the adoption of one of Merton's adaptations.

Agnew (1995, 1997) has taken further reductionist steps in the process, citing criticism that anomie/strain theory does not speak to developmental aspects of delinquency. In particular, he asserts that the theory should predict an increase in antisocial behavior in late adolescence because that is when individuals seriously enter the job market, but what we actually observe is a decrease in antisocial behavior during this period (1997:101). Just when the central issue of the theory (perceptions about the possibility of achieving the American dream via occupational success) becomes most salient, instead of a dramatic increase in delinquency as people begin to perceive Merton's "disjunction," offending actually decreases.

Responding to this potentially fatal criticism, Agnew reflects on the fact that developmental psychologists talk about two general types of offenders, those who begin offending before adolescence and continue long into adulthood, and those who limit their offending to adolescence (1997:103). He wants to differentiate the reactions to strain between the two types of offenders by adding aggressiveness, which he views as an umbrella term covering hyperactivity, attention deficit disorder, impulsivity, and insensitivity, to his previous list of individual differences affecting delinquency and criminality. Agnew indicates that individuals high on these traits are likely to evoke negative reactions from others and are "less likely to form close attachments to conventional others—such as parents, teachers, and spouses; learn prosocial beliefs and behaviors; do well in school; and obtain rewarding jobs" (1997:106).

There are hints at evocative rGE in the preceding passage, but faithful to his discipline, Agnew appears to claim these traits are the products of early childhood socialization (1997: 107), although he does indicate that others have pointed out that they are heritable (1997:109). The next logical step in anomie/strain theory is to review the genetic and environmental contributions to the development of the traits important to the theory. Agnew's work over the past two decades points to the necessity of reducing the level of analysis in any science as new issues arise.

Agnew has recently narrowed down further his search for traits conducive to criminal behavior in his new general theory of crime (2005), although there is little resemblance to any form of anomie/strain left in it. In his new theory, Agnew identifies five *life domains* that contain possible crime-generating factors—personality, family, school, peers, and work that interact and feed back on one another across the life span. He further suggests that personality traits set individuals on a particular developmental trajectory that influence how other people in the family, school, peer group, and work domains react to them in evocative rGE fashion. In other words, personality variables "condition" the effect of social variables on criminal behavior.

The traits of *low self-control* and *irritability* are identified by Agnew as "super traits" that are composites of many endophenotypes such as sensation seeking, impulsivity, poor problem-solving skills, and low empathy. People with low self-control and irritable temperaments are likely to evoke negative responses from family members, schoolteachers, peers, and workmates that feed back and exacerbate those tendencies (the multiplier effect in evocative rGE). Agnew is more forthright than heretofore about the role of biology when he states that "biological factors have a direct affect on irritability/low self-control and an indirect affect on the other life domains through irritability/low self-control" (2005:213). He also notes neuroscience claims that the immaturity of adolescent behavior mirrors the immaturity of the adolescent brain (discussed in Chapter 6), and that teens experience massive hormonal surges that facilitate aggression and competitiveness. Thus, says Agnew, neurohormonal changes during adolescence *temporarily* increase irritability/low self-control among adolescents who limit their offending to that period, while for those who continue to offend irritability and low self-control are *stable* characteristics. Agnew has neatly integrated concepts and research from genetics, evolutionary psychology, and neuroscience into his theory, as well as showing how the various components (domains) influence each other developmentally across the life span.

THE PSYCHOLOGICAL PILLARS OF STATUS ATTAINMENT

All the traits Agnew lists as important for understanding antisocial behavior will be addressed later; for now I want to briefly review the literature on the two major individual-level factors that he identifies as important to successful coping and the attainment of legitimate success: temperament and intelligence. Temperament and intelligence have been called "the two great pillars of differential psychology" by Chamorro-Premuzic and Furman (2005:352), who go on to add that these two constructs are vital to predicting all kinds of life outcomes.

Agnew remarks on a number of occasions that (1) temperament and intelligence bear a strong relationship to problem-solving skills, and (2) the lower classes feel strain most acutely (1997:111–114). Unfortunately, even though SES is central to anomie /strain theory, he never attempts to make the connection between SES and the temperamental and cognitive correlates of problem solving (assuredly an important coping resource), although he is more than willing to state that such traits are a function of both biological and social factors (1997:105).

Making a connection between individual traits and SES is one most sociologists are reluctant to make. Sociology tends to treat SES as a variable that explains all sorts of other things but which needs little explanation itself. In effect, SES is generally viewed as self-perpetuating, so if offspring SES is caused by anything other than an unfair social system, it is caused by parental SES via modeling and the transmission of values and attitudes. Any attempt to predict a person's SES from his or her parental SES, however, is hopelessly confounded by genetics. As we shall see, in an open society offspring IQ predicts offspring's SES more strongly than parental SES or any other single variable for that matter.

Intelligence and SES

Given the depth and breadth of research findings (twin and adoption studies, fMRI, PET and molecular genetic studies) touched on in Chapter 2, few scientists familiar with the literature seriously doubt the importance of genes in explaining variation in IQ. Intelligence, as operationalized by IQ tests, is an obvious one of a number of determinants of both a person's occupational success and coping strategy, yet it is one that is conspicuously absent in sociological discussions of social status. It is as though to discuss intelligence as a possible determinant of SES amounts to sociological heresy (the stance reminds us of the aristocratic lady's famous reaction to Darwin's theory of evolution: "Let us hope that it is not true; but, if it is true, let us hope that it will not become generally known"). For instance, a densely packed 750-page book of readings on social stratification (Grusky 1994) amazingly does not even have the terms *IQ* or *intelligence* listed in the index. More contemporary social stratification textbooks have not improved on this, although IQ does sometimes appear (mostly to rail against *The Bell Curve*), but never as a factor predictive of SES.

Lee Ellis (1996:28) has also commented on this strange absence on any discussion of intelligence in sociological theories of the origins of SES: "Someday historians of social science will be astounded to find the word intelligence is usually not even mentioned in late-twentieth-century text books on social stratification." Ellis is aware of the inflammatory effects of making the obvious link between SES and IQ, but also that the link is in desperate need of dispassionate consideration. After all, IQ test were specifically designed to be a class-neutral measure of aptitude to turn schools into capacity-catching institutions to provide the ever increasingly complex state and economic machinery with competent workers (Nettle 2003; Pinker 2002).

The litmus test for any assessment tool is its criterion-related validity—its ability to predict outcomes. IQ tests do a particularly good job in this regard. An examination of 11 meta-analyses of the relationship between IQ and occupational success found that IQ predicted success better than any other variable in most occupations, particularly in higher status occupations, and that it predicted equally well for all classes and racial or ethnic groups (Gottfredson 1986). One study of over 32,000 workers in 515 different occupations revealed that the correlations between IQ and job performance rose from .23 for "low-complexity" jobs such as cannery worker, to .58 for "high-complexity" jobs such as circulation manager (Gottfredson 1997). Correlations would doubtless be higher for even more complex occupations not included in the study, such as engineer, physician, lawyer, or scientist.

Intelligence is particularly important in technologically advanced societies in which low-complexity occupations become mechanized and high-complexity occupations become more prevalent. Industrial and postindustrial economies are incompatible with a closed caste-like society in which occupations are assigned by accident of birth. Rather, modernization requires open competition for the choicest occupations, with class, gender, and race/ethnicity very much playing second fiddle to talent, the engine driving any modern economy. Employers compete for talented employees, and IQ testing has been *the* major tool in capitalist (and surreptitiously, in communist societies) to locate them from all segments of society (Eysenk 1982). This is not to deny that IQ has been a tool of exclusion also. For instance, because of high training costs and levels of failure among low-IQ recruits, the U.S. Army is forbidden by law to enlist anyone with an IQ below 80 (Gottfredson 1997). Whether we choose to view the role of intelligence and intelligence testing in modern societies positively or negatively, the fact remains that "in open societies with high degrees of occupational mobility, individuals with high IQs migrate, relative to their parents, to occupations of higher SES, and individuals with lower IQs migrate to occupations of lower SES" (Bouchard and Segal 1985:408).

This was true even in the 1930s, although given the discrimination against blacks and women back then, probably only for white males (Merton was the son of Jewish immigrants and was raised in the slums, but he achieved the American dream [Lilly, Cullen, and Ball 2007]). Ironically, the article following Merton's *Social Structure and Anomie* in the *American Sociological Review* examined the IQ/status relationship (Clark and Gist 1938) and found that they were highly correlated, and that IQ served to funnel people into their various occupations. Clark and Gist were not saying that IQ is the only cause of SES any more than I am, although they did conclude that it was perhaps the most important cause. It is without question that being born into an upper- or middle-class family confers many advantages, and that being born into a lower-class family brings with it many disadvantages, but let us not forget that those advantages and disadvantages are both genetic and environmental (Gordon 1997; Nielsen 2006).

In primitive hunting-and-gathering societies genetic influences were doubtless large, as one's status depended on a variety of talents related to exploiting the environment. As societies became larger and organized along agrarian lines, casts and classes emerged and the role of individual differences (and hence genes) declined precipitously as social status was largely ascribed by accident of birth. Individual differences became important again with industrialization, and even more so in postindustrial modern societies (Guo 2006). Although there is no such thing as a totally open society, unlike the rigid caste-like societies of the past, modern "class attainments do not represent environments imposed on adults by natural events beyond their control" (Rowe 1994:136). Genes become important to determining status in rough proportion to the equalization of environments. While this may seem paradoxical, it is a basic principle of genetics: The more homogeneous or equal in terms of access to opportunities the social environment (as in hunter/gather and modern capitalist societies), the greater the heritability of a trait; the more heterogeneous (or unequal) the social environment (as in agrarian societies), the lower the heritability of a trait (Plomin 2005). High heritability coefficients for socially important traits tell us that the society is doing a good job of equalizing the environment with respect to the traits in question (Nettle 2003).

Just how mobile is the American occupational structure? According to one major study, it is quite substantial (Hurst 1995). This study found that 48% of sons of upper-white-collar-status fathers had lower status occupations than their fathers, with 17% falling all the way to "lower-manual" status, and that 51% of sons of lower-manual-status fathers achieved higher status, with 22.5% achieving "upper-white-collar" status (Hurst 1995:270). Hurst (1995:276) concluded, "There was a great deal of movement both within and between generations. Well over half of the sons moved out of the occupational strata of their fathers and out of the strata of their own first jobs." Given this degree of upward and downward social mobility, and the degree to which IQ predicts it equally for all races and social classes, it is difficult to maintain that any group is systematically denied (i.e., denied by "the system") access to legitimate opportunities to attain middle-class status.

Indeed, far from being denied by a discriminatory and racist socioeconomic system, studies have found that because of programs such as affirmative action the occupational mobility of African Americans (the group most often considered discriminated against) into the most prestigious positions has increased at a more rapid rate than whites (Farley 1996; Wilson, Sakura-Lemessy, and West 1999). As Vold, Bernard, and Snipes (1998:177) conclude: "It is not merely a matter of talented individuals confronted with inferior schools and discriminatory hiring practices. Rather, a good deal of research indicates that many delinquents and criminals are untalented individuals who cannot compete effectively in complex industrial societies."

ASCRIPTION VERSUS ACHIEVEMENT

When compelled to confront the correlation between IQ and SES, social scientists tend to consider a child's IQ to be an effect of his or her parents' SES; i.e., a reification of class advantage or disadvantage. If this is the case, then any occupational success someone achieves which he or she claims to have meritoriously achieved is actually no more than ascription—class origin = class destination. A number of studies have found the correlation between parental SES and children's IQ to be within the .30 to .40 range (which is predictable from polygenic transmission models). The real test, however, is not the correlation between parental SES and the individual's IQ, but rather the correlation between the *individual's* IQ and his or her attained adult SES. This correlation has been found across a number of studies to be in the .50 to .70 range (Jensen 1998). As Jensen remarks: "If SES were the cause of IQ [rather than the other way around], the correlations between adults' IQ and their attained SES would not be markedly higher than the correlation between children's IQ and their parents' SES" (1998: 491). Offspring IQ is thus a much more powerful predictor of offspring SES than is parental SES.

One study tackled the SES/IQ issue head on as it relates to class by controlling for the entire complex of variables that constitute the environment of rearing by comparing siblings growing up in the *same home* with the *same parents*, but who had different IQs (Murray 1997). The 5,863 subjects came from the National Longitudinal Study of Youth, and were divided into "very bright" (IQ = 125+), "bright" (110–124), "normal" (90–109), "dull" (75–89), and "very dull" (< 75). Among the findings were that the "very brights" earned an average of $26,000 annually more than their "very dull" siblings, were significantly more likely to be married, and less likely to have children out of wedlock. Thus in this study that "perfectly" controlled for SES of rearing, very large differences in many areas of life were predicted by IQ level.

Similarly, DiRago and Vailant (2007) studied the original nondelinquent control group used in the famous studies of delinquency carried out by Sheldon and Eleanor Glueck (1950) 60 years after the original study began. All subjects were born between 1925 and 1932 and lived in poor high-crime neighborhoods in Boston. These men were interviewed about their occupational status (among many other things) when they were 25, 32, 47, and 65 years of age. Attrition whittled the original 500 males in the sample down to 345 at age 65. Childhood social class was significantly related to occupational status (r = .17) at age 25, but the correlation progressively dwindled to a nonsignificant .079 at age 65. At age 65 none of the measured environmental factors were related to occupational status. Years of education (–.410) and IQ (–.347) were related to occupational success at age 65 (correlations are negative because occupational success was coded 1 = professional down to 7 = unskilled). Although not a behavior genetic study, the results support the behavior genetic "law" that states that as we age the effects of shared environment (in this case, childhood SES, etc.) on phenotypic traits fade to insignificance while genes and nonshared environments become more salient.

Moving across the Atlantic, Nettle's (2003) study of all children born in Britain in one week in March 1958, followed to the age of 42, found that childhood IQ is associated with class mobility in adulthood uniformly across all social classes of origin. Nettle found an IQ difference of 24.1 points between those who attained professional class and those in the unskilled class, *regardless of the class or origin*. He concluded that "intelligence is the strongest single factor causing class mobility in contemporary societies that has been identified" (2003:560).

Another British longitudinal study (Bond and Saunders 1999) directly tested what they called the "class structuralist" position (the position that the social advantages/disadvantages of childhood SES largely determine adult occupation) with the "status attainment" position (the position that individual ability and motivation largely determine adult occupation). The study found that individual meritocratic factors (assessed when subjected were 7 years old) accounted for 48% of the variance in occupational status at age 33. All measured background variables (including parental SES) combined accounted for only 8%. Based on this sixfold difference in the proportion of variance explained, Bond and Saunders concluded that: "occupational selection in Britain appears to take place largely on meritocratic principles" (1999:217).

Finally, Nielsen's (2006) American behavior genetic longitudinal study of 1,072 sibling pairs (MZ and DZ twins, full siblings, half siblings, cousins, and nongenetically related), designed to pit the ascription thesis against the achievement, thesis found strongly in favor of the latter. The study looked at verbal IQ (VIQ), grade-point average (GPA), and college plans (CPL). Partitioning the variance of all three measures into genetic, shared environment, and nonshared environment components, the heritability coefficients (h^2) were VIQ = .536, GPA = .669, and CPL = .600; the shared environment coefficients (c^2) were VIQ = .137, GPA = .002, and CPL = .030, and the nonshared environmental (e^2) variances were .327, .329, and .370, respectively. Shared environment, of course, is everything shared by siblings, including SES, as they grow up. The proportions of variance explained by class origin across all measures are minuscule compared with the proportions explained by genes and nonshared environment.

None of these studies make the claim that class of origin or the family does not matter. A loving and caring family is preferable to an unloving and neglectful one regardless of what effect it may or may not have on offspring IQ, or any other trait. Families obviously have lasting effects on offspring, but not effects that make offspring similar on cognitive and personality traits. The lack of shared environmental effects in the Nielsen (2006) study, for instance, only shows that common SES background has not made siblings similar in terms of VIQ, GPA, or college plans; i.e., SES has little uniform effect on these variables for children of various degrees of genetic relatedness raised within the same households.

There does remain a small but persistent association between class of origin and class of destination. In the Nettle (2003) British cohort, 39.5% remained in their father's class, but this figure is inflated by the fact that those in the lowest class can only stay the same or go up, and those in the highest class can only stay the same or go down. Even then, class stability cannot all be attributed to class per se because class of origin reflects the same genetic effects in the parental generation that class of destination does in the offspring generation (Nielsen 2006).

IQ and Criminality

If IQ predicts adult SES, and if the lack of success leads to criminal adaptation, then IQ must itself be a predictor of criminal behavior. When evaluating the relatively small (8 to 10 IQ points) difference said to separate criminals and noncriminals (Wilson and Herrnstein 1985), we must remember that researchers do not typically separate what Moffitt (1993) calls *adolescent-limited* (AL) from what she calls *life-course persistent* (LCP) offenders. As statistically normal individuals temporarily responding to the contingencies of their environments in antisocial ways, we would not expect AL offenders to be significantly different from nonoffenders on IQ, and they are not. Moffitt (1993) reports a 1-point mean IQ deficit between AL offenders and nonoffenders but a 17-point deficit between LCP offenders and nonoffenders. Aggregating temporary and persistent offenders on IQ creates the erroneous perception that IQ has minimal impact on antisocial behavior. Other studies find similarly large gaps when subjects are disaggregated. A longitudinal study that looked at high- and low-sensation seekers who were and who were not seriously involved in delinquency found some interesting results (Gatzke-Kopp et al. 2002). High-sensation-seeking delinquents were significantly more impulsive, had significantly lower IQs, and were lower in socioeconomic status than high-sensation-seeking control subjects. Among the low-sensation seekers, only IQ differentiated delinquents (average IQ = 85.5) from nondelinquents (average IQ = 102.8)—the same 17-point difference that differentiated Moffitt's AL and LCP offenders.

While these studies separated temporary and persistent offenders, they did not separate IQ subtest scores, which also leads to an underestimation of the effects of cognitive abilities by pooling verbal IQ (VIQ), which uniformly shows a significant difference between offenders and nonoffenders, and performance IQ (PIQ), which does not. The most serious and persistent criminal offenders tend to have a PIQ score exceeding their VIQ score (P > V) by about 12 points. Averaged across a number of studies, the odds ratio comparing V > P boys (2.6:1 against becoming delinquent) with P > V boys (2.2:1 in favor) was 5.6 (Walsh 2003). Low VIQ indexes poor abstract reasoning, judgment, and school performance, as well as high impulsiveness and low empathy. These traits are not conducive to occupational success, but they are conducive to antisocial behavior, especially if combined with a disinhibited temperament (Farrington 1996).

The most frequently explanation for the IQ/delinquency relationship is that low IQ impacts delinquency and criminality via poor school performance (Ward and Tittle 1994), which is no doubt true. However, a national cohort study found that IQ scores at age 4, long before the children could accumulate school experiences, predicted later delinquency (Lipsitt, Buka, and Lipsitt 1990), and other research has shown that early childhood conduct problems predict adolescent delinquency better than school performance per se (Fergusson and Horwood 1995). In other words, poor cognitive skills and antisocial behavior are evident before children enter school, and poor school performance is just another manifestation of this disability.

Another interpretation of the relationship between IQ and crime is that it is a spurious consequence of the possibility that only the less intelligent criminals get caught. Hirschi and Hindelang (1977) surveyed the available evidence pertinent to this "differential detection hypothesis" and concluded that it was not empirically supported. Moffitt and Silva (1988) also tested the hypothesis with an elegantly designed study of a birth cohort. Dividing the cohort into three groups (self-reported delinquents with police records, self-reported delinquents unknown to the police, and self-reported nondelinquents), they found that the first two groups (detected and undetected delinquents) did not differ from one another; i.e., undetected delinquents are no more intelligent than detected delinquents. Both delinquent groups, however, differed significantly from the nondelinquent group on verbal, but not performance, IQ.

TEMPERAMENT AND SES

Temperament is the second "pillar of differential psychology" identified by Agnew (1992) as important to understanding prosocial/antisocial responses to strain. It is also important for understanding occupational success. Intelligence alone is not sufficient for occupational success; one must also have the requisite temperamental qualities, such as agreeableness, perseverance, patience, and sense of responsibility. It was noted earlier that Nielsen's (2006) behavior genetic study found a higher heritability coefficient for GPA than for VIQ. This is to be expected since GPA is a more "extended" phenotype than VIQ, encompassing such things such as the studious and dutiful application of one's cognitive abilities to the tasks at hand.

Temperament is a phenotypic trait identifiable early in life that constitutes an individual's habitual mode of emotionally responding to stimuli. Temperamental endophenotypes include *mood* (happy/sad), *sociability* (introverted/extraverted), *activity level* (high/low), *reactivity* (calm/excitable), and *affect* (warm/cold), among others. These various components make it easy or difficult for others to like us and to get along with us. Temperamental variation is largely a function of heritable variation in central and autonomic nervous system arousal patterns (Kagan and Snidman 2007; Kochanska 1991; Lemery and Goldsmith 2001). Heritability estimates for the various endophenotypes of temperament range from about .20 to .80, which indicates a substantial environmental contribution (Kagan 2007; Saudino 2005). Although reasonably stable across the life course, environmental input can strengthen or weaken innate temperamental propensities, and different temperamental components emerge at different life junctures as neurological and endocrine arousal systems are fine-tuned by experience, perhaps via epigenetic mechanisms. In other words, "Temperament develops, that is, emotions and components of emotions appear at different ages" (Rothbart, Ahadi, and Evans 2000:124).

It is largely temperamental differences that make children differentially responsive to socialization via evocative rGE. Temperamental unresponsiveness to socialization is exacerbated by the fact that the temperaments of parents and children are typically positively correlated. Parents of children with difficult temperaments tend to be inconsistent disciplinarians, irritable, impatient, and unstable, which makes them unable or unwilling to cope constructively with their children, thus saddling their children with both a genetic and an environmental liability (Lykken 1995; Saudino 2005). Children's temperamental irritability may also adversely affect the quality of parent–child interactions regardless of parent temperament, and thus lead to a situation of non-attachment and all the negative consequences that implies (Rothbart, Ahadi, and Evans 2000). Numerous studies has shown that children with difficult (disinhibited, irritable) temperaments are responded to negatively not only by parents, but also by teachers and peers, and that these children find acceptance only in association with peers with similar dispositions (reviewed by Ellis 1996b). As Caspi (2000:170) summarizes this literature:

> In the early years of life, person-environment covariation occurs because of the joint transmission of genes and culture from parents to offspring. Given that parents and children resemble each other in temperamental qualities, children whose difficult temperament might be curbed by firm discipline will tend to have parents who are inconsistent disciplinarians, and the converse is also true: Warm parents tend to have infants with an easy temperament. Later in life, person-environment covariation occurs because people choose situations and select partners who resemble them, reinforcing their earlier established interaction style.

Conscientiousness

Temperament is the biological foundation upon which personality is built. Personality refers to the relatively enduring, distinctive, integrated, and functional set of psychological traits that result from an individual's temperament interacting with his or her environment. These traits can help or hamper the individual in seeking and keeping a job, and performing well or poorly at it. Conscientiousness is one of the "big five" factors of personality and is particularly important to success in the workforce and climbing the class ladder, and, thus, if the anomie/strain argument has merit to criminal behavior. Conscientiousness is a dimension ranging from well-organized, disciplined, scrupulous, orderly, responsible, and reliable at one end of the continuum, and disorganized, careless, unreliable, irresponsible, and unscrupulous at the other (Lodi-Smith and Roberts 2007). Conscientiousness has been called the "will to achieve," and is highly linked with upward mobility (Kyl-Heku and Buss 1996:49). An analysis of 21 behavior genetic studies of conscientiousness found a median heritability of .66 (Lynn 1996), which still leaves considerable room for environmental influences on trait variance.

Employers can be strongly expected to favor high levels of conscientiousness in their employees and perspective employees, and they do. Conscientiousness is more important in high-autonomy jobs than in low-autonomy jobs because it "affects motivational states and stimulates goal setting and goal commitment" (Schmidt and Hunter 2004:169). In an intergenerational study following subjects from early childhood to retirement, Judge and his colleagues (1999) found that conscientiousness measured in childhood predicted adult occupational status (r = .49) and income (r = .41) in adulthood, which were only slightly less than the correlations between "general mental ability" (GMA) and the same variables (.51 and .53, respectively). Schmidt and Hunter's (2004:170) analysis of GMA and personality variables in attaining occupational success found that "the burden of prediction is borne almost entirely by GMA and conscientiousness."

Another longitudinal study having to do with the influence of temperament on occupational success was conducted by Caspi, Bem, and Elder (1989). This study identified males with a history of temper tantrums in childhood and traced them for 30 years investigating multiple areas of their lives. The majority of bad-tempered boys from middle-class homes ended up in lower status occupations than their fathers, had erratic work histories, and experienced more unemployment than other males with more tranquil temperaments. They were also more than twice as likely as other men to be divorced by age 40, which illustrates the heterogeneity of negative outcomes that can arise from a single temperamental dimension. It also illustrates that individuals with a tendency to be disagreeable tend to take it with them wherever they go, from home, to school, to work, to marriages, as Agnew's (2005) "super traits" theory predicts.

In short, individuals with certain types of temperament tend to develop certain types of personalities, but this innate tendency may be weakened or strengthened by cultural and experiential variables. When we survey the range of subtraits conscientiousness envelops, it is reasonable to assume that this is a trait that almost all cultures and families would expend major effort in trying to inculcate in its members (Roberts, Walton, and Viechbauer 2006). However, persons with disinhibited, irritable, and irresponsible temperaments do not develop the personal qualities they need to apply themselves to the long and arduous task of achieving legitimate occupational success, even if they possess adequate mental abilities. They will become "innovators" and/or "retreatists," not because a reified social structure has denied them access to the race, but rather because they find the race intolerably boring and busy themselves with more "exciting" pursuits instead.

STATUS IN A MERITOCRATIC SOCIETY

Figure 5.1 illustrates the model of the pathway to SES achievement presented here. Some people applaud the meritocratic process; others do not. For instance, in Steven Messner and Richard Rosenfeld's *Crime and the American Dream* (2001), in which they advanced their institutional anomie theory, they point out that America's merit-based competition for success *requires* inequality of outcomes because "winning and losing have meaning only when rewards are distributed unequally" (2001:9). They view this as both criminogenic and unfair, as do Sawhill and Morton (2007:4), who write: "people are born with different genetic endowments and are raised in different families over which they have no control, raising fundamental questions about the fairness of even a perfectly functioning meritocracy."

Fairness is a philosophical issue suffused with contradictory notions. With Sawhill and Morton, one cannot help feeling sorry for individuals burdened with genetic and environmental crosses not of their own making, but how does the moral issue of fairness fit into this sentiment? Those with Thomas Sowell's constrained vision view fairness as an equal opportunity *process*—a nondiscriminatory chance to enter the race—which governments can guarantee via law. Those with Sowell's unconstrained vision view fairness as equality of *outcome*—which implies that all should cross the finishing line at the same time—which no power on earth can guarantee. "Fairness" can only be obtained from those in full control of the resources that are allegedly unfairly distributed. Only governments can begin to approach the kind of control over economic outcomes pleasing to unconstrained visionaries, but who wants the state to have such power?

For constrained visionaries, all are treated fairly if all are subject to the same rules and judged by the same standards. What unconstrained visionaries cannot get over is that life deals us all a certain hand of cards that we must play the best way we can within the rules of the game. We bring to the game advantages and disadvantages, but these are not the fault of the game. I played soccer for 40 years under the same rules and standards of judgment as David Beckham, but our different talents brought Beckham a $250 million contract and me a series of aches and pains, a concussion, and a bum knee. In Messner and Rosenfeld's terms, the rules of the game practically guaranteed that I would fail and Beckham succeed. Beckham had "what it takes" and I didn't—it's really quite that simple. It is not fair that Arabs can't ice fish or Eskimos grow oranges, and surely the dull siblings in Murray's (1997) study felt it grossly unfair that their smarter siblings did so much better in the SES game simply because they inherited different snippets of their parent's DNA. Criminologists (and other social scientists) must shake free of their habit of automatically thinking "discrimination," "prejudice," "bias," and "unfairness" when they observe differences in

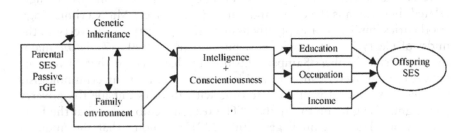

Figure 5.1 Pathway from parental to offspring SES in a meritocratic society.

valued outcomes until they have examined what those who finished the race at the tail end (or did not even get into it) brought with them to it.

The alternatives to meritocracy are either caste-based societies in which one can simply draw a straight arrow from parental SES to offspring SES or failed social engineering experiments exemplified by the old Soviet Union that were supposed to produce equality and "fairness." Note how the pro-meritocratic words of Federal Reserve chairman Ben Bernanke (cited in Sawhill and Morton 2007:1) echo Messner and Rosenfeld's with respect to the system requiring unequal outcomes, but from which he draws the opposite conclusion:

> Although we Americans strive to provide equality of economic op-portunity, we do not guarantee equality of economic outcomes, nor should we. Indeed, without the possibility of unequal outcomes tied to differences in effort and skill, the economic incentives for productive behavior would be eliminated and our market-based economy—which encourages productive activity primarily through the promise of finan-cial reward—would function far less effectively.

Of course, we could temper the intensity of American competitiveness somewhat by recognizing that some are disadvantaged simply by accident of birth. Institutional anomie theory recommends taming the power of the market via decommodification. For instance, the decision to have children could be freed from economic considerations by granting government-guaranteed maternity leave benefits and family allowances/income support, and higher education could be accessible to all people with talent without regard for the financial ability to pay. In other words, policies that ensure an adequate level of material well-being that is not so completely dependent on an individual's ability-based performance in the marketplace.

CRIME AS A CAUSE OF POVERTY

The link between SES and crime often boils down to the assertion that poverty causes crime. There is no doubt that poverty and crime are strongly related, but which is the cause and which is the effect? Many criminologi-cal theories implicitly or explicitly assume that poverty, or "poverty in the midst of plenty," as one sociologist put it (Farley 1990:217), is *the* major cause of crime. Robert Sampson (2000:711) has written that "Everyone believes that 'poverty causes crime' it seems . . . [and many criminologists wonder why anyone] would waste time with theories outside the poverty paradigm." He goes on to say that "The reason we do . . . is that the facts demand it." Frank Schmalleger (2004:223) also notes that the underly-ing assumption of all structural theories is that the "root causes" of crime are poverty and various social injustices. But as he also notes: "Some now argue the inverse of the 'root causes' argument, saying that poverty and what appear to be social injustices are produced by crime, rather than the other way around." Sampson and Schmalleger have committed sociological heresy, but there is much truth in what they say; moral poverty, not economic poverty, is the prime mover of criminal behavior.

Take two hypothetical groups of individuals, one group consisting of individuals born into poverty but who do not engage in or criminal behavior and who have at least average levels of intelligence and conscientiousness, and the other consisting of individuals born to middle-class parents, who embark on a life of crime. It is not difficult to predict which group will have the highest average level of financial success in adulthood, regardless of race or parental class. Barring major accidents or illnesses, a male of any race in the United States is virtually guaranteed to avoid poverty if he does just two things: finishes high school and secures and maintains employment. For females of any race, avoiding unwed motherhood is an added proviso. In 1995, fewer than 3% of white *or* black males who worked full-time year-round were in poverty. Among black females who worked full-time year-round, 7.5% were in poverty compared with 2.2% of similarly situated white females, a difference that probably reflects the greater likelihood of black females having to support young children without the benefit of a male partner or of child support (Thernstrom and Thernstrom 1997:242).

The "crime causes poverty" thesis runs something like this: Individuals who drop out of school, do time in a juvenile detention center, and acquire a criminal record severely compromise their opportunities to gain meaningful employment, to form prosocial networks, to become attractive as a marriage partner to prosocial females, and to lead a "straight" life. Although there are perhaps a few exceptionally talented career criminals who enter middle and old age financially secure, and a few who manage a decent life despite their bad starts, the great majority live their lives in poverty (Raine 1993). Economic studies have found that incarceration reduces employment opportunities by about 40%, reduces wage by about 15%, and wage growth by about 33% (Western 2003).

As well as foreclosing on personal opportunities for middle-class monetary success, crime may also cause poverty by stripping communities where crime is most prevalent of its businesses and its jobs. Why would a small business want to stay in an area where robberies, thefts, break-ins, and muggings are an everyday occurrence? In such areas the cost of doing business is sharply increased (increased insurance premiums and additional expenses for security guards and alarm systems, etc.). After building costs and space constraints, crime is the most important reason for companies either not moving into inner-city areas or for moving out of them if they are already there (Porter 1995). When teachers and children are more concerned about rapes, dope dealing, and assaults, how can any meaningful learning take place in schools serving high crime area? How can people be proud of their homes and hope to accrue equity when all other houses around them are run down and bullet marked? "Poverty is endemic where crime is endemic" is just as plausible a proposition as the proposition "crime is endemic where poverty is endemic." Democratic Congressman John Lewis sums up this argument well: "It is not only poverty that has caused crime. In a very real sense it is crime that has caused poverty, and is the most powerful cause of poverty today" (in Walinsky 1997:11).

STATUS STRIVING AND THE INTRINSIC
REWARDS OF CRIMINAL BEHAVIOR

How does temperament impact the probability of antisocial behavior, and how are these two constructs tied to status striving? Anomie/strain theory shares its deep interest in status striving with evolutionary psychology and views status concerns as fundamental motivating factors behind much of human behavior, both deviant and conforming. They differ on the origins of status striving, however, with anomie/strain theorists viewing it as behavior learned in conformity with a particular set of cultural values and evolutionary psychologists viewing it as a cross-cultural universal. The desire for status is an unlearned universal behavior-motivating feature of all social animals because of its ubiquitous role in reproductive success (Barkow 2006; Buss 2005; Wrangham and Wilson 2004). Of course, evolutionary psychologists are well aware that the specific actions and activities that confer social status are socially determined. That is, although our taste for status runs deep, it means different things in different cultures and subcultures. However, whatever form it takes it is something that social animals, most particularly males, are designed by nature to seek. The status sought by criminals may be 180 degrees askew from that sought by the rest of us, but they seek it for the same reasons.

Just how important are middle-class indicators of status to young males in economically deprived environments relative to other indicators? David Rowe denies that young males compare themselves with "distant middle-class standards," and asserts that the "main source of strain is an attempt by males to win in male–male encounters a high level of social prestige in local peer groups" (1996:305). A life of crime is not merely the default option of those unable to meet the intellectual and temperamental demands of occupational success. Chronic criminals not only lack the requisite characteristics for middle-class success; evidence suggests that their personalities are such that the straight life is unappealing. Studies of criminals in the witness-protection program show that even when provided with such a ready-made straight life and the resources to go with it, approximately 21% of them are arrested under their new identities within two years of entering the program (Albanese and Pursley 1993:75).

Jack Katz's (1988) seminal study of inner-city criminals draws the same conclusion as Rowe: chronic criminals are seduced by a life of action and value their "badass" reputations and the dominance and status that accompany it more than any middle-class American dream. Crime's primary appeal, says Katz, is the intrinsic rewards that accompany it, the thrills, the rush of taking risks and getting away with it, not the frequently negligible material rewards of such activities. This is a reiteration of Cohen's (1955) view of delinquency as motivated by short-run hedonism rather than material gain and as malicious and destructive rather than instrumental.

Intrinsic reward refers to internally derived reinforcements that lead to a greater probability of repeating whatever activity led to that reward. When we survey the meager monetary gains and dismal long-term consequences of crime, the claim that it is an alternative way to achieve monetary success does not cohere well with phenomenological accounts of crime presented to us by the works of theorists such as Katz (1988) and Anderson (1999). There must be something that makes crime appealing for its own sake independent of any extrinsic rewards that may accrue from it. Chronic criminals are certainly motivated by the need for fast cash to feed their "every night is Saturday night" lifestyles, but interviews with criminals find that the internal rewards of committing the crime itself are powerful motivators (Wood et al. 1997). As Jock Young (2003:391) put it: "[t]he sensual nature of crime, the adrenaline rushes of edgework—voluntary illicit risk-taking and the dialectic of fear and pleasure . . . all point to a wide swath of crime that is expressive rather than narrowly instrumental." Because expressive crimes tend to be violent and often committed to gain status ("juice"), an exploration of the biosocial mechanisms underlying the intrinsic pleasure derived from committing them is in order.

Arousal Theory

It has been suggested that the "high" achieved when committing crimes is analogous to that obtained from illicit drugs and from other kinds of risky behavior (Gove and Wilmoth 2003; Sampson and Laub 2005). As the infamous Depression-era bank robber Willie Sutton put it: "I was more alive when I was inside a bank robbing it than at any other time in my life" (Sutton and Linn 1976:120). The common link between all kinds of sensation-seeking behavior is low arousal. Arousal theory is based on the well-established finding that different levels of physiological arousal correlate with different personality and behavioral patterns. In identical environmental situations, some people are suboptimally aroused, and other people are superoptimally aroused. Sub- and superoptimal arousal levels are opposite tails on a normal distribution, with most people being optimally aroused under the normal range of environmental conditions (neither too constant nor too varied). What is optimal level of environmental stimulation for most of us will be stressful for some and boring for others. In other words, many chronically underaroused criminals may engage in crime simply to "feel alive."

Numerous studies have shown that, relative to the general population, serious criminals are chronically underaroused as determined by EEG brain wave patterns, resting heart rate and skin conductance, histories of ADHD, and low levels of the stress hormone cortisol (Ellis 1996b; Scarpa and Raine 2003; van Goozen et al. 2007). Individuals who are chronically bored and who continually seek intense stimulation are not likely to apply themselves to school or endear themselves to employers, regardless of their IQ. Yet these individuals have the same needs as their less frenetic peers for status, and this is what may lead them into trouble. We know that suboptimal arousal is linked to the 7-repeat polymorphism of the DRD4 allele. Several molecular genetic studies have linked the DRD4 polymorphism with lack of behavioral control and addiction (Kreek et al. 2005). Substance abuse is frequently considered an effort to raise one's level of arousal to optimal levels (Vaughn 2009).

Recall that the basic regulator of neurological arousal is the reticular activating system (RAS), the brain's filter system that determines to what stimuli higher brain centers will pay attention. Whatever the influences of the DRD4 allele (or any other allele) on behavior may be, they are mediated by environmental input, and environmental input has to be funneled to the brain via the RAS. Some individuals are *augmenters*; that is, they possess a RAS that is highly sensitive to incoming stimuli (more information is taken in and processed); others are *reducers,* possessing a RAS that is unusually insensitive. No conscious attempt to augment or reduce incoming stimuli is implied. Augmentation or reduction is solely a function of the differential physiology of the RAS. Augmenters prefer more constancy than variety in their world, and seek to tone down environmental stimuli that most of us find to be "just right." Such people are rarely found in criminal populations. Suboptimally aroused people, on the other hand, are reducers who are easily bored with "just right" levels of stimulation, and continually seek to boost stimuli to more comfortable levels for them (Raine 1997; Zuckerman 2007).

Gove and Wilmoth (2003) have suggested that the risky low-payoff crimes produce positively reinforcing neurological events involving dopamine arising in the ventral tegamental area and terminating at the nucleus accumbens. According to Gove and Wilmoth, the stress and anxiety accompanying the anticipation and commission of high-risk crimes activates the endogenous opioids that exert opiate-like effects by interacting with central nervous system opioid receptors (*opioid* refers to naturally occurring endogenous chemicals such as the endorphins; *opiate* refers to exogenous drugs such as heroin that bind with opioid receptors). The activation of these opioids, in turn, results in activation of dopamine synapses in the nucleus accumbens to induce the neuropsychological high after the successful completion of the act. The high is more intense the more arduous and risky the behavior is.

Why would humans (and other animals) have evolved a neurological reward system for engaging in risky behavior that may afford them little or no tangible reward? Gove and Wilmoth suggest that the major function of the opioids is to counteract the negative effects of stress. If stress (such as the anxiety accompanying burglarizing a house) were not negated somewhat, then we may never engage in any kind of risky behavior, legal or otherwise. Noting that sensation-seeking is much more prevalent in males than in females, and that much of sensation-seeking involves risky behavior, it has been speculated that sensation-seeking, despite its many dangers, may have contributed to male reproductive success in ancestral populations (Crawford 1998; Zuckerman 2007). Risky behavior may have often been necessary to obtain scarce resources in a competitive environment, and sensation-seeking may be the proximate mechanism that aided immature and disadvantaged males in ancestral environments to pursue contingent cheating strategies that improved their reproductive success (Crawford 1998). In modern environments, this search for amplified excitement often leads the underaroused person into conflict with the law, particularly if the person lacks the intellectual and temperamental resources to pursue socially acceptable ways of seeking thrills.

NEUROHORMONAL UNDERPINNINGS OF STATUS HIERARCHIES

A large body of literature indicates that testosterone (T), serotonin (5-HT), and cortisol are related to dominance and status across animal species from fish to humans (Anestis 2006; Archer 2006; Mehta and Josephs 2006). Although there is a complex interacting stew of other chemicals and a number of different receptor and transporter molecules for them, twisted feedback loops, and social-environmental contexts to consider (see Summers 2002), the basics are that high levels of T foster dominance and extraversion, low cortisol reflects low levels of anxiety, stress, and fear, and high 5-HT fosters confidence and self-esteem (Figure 5.2). While basal levels of these substances are heritable, their secretion patterns are highly dependent on environmental context, rising and falling according to the organism's experiences. In well-established status hierarchies, leaders may best assert their dominance by a subtle mix of coercion and bonhomie, while in chaotic hierarchies more aggressive strategies may be needed to maintain control, with a shift in strategies perhaps favoring a different mix in the levels of these chemicals (Anderson and Summers 2007).

For instance, although T appears to be the best predictor of status/dominance attainment among most animal species, there is no uncomplicated linear causal relationship between the two variables. A high T male with a hyperactive ANS and/or HPA axis will encounter anxiety and fear, and thus elevated cortisol, in the face of status challenges. Cortisol suppresses T, and so the combination of anxiety and the loss of T driven surgency may lead him to withdraw from future status competitions. However, although chronically high levels of cortisol are associated with subordinate

High Status Males		Low Status Males	
Substance	Major Characteristics	Substance	Major Characteristics
High T	Dominance, Extroversion	Low T	Subordination, Introversion
High 5-HT	Calm, Confident	Low 5-HT	Impulsive, Insecure
Low Cortisol	Low fear and anxiety	High Cortisol	High fear and anxiety

Figure 5.2 Basic neurohormonal features of high- and low-status males. T= testosterone. 5-HT = Serotonin.

social status and with aggression inhibition, acutely elevated levels and/or chronic extreme low levels tend to promote aggression (Kruk et al. 2004; van Goozen et al. 2007).

Experiments with nonhuman primates suggest that serotonergic mechanisms may calibrate or modulate hormonal (cortisol and T) responses in status competitions (Anderson and Summers 2007; Zuckerman 1990). Artificially augmenting 5-HT in male vervet monkeys typically results in them attaining high dominance status in the troop (Raleigh et al. 1991). In naturalistic settings, the highest ranking males typically have the highest levels of 5-HT, which is indicative of a confident and mellow leadership style (Anestis 2006) and the lowest ranking generally have the lowest levels. In established dominance hierarchies, low-ranking males defer without much fuss to higher ranking males over access to females and other resources. When the hierarchy is in flux, low-status, low 5-HT males usually become the most aggressive in status competitions. Serotonin levels of newly successful males rise to levels commensurate with their new status (Brammer, Raleigh, and McGuire 1994), which indicates that environmental events strongly influence serotonin's secretion patterns.

Criminologists are most concerned with status competitions that take place in chaotic environments. Risky violent confrontations among males over status issues are most often observed in environments lacking firmly established dominance hierarchies and in which social restraints have largely dissolved. These environments have been termed "subcultures of violence" or "honor subcultures" in which taking matters into one's own hands is seen as the only way to obtain the all-important "juice" (status) on the street (Anderson 1999; Mazur and Booth 1998). Serious assaults and homicides among members of honor subcultures are generally the result of trivial altercations over matters of honor, respect, and reputation in the context of a culture where the violent defense of such intangibles is a major route to status (Bernard 1990; Mazur and Booth 1998). Assaults and homicides tend to take place in front of an audience composed of friends of both the killer and the victim, thus squeezing the maximum amount of "juice" from the incident (Buss 2005; Wilson and Daly 1985). Reacting violently to even minor threats to one's status lets others know that "You can't push *me* around!" (Fessler 2006)

The cost/benefit ratio attending the violent behavior of culturally disadvantaged males, while seeming to defy rational choice assumptions, is quite understandable when viewed by the light of evolutionary theory. According to Daly and Wilson (1988:129), killing has been "a decided social asset in many, perhaps most, prestate societies," and dueling over trivial matters of "honor" was ubiquitous among the aristocracy of Europe and the American South until fairly recently (Baumeister, Smart, and Boden 1996). Status, prestige, and respect are so highly valued by males because these things mattered socially more than just about anything in ancestral environments in terms of reproductive success. We should not be too surprised that similar duels over trivial matters of honor still occur in subcultures where there is not likely to be much respect for laws forbidding it, and where alternate forms of status striving are perceived to be unavailable. Although seeking status through violence is maladaptive in modern societies, our psychological mechanisms were crafted to solve status problems faced by males in environments far different from those faced by young males in the inner cities of the modern world. From an evolutionary perspective, the more young males come to devalue the future, the more risks they are willing to take to obtain their share of street status (see the Wilson and Daly [1997] study of Chicago neighborhood homicide rates in Chapter 3).

The same kinds of relationships between serotonin levels and self-esteem, status, impulsivity, and violence are consistently found among human males, indicating a common set of fitness concerns among all primates (van Goozen 2007; Virkunen, Goldman, and Linnoila 1996). As previously noted, rising to a dominant position in a status hierarchy among primates is not a matter of individual combativeness. Confident and ambitious individuals form alliances, coalitions, and "gangs" within the troop to help them to achieve their aim just as aspiring human leaders do (Raleigh et al. 1991). Given the bidirectional relationship between social status and the various neurohormonal mechanisms, it may well be that natural selection has equipped us to adjust ourselves to the social statuses we find ourselves in within well-ordered groups. This is not social Darwinism asserting that the status quo exists because it is naturally ordained, and therefore good. These same mechanisms also equip those with little to lose with the necessary biological wherewithal to attempt to elevate their status by taking violent risks when social restraints are weak (Anderson and Summers 2007; Brammer, Raleigh, and McGuire 1994).

The relationship between low serotonin and impulsive violent behavior is consistently found among humans, and has been called "perhaps the most reliable findings in the history of psychiatry" (Fishbein 2001:15).

6 The Social Learning Tradition and Adolescence

The social learning tradition in criminology is a social process perspective that has its theoretical home in symbolic interactionism, which in turn has its home in philosophical idealism (a system that avers that we only experience reality through our subjective experience of it). Unlike social structural theorists who sometimes write about social structures as if they exist apart from human thought and activity, symbolic interactionists focus on how people interpret their social reality, and how they create, sustain, and change it. The perspective is summed up in the *Thomas theorem*: "If men define situations as real, they are real in their consequences" (in Nettler 1978:272). Social learning is thus about how people define social reality and how they learn these definitions as they are socialized into the groups of which they are a part and is the other face of the gene/environment whole.

Humans are exquisitely designed to respond to the environment and to incorporate it into their genomes and neural circuitry; i.e., to learn. Learning is some change in behavior, belief, or ability in response to experience, and is a way for organisms to adapt to their immediate environment in ways that could not possibly be specified a priori by their genes: "The extended period of neuroplasticity is an aspect of human nature that allows and *requires* environmental input for normal human development" (Wexler 2006:16). The most important feature of the environment for humans is other humans, and it is the importance of other humans in the environment that social learning theorists emphasize. Biosocial and social learning theorists are thus in agreement about the importance of social learning, but like so many other things, the devil is in the details. Social learning and biosocial criminologists differ on at least four interrelated points:

- Social learning theorists do not explicitly recognize the role of biology in learning.
- Social learning theorists do not explicitly recognize individual differences in ability to learn.
- Social learning theory does not explicitly recognize the people in identical environments may learn vastly different lessons because of those differences.

- Social learning theorists explicitly share sociology's unconstrained vision that assumes antisocial behavior is something that is learned rather than something that will emerge in the absence of moral socialization.

Differential Association Theory (DAT)

DAT is the modern progenitor of social learning theories and is the brain-child of Edwin Sutherland. Sutherland's ambitious agenda was to develop a theory that could identify conditions that must be present for crime to occur and that are absent when crime is absent (i.e., a necessary and sufficient cause of criminal behavior) without reference to psychology or biology, to which he was openly hostile. Although Sutherland often used *crime, criminality*, and *delinquency* interchangeably, his theory is clearly couched in terms of juvenile delinquency. This chapter therefore concentrates on anti-social behavior in childhood and adolescence. It is important to understand the biological mechanisms associated with antisocial behavior surrounding the adolescent years because such behavior is so prevalent during this period. Almost all adults arrested for felony crimes have juvenile records, and longitudinal studies show very low rates of adult-onset criminal behavior of about 4% (Elliot, Huizinga, and Menard 1989). A more recent study (Gomez-Smith and Piquero 2005) found that 7.9% of a sample of offenders was adult-onset, although it is possible that these individuals were unde-tected juveniles offenders.

DAT asserts that humans, like chameleons, take on the hues and colors of their environments, blending in and conforming with natural ease. Most Americans probably like baseball, hot dogs, apple pie, and Chevrolets, as a Chevrolet commercial used to remind us. But do we prefer these things over, say, soccer, bratwurst, strudel, and Volkswagens because the former are demonstrably superior to the latter, or simply because we are Americans and not Germans? We view the world differentially according to the attitudes, beliefs, and expectations of the groups around which our lives revolve; it could hardly be otherwise, particularly in our formative years. Sutherland's basic premise is that delinquent behavior is learned just as readily as we learn to play the games, enjoy the food, and drive the cars that are integral parts of our cultural lives.

Sutherland was more interested in the process of learning delinquent attitudes than in their alleged structural origins. His theory takes the form of nine propositions outlining the process by which individuals come to acquire attitudes favorable to criminal or delinquent behavior. His first three propositions assert that criminal behavior is learned in the process of social interaction, particularly within intimate personal groups. These propositions also assert that criminal and delinquent behavior are not biologically inherited, the result of psychological abnormalities, invented anew by each criminal, nor learned from impersonal communication (i.e., from movies, magazines, and other such distant teachers). The learning of delinquent behavior involves the same mechanisms involved in any other learning, and includes specific skills and techniques, as well as motives, rationalizations, justifications, and attitudes.

The key proposition in the theory is "A person becomes delinquent because of an excess of definitions favorable to violations of law over definitions unfavorable to violations of law" (Sutherland and Cressey 1974:77). Learning delinquent conduct is not a matter of simple imitation; it is a process of modeling the self after, and identifying with, individuals who hold an excess of procriminal definitions over anticriminal definitions, and whom we respect and value. "Definitions" refer to meanings that experiences have for us, how we see things, our attitudes, values, rationalizations, and habitual ways of viewing and responding to the world. In short, DAT asserts that how we think determines how we behave—cognition causes conduct.

Associations with others holding definitions favorable to violation of the law vary in *frequency, duration, priority,* and *intensity.* That is, the earlier in life we are exposed to criminal definitions, the more often we are exposed to them; the longer those exposures last, and the more strongly we are attached to our mentors who supply us with them, the more likely we are to internalize them. DAT thus shares with evolutionary psychology the notion that crime and delinquency are normal products of normal individuals engaging in normal social interactions.

Sutherland agreed that criminal behavior is to some extent an expression of general needs and values (e.g., the need for status), but he insisted that they are not explanations of crime since noncriminal behavior is also an expression of the same needs and values (Sutherland and Cressey 1974). Thus he took issue with strain theorists' emphasis on such things as causes of antisocial behavior. Sutherland was saying that since these needs are constants, they cannot be used as explanations for behavior that varies between criminal and noncriminal. Sutherland seemed oblivious here to the distinction between ends and means. It is the means one uses to meet one's needs that anomie/strain theory (and almost all other theories) seeks to explain, not the ends. Be that as it may, for Sutherland criminal definitions are the real causes of crime.

Sutherland's excess of "criminal definitions" proposition has been criticized as both "true and trivial," because all it essentially says is that people are apt to engage in acts contrary to the law when they don't respect the law (Hirschi 1969:15). Why some people hold these definitions and others do not is a question that needs answering. The response from DAT theorists would be that people acquire them through association with delinquent peers, and they may point out that the correlations found between association with delinquent peers and delinquent behavior are among the strongest found in social science. Because antisocial definitions flourish in delinquent groups, proponents of the theory interpret these correlations to mean that association with delinquent peers *causes* delinquency, or at least precedes it (Warr 2002). Associating with delinquent peers is certainly a very good predictor of a person's own delinquency, but is it causal? What is in need of explanation is why people have the associations they do, and this is where the theory can benefit from biosocial insights.

Sutherland recognized that individual differences affect relationship patterns in his1939 formulation of DAT when he wrote: "Individual differences among people in respect to personal characteristics or social situations cause crime only as they affect differential association or frequency and consistency of contacts with criminal patterns" (1939:8). Sutherland was specifying a path model in this proposition in which differential association is clearly an *intervening* variable between individual differences and crime as follows:

Individual differences → Contact with criminal patterns → Crime.

Sutherland continued to claim, however, that the antisocial attitudes and values learned in association with criminal peers was *the* cause of crime. It is difficult to see from the form of his 1939 proposition why he failed to recognize that if individual differences sort people into different relationship patterns, it may be the characteristics of people that "cause" the activities they engage in, that their associations merely facilitate and accentuate them, and that their "definitions" merely grant them permission. Perhaps he did, and perhaps the realization became an embarrassment to his goal of developing a pure sociological theory. In any case, he dropped any reference to individual differences in later versions of the theory.

Consistent with the claim that individual differences lead to different associations, various reviews of friendship patterns (e.g., Rodkin et al. 2000) have shown that the propensity for a given pattern of activity (including antisocial activity) precedes association with like-minded individuals. "Birds of a feather flock together" and become more alike on the basis of association. Reviews of the onset of delinquent behavior do find that such behavior typically *precedes* gang membership (Matsueda and Anderson 1998). Association with delinquent peers acts more like a catalyst speeding up and enhancing antisocial conduct among the predisposed than as a stimulator of uncharacteristic behavior among the innocent. As Gottfredson (2006:92) summarized a number of studies addressing this issue: "The evidence is consistent with the proposition that much of the variance in peer effects on delinquency is attributable to the selection effect of like individuals associating together."

Although we see no mention of it in the writings of social learning theorists, behavior genetics has been informing us for some time that Gottfredson is correct. An early study (Rowe and Osgood 1984) showed that about 60% of the correlation between antisocial behavior and association with delinquent peers was due to common genetic factors; i.e., individuals select (active rGE) or are select by (evocative rGE) delinquent peers because of the traits that the individual and the peer group have in common. A later study of 1,036 sibling pairs (Cleveland, Wiebe, and Rowe 2005) found that genetic factors accounted for 64% of the variance in delinquent (defined in terms of substance abuse) peer affiliation.

Two more recent studies highlight the involvement of both genes and environment in the peer group choices people make and in peer group delinquency. The first studied 533 MZ and 558 DZ twin pairs with a mean age of 15.11 years (Button et al. 2007). This study showed that, overall, genes accounted for 21% of the variance in antisocial peer affiliation, with the shared and nonshared environment accounting for 40% and 39%, respectively. Shared environment is thus a very important source of variance in peer affiliation in mid-adolescence. Looking at the patterns of G x E interaction in the sample, Button and her colleagues found that the heritiabilities of "conduct problems" increased as the frequency and intensity of delinquent peer group affiliation escalated. Otherwise stated, the magnitude of the genetic effects on conduct problems increased as the level of association with delinquent peers increased.

A longitudinal study of peer group deviance using data from 469 MZ and 287 DZ twin pairs followed from age 8 to 25 found that as twins matured and created their own mini-worlds (active rGE), genes played an increasingly larger role in peer choice (Kendler et al. 2007). As twins left their parental homes, shared environmental effects (an important determinant in childhood) diminished and nonshared environment became more important. The heritabilities of peer group deviance across five age categories rose from .138 in the 8–11 age group to .462 in the 22–25 age group. Shared environment effects declined from .27 in the youngest group to .143 in the oldest group, and nonshared effects stayed relatively stable at .342 and .352, respectively. These studies bring home the fact that criminologists can ill afford to ignore the shifting patterns of genetic and environmental effects on peer affiliations and deviance across the life span. Even molecular genetics is getting into the act, with one recent study demonstrating a significant effect of the 10-repeat allele of the dopamine transporter (DAT1) on peer group affiliation, controlling for a number of other putatively causal variables (Beaver, Wright, and DeLisi 2008).

As the preceding studies show, the extent of peer group association is context as well as genetically contingent. Even individuals who persist in offending well past adolescence may not have joined an antisocial peer group because they had an affinity for antisocial behavior, as those who join judo clubs and participate with gusto presumably have an affinity for the martial arts. Peer choices in some neighborhoods are not as varied as they are in others (e.g., the neighborhoods in which most of the subjects in the aforementioned genetic studies lived), nor are the consequences of those choices similar across all environments. Like the impact of a few rotten apples in a large barrel of good ones, the moral ambiance in inner-city neighborhoods tends to be determined by its drug dealers and hustlers, not its decent people who often feel it imperative to abide by the pernicious "code of the street" or become subject to reprisals (Anderson 1999).

These facts lead defenders of DAT to insist that the concept of differential social organization accounts for the associations people have (Akers and Jensen 2006). Children associate, play, and become friendly with individuals in the neighborhoods their parents provide. If the neighborhood is so densely populated that "good" and "bad" kids live on the same streets or in the same buildings, there is little element of choice as to the company children keep. In certain neighborhoods, delinquent peers may indeed "cause" delinquency among youths who are otherwise insulated from it, as well as facilitate and accelerate it among others who have an affinity for it. Thus, while the causal-order criticism may be valid for children growing up in better neighborhoods with roughly equal access to both pro- and antisocial peers, it may not be valid for kids growing up in the urban slums where prosocial peers and legitimately successful adult role models are rare. Ronald Akers (1999:480) responded to the "birds of a feather" adage with the equally pithy riposte: "If you lie down with dogs you get up with fleas."

Learning, Attitudes, and Behavior

Criminologists in the social control tradition take issue with DAT's assumption that antisocial behavior is learned, and stress that antisocial behavior comes naturally to the unsocialized individual (Gottfredson 2006; Sampson 1999). "What is there to be learned about simple lying, taking things that belong to another, fighting and sex play," asked an early critic (Glueck 1956:94). Individuals certainly learn to get better at doing these things through their associations with other like-minded individuals, but they hardly have to be taught them. What they have to be taught is how to curb these natural behaviors, what constitutes moral behavior, and how to consider the rights and feelings of others.

If antisocial behavior "comes naturally," the implication is that behavior precedes attitudes, and thus behavior causes definitions favorable or unfavorable to law violation rather than the other way around. The primacy status of attitudes versus behavior depends on a large number of factors such what attitudes and behaviors we are talking about, the strength and emotional content of the attitude, and whether or not an attitude may conflict with self-interest, as well as the kind of people we are talking about. Studies consistently find that behavior has a stronger impact on attitudes than attitudes have on behavior when self-interest is involved (Crano and Prislin 2006). This would mean that Sutherland's "definitions favorable" are nothing more than the morality of expediency, a set of post hoc rationalizations that are caused by, rather than causes of, behavior.

Mark Warr (2002) believes that at some level certain individuals come to accept antisocial definitions as righteous and makes a distinction between compliance and private acceptance of group attitudes, values, and behavior. *Compliance* refers to individuals who conform to group attitudes and behavior without privately agreeing with the underlying morality of what they are doing. *Private acceptance* refers to the acceptance of the beliefs, attitudes, and values expressed by the delinquent group as right and proper. Warr takes DAT to task because it "was built squarely on the idea of private acceptance" (Warr 2002:7), probably because of the allure of the idea of attitude/behavior consistency and the assumption of the priority of attitudes. We know that even if there is some level of private acceptance, the majority of juveniles who run with antisocial peers during adolescence desist by their early to mid-20s (Caspi and Moffitt 1995).

Social Learning Theory

The sum of these criticisms is that it is not necessary for attitudes to be transferred in intimate social groups before offending becomes probable, and the transmission of antisocial attitudes seems less important than other social learning mechanisms such as peer pressure, modeling, imitation, and vicarious reinforcement. DAT has been revised across a number of dimensions in response to such criticisms, the most thorough being social learning theory (SLT). The theory originated with Robert Burgess and Ronald Akers (1966), and was designed to discover the mechanisms (beyond looking at frequency, priority, duration, and intensity) by which antisocial behavior is learned. Burgess and Akers applied the powerful mechanisms of operant psychology to the vague "definitions favorable" concept of DAT.

Operant psychology is a perspective on learning that asserts that behavior is governed by its consequences. *Operants* are behaviors that operate on the environment with the intention of bringing about some positive outcome for the actor. Behavior has two general consequences: it is reinforced or it is punished. If it is reinforced it is likely to reoccur in similar situations in the future, and if it is punished it is less likely to reoccur. For instance, acting like a "badass" is an operant, and if a gang member is rewarded with the approval of other gang members for acting that way, the behavior is said to have been reinforced, and thus likely to be repeated. Expressed attitudes to acting bad are likewise operants that are reinforced by the approval of peers who think such attitudes are "cool." Behavior may emerge initially by imitation, modeling, or even spontaneously, but how well that behavior is learned, and whether it is repeated, depends on how others in the immediate social environment react to it (Akers and Jensen 2006).

Behavior is reinforced either positively or negatively. Positive reinforcement occurs when a reward (all reinforcers are rewards, but not all rewards are reinforcers) is received, and negative reinforcement occurs when some aversive condition is avoided or removed. Examples of positive reinforcement might be the tangible proceeds obtained in a burglary or the intangible but immensely desired status a gang member might gain by beating or killing a member of a rival gang. Beating up people then becomes conditioned behavior for this person due to its rewarding consequence. Examples of negative reinforcement might include the freedom from boredom a suboptimally aroused delinquent obtains by running with a gang or the removal of an unwanted reputation after demonstrating some act of bravado.

Punishment can also be positive or negative. Positive punishment is the application of some aversive stimulus, such as a prison term or disapproval from some valued source. Negative punishment is the removal of a pleasant stimulus, such as withdrawal of parental love, or the loss of status in the gang in response to some disapproved behavior such as "punking out."

SLT accepts that Sutherland's "definitions" are normative meanings defining behavior as right or wrong, but views them as verbal behaviors that are exhibited because they and the behaviors they refer to have been reinforced. The acquisition of definitions favorable to delinquency or conformity depends on the individual's history of reinforcement and punishment attached to his or her actions. This is not a simple process of adding and subtracting rewards and punishments. Rewards and punishments are differentially valued and influential according to their source, and according to the meaning they have for the person experiencing them. A dressing down by a teacher, for instance, may be experienced by some as punishment because they value the teacher's approval, and for others it may act as a reinforcer of the behavior that elicited the teacher's response because it served as a source of levity and approval among valued classmates. Thus, the social context is an extremely important component of SLT: "Most of the learning relevant to deviant behavior is the result of social interactions or exchanges in which the words, responses, presence, and behavior of other persons makes reinforcers available, and provide the setting for reinforcement" (Akers 1985:45).

Another important concept in operant psychology is *discrimination*, which means being able to discriminate between stimuli based on prior experiences. The process of socialization is one of becoming "fine-tuned" to the nuances of social life so that we can operate more efficiently in a variety of contexts by making progressively finer and finer distinctions among the stimuli we experience. Discriminative stimuli are thus signals or clues transmitted by others indicating the kinds of behaviors that will be rewarded or punished in a particular social context. Consider the different social signals presented to a person stopped in the street by a little girl, a clergyman, a Hell's Angel, a police officer, an aggressive wino, or the person's mother, and then consider his or her possible responses. The person's response will represent what he or she has previously learned (personally or vicariously) about those people or others like them.

Unlike the anomie/strain tradition, which has shown a slow but steady reductionist trajectory over the decades, the social learning tradition has changed very little. Akers himself has written that his social learning theory is not in competition with DAT, but rather that it is "a broader theory that retains all the differential association processes in Sutherland's theory" (1994:94). Cao (2004:97) points out that both DAT and SLT begin with a blank-slate concept of human nature and assume "a passive and unintentional actor who lacks individuality . . . Both are better at explaining the transmission of criminal behavior than its origins" and because of their "limited conception of human nature learning theories generally also ignore the differential receptivity of individuals to criminal messages."

By couching these processes in terms of operant psychology, Akers (Burgess has long abandoned this line of thinking for biosocial science) did help to clarify some of the murkier aspects of DAT, but he does not venture into the realm of individual differences in the ease or difficulty with which persons learn prosocial versus antisocial behavior. Because of different constitutions, some people naturally find hell raising more exiting (and thus more reinforcing) than others; some are more susceptible to short-term rewards because they are impulsive; some are better able to appreciate the long-term rewards of behaving well; and some are more ready to engage in aggressive behavior than others because of their temperaments and will find such behavior reinforcing. Gwynn Nettler (1984:295) puts it in classical G x E interaction terms: "Constitutions affect the impact of environment. What we learn and how well we learn it depends on constitution . . . The fire that melts the butter hardens the egg."

In response to Sampson's (1999) criticism that his SLT neglects individual differences, Akers replied that he did not neglect them because he wrote that "An individual in a low crime group or category who is nevertheless more exposed to criminal associations, models, definitions, and reinforcement than someone in a high crime group or category will have a higher probability of committing criminal or deviant acts" (1999:482). This is an explanation in terms of different *environments,* not in terms of different *individuals.* It reinforces Cao's (2004) point that SLT reflects the typical sociological appetite for positing equipotential automaton-like individuals entering different environments at one end and emerging out the other as criminal or not criminal based solely on what they are exposed to, and once again demonstrates why criminology is sorely in need of the biosocial sciences.

THE AGE-CRIME CURVE AND ADOLESCENCE

Figure 6.1 presents a graph showing crime rates by age in 1999. Although it represents just one year, the pattern is so law-like that it stands as a close approximation across time and cultures (Ellis and Walsh 2000). The kurtosis of the curve varies by time and culture, but it is invariably present. Noting the constancy of the age-crime distribution, Charles Goring concluded in 1913 that it conformed to a "law of nature" (in Gottfredson and Hirschi 1990:124). Laws of nature describe regularities of nature; they do not explain why these regularities happen. Sociologically trained criminologists have been unable to explain why this pattern is consistently

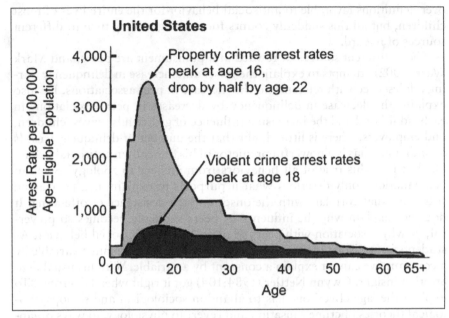

Figure 6.1   Arrest rates per 100,000 in the U.S. by age in 1999. Source: Ellis &
Walsh, *Criminology: A Global Perspective* (2000). Reprinted with permission Ellis
and Walsh.

present. For instance, Hirschi and Gottfredson (1983:554) have stated that
"the age distribution of crime cannot be accounted for by any variable or
combination of variables currently available to criminology." Shavit and
Rattner (1988:1457) share this opinion when they write that delinquency
remains "unexplained by any known set of sociological variables."

Other criminologists disagree, but in doing so they offer some very obtuse
explanations. For instance, David Greenberg (1985) offers us a "strain"
explanation and wants us to believe that because the pocket money parents
give to their children is insufficient for their needs, they rob and steal to
make up the difference! It escapes him that adults are subjected to considerably more strain with families to support and mortgages to pay with
salaries that may be inadequate without resorting to crime to make up the
difference. It is this kind of desperate explanation that so infuriates those
who want to move beyond a pure sociological criminology.

Then there is Ronald Akers (1998:338), who believes that "Age-specific
[crime] rates differ because individuals are differentially exposed to the
learning variables at different ages." Robert Sampson (1999:447) replies
that such an explanation "seems rather incongruous [in that] teens are suddenly able to reverse course after more than a dozen years of anti-delinquent
learning ratios." In other words, definitions favorable to prosocial behavior have presumably enjoyed priority, frequency, duration, and intensity

over definitions favorable to antisocial behavior for the entire lives of most children, but all this suddenly counts for nothing as they turn to different sources of reward.

These different sources of reward and punishment are peers, and Mark Warr (2002) attempts to explain the onset and increase in delinquency during adolescence with reference to the increase in peer associations. He also explains the decrease in delinquency by decreases in peer associations in early adulthood and the increasing influence of girlfriends, wives, children, and employers. There is little doubt that the number of delinquent friends an individual has is a powerful predictor of his or her own antisocial behavior; but isn't this true of all behaviors from archery to zoology? Such an "explanation" only describes what it purports to explain; i.e., portraying situations that correlate with the onset and the desistence of offending. It does not explain why the influence of peers suddenly becomes so powerful, or why association with peers so often leads to antisocial behavior. As students in Statistics 101 are aware, just as we cannot explain a variable by a constant, we cannot explain a constant by a variable. With his usual penetrating insight, Gwynn Nettler (1984:104) got it right when he wrote: "To explain the 'age effect' one has to abandon sociological and sociopsychological theories of crime causation and revert to physiology, to ways organisms differentially function with age." In other words, the invariance of the age-crime curve must be explained by something which itself is invariant (the relative peakedness of the curves, indicating changes in prevalence, are explicable in terms of social, political, and economic variables, but not their ubiquitous presence), and that is age-specific developmental physiology.

BIOSOCIAL EXPLANATIONS FOR THE AGE-CRIME CURVE

Social influences on behavior during adolescence are experienced in the context of the profound neurological and endocrinal changes that occur during the second decade of life. The inaugural event for all these changes is the onset of puberty, the invariant age-dependent physiology across time and place with which Nettler was concerned. Puberty is the developmental stage that occurs around 11 years of age for girls and 12 for boys in the modern Western world marking the onset of the transition from childhood to adulthood and preparing us for procreation (Sisk and Zehr 2005). Puberty initiates a cascade of physical, hormonal, and neurological events that sometimes dramatically affect the behavior of adolescents.

Adolescence is a period of social limbo in which individuals no longer need the same level of parental care as children but are not yet ready to take on the roles and responsibilities of adulthood. Socially defined adulthood means taking on socially responsible roles such as acquiring a steady job and starting one's own family that mark individuals as independent members of society. Unfortunately, the legal definition of adulthood as 18 years of age rarely matches socially defined adulthood today. The decreasing age of puberty combined with the increasing time required to prepare for today's complex workforce has led to a large "maturity gap" which provides fertile soil for antisocial behavior (Moffitt 1993).

Adolescence is a normal and necessary period in the human life span (Rosenfeld and Nicodemus 2003). There is much to learn about being an adult, and adolescence is a time to experiment with a variety of social skills before putting them into practice in earnest, and a time of sorting out what aspects of the previous generation to retain and what to discard. Although rebellious teens are often a source of parental despair, if teens are to become capable of adapting to new situations it is necessary to temporarily strain the emotional bonds with parents that served their purpose well in childhood. For adolescents not to assert themselves would be to hinder their quest for independence. Adolescents must leave their childhood nests and bond and mate with their own generation and explore their place in the world. Leaving the nest is fraught with risk, but it is an evolutionary design feature of all social primates as males seek out sexual partners from outside the rearing group. Fighting with parents and seeking age peers "all help the adolescent away from the home territory" (Powell 2006:867). This parent/child conflict rarely results in permanent fracturing of the bond, and it has often been found that moderate conflict with parents leads to better postadolescent adjustment than either the absence of conflict or frequent conflict (Smetana, Campione-Barr, and Metzer 2006).

Natural selection has provided adolescents with the necessary tools to engage in all this experimentation, such as the huge increase in testosterone (T) that accompanies puberty (Felson and Haynie 2002). Figure 6.2 illustrates T fluctuations across the life span. T organizes the male brain during the second trimester of pregnancy so that it will respond to the pubertal surge of T in male typical ways. After T *organizes* the male brain there is little difference between the sexes in levels of T until puberty when the second surge of T—dramatically higher for males than for females—*activates* the male brain to engage in male-typical behavior (Ellis 2005).

In addition to preparing us for procreation, the evolved purpose of the pubertal T surge is to facilitate the behaviors—risk taking, sensation seeking, sexual experimentation, dominance contests, self-assertiveness, and so on—that cause so much consternation among parents. None of these behaviors are necessarily antisocial, although they can easily be pushed in that direction in antisocial environments. Nor do biosocial criminologists claim a causal role for T in these behaviors. Although T levels are highly heritable, at least 40% of the variance is attributable to environmental factors (Booth et al. 2006). T levels rise and fall to help organisms meet the challenges they confront. The "need" to conform to risky behavioral patterns, to seek dangerous sensations, and to engage in dominance competitions with other males certainly qualifies as challenges that would require raising T levels to meet them (Mazur 2005).

Adolescent risk taking does not mean that teenagers lack the reasoning capacity to evaluate danger. In purely rational terms, the reasoning of 15-year-olds is on par with adults in that they are just as able to perceive risk and estimate vulnerability (Reyna and Farley 2006). The problem is that teens tend to discount their perceptions and risk estimates because of poor impulse control and poor emotional regulation, not poor logical reasoning. Adolescents can act just like adults under nonstressful conditions, but they fall back on emotions and on cues from their peers when under stress. Logical decisions relating to risky behavior (including antisocial behavior) are easily undermined among adolescents when emotions are in the driver's seat and similarly emotion-driven peers occupy the passenger seats.

THE ADOLESCENT BRAIN

While the T surge explains some of the changes in adolescent behavior, the problem of impulse control and emotional regulation suggests that teens are not fully able to access areas of the brain that allow adults to react to stress and danger in a more controlled ways. The changes that occur in the brain during adolescence are more salient than the hormonal surge to explaining the teenage jump in antisocial behavior. Ernst, Pine, and Hardin (2006:299) explain risky adolescent behavior in terms of an imbalance between areas in the brain associated with approach/avoidance behaviors (the BAS/BIS systems): "The propensity during adolescence for reward/novelty seeking in the face of uncertainly or potential harm might be explained by a strong reward system (nucleus accumbens), a weak harm-avoidance system (amygdala), and/or an inefficient supervisory system (medial/ventral prefrontal cortex)." Similar conclusions relating to the adolescent brain emerged from the 2003 neuroscience conference of the New York Academy of Sciences (NYAS) as summarized Aaron White (2004:4):

1. Much of the behavior characterizing adolescence is rooted in biology intermingling with environmental influences to cause teens to conflict with their parents, take more risks, and experience wide swings in emotion.
2. The lack of synchrony between a physically mature body and a still maturing nervous system may explain these behaviors.
3. Adolescents' sensitivities to rewards appears to be different than in adults, prompting them to seek higher levels of novelty and stimulation to achieve the same feeling of pleasure.

Fortunately, these things are only temporary bends in a road that will straighten out in time; which is the fourth conclusion coming from NYAS conference:

4. With the right dose of guidance and understanding, adolescence can be a relatively smooth transition.

The neurobiological data show the physical reasons for the adolescent surge in sensation seeking and the propensity to favor short-run hedonism over more reasoned long-term goals. Functional magnetic resonance imaging (fMRI) studies have shown that relative to children and adults, adolescents have exaggerated nucleus accumbens activity feeding into the prefrontal cortex (PFC) (Eshel et al. 2007; Galvan et al. 2006). As we have seen, the nucleus accumbens is implicated in reward-seeking behaviors and the PFC is an inhibitor and guide of impulses.

Adolescence is accompanied by changes in the ratios of excitatory and inhibitory neurotransmitters. The excitatory neurotransmitters dopamine and glutamate peak during adolescence and the inhibitory neurotransmitters gamma-aminobutyric acid and serotonin are reduced (Collins 2004; Walker 2002). Adolescents are thus provided with all the biological tools— increased T and excitatory chemicals and decreased inhibitory chemicals— needed to increase novelty seeking, sensation seeking, status seeking, and competitiveness. All these changes strongly suggest that the mechanisms of adolescence are adaptations forged by natural selection (Spear 2000; White 2004). Mid-adolescence and early adulthood is a period of intense competition among males for dominance and status among many primate species aimed ultimately at securing more mating opportunities than the next male. As Martin Daly (1996:193) put it: "There are many reasons to think that we've been designed to be maximally competitive and conflictual in young adulthood."

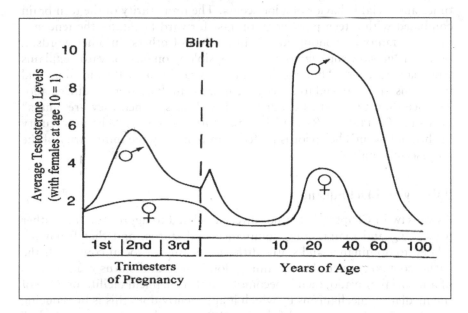

Figure 6.2 Testosterone levels of human males and females from conception to old age. Source: Ellis and Walsh, Criminology: A Global Perspective (2000). Reprinted with permission Ellis and Walsh

In addition to the changes in neurotransmitter ratios, the pubertal hormonal surge prompts the increase of gene expression in the brain to slowly refine the neural circuitry to its adult form (Sisk and Zehr 2005). The PFC undergoes a wave of synaptic overproduction just prior to puberty followed by a period of pruning during adolescence and early adulthood (Giedd 2004; Sowell et al. 2004). As Laurence Steinberg (2005:70) points out, these changes are particularly significant in the PFC: "Significant changes in multiple regions of the prefrontal cortex [occur] throughout the course of adolescence, especially with respect to the processes of myelination and synaptic pruning." Myelination of the PFC is not complete in the adolescent brain, which accounts for the "time lapse" between the adolescent perception of an emotional event in the limbic system and the rational judgment of it in the PFC. Functional MRI studies (Monk et al. 2008; Nelson et al. 2003) have shown that adolescents exhibit greater limbic activation in response to angry and fearful faces than to happy and neutral faces, whereas adults show the opposite. These data indicate that adolescents pay greater attention to negative emotional states than adults. The incompleteness of PFC myelination has also been implicated in the weaker decision-making skills of adolescents (Beckman 2004). The brain areas most often used show increased myelination; myelination thus depends at least in part on activity-dependent interaction between neurons and glial cells (Fields 2005).

The conclusion that is reached from this brief survey of the neurobiology of adolescence is that there are *physical* reasons for the immature and sometimes antisocial behavior of adolescents. The immaturity of the teen brain combined with a teen physiology on fast forward facilitates the tendency to assign faulty attributions to the intentions of others. In other words, a brain on "go slow superimposed on a physiology on fast forward" explains why many teenagers "find it difficult to accurately gauge the meanings and intentions of others and to experience more stimuli as aversive during adolescence than they did as children and will do so when they are adults" (Walsh 2002:143). As Richard Restak (2001:76) put it: "The immaturity of the adolescent's behavior is perfectly mirrored by the immaturity of the adolescent's brain."

Puberty and Delinquency

As we saw in Chapter 5, T is not directly related to aggression, but rather to seeking and maintaining dominance and status, and that behaviors related to dominance and status differ according to social context. If the social context is such that obtaining dominance and status requires acts of antisocial bravado, then T becomes a useful chemical facilitator. One of the mediating mechanisms by which it apparently does this is by reducing fear (van Honk, Peper, and Schutter 2005), which explains why high T levels are found in high-sensation seekers (Zuckerman, 2007). But let me be absolutely clear; T is only one link in a chain of interacting biological and social variables which push behavior in one way or another. To make a somewhat prurient but useful analogy, T is like Viagra; it doesn't make the turgidly challenged want to make love; rather, it helps them to do so once they have decided that's what they want. T is only an accomplice in an aggressive act; a necessary one perhaps, but it is not the perpetrator. The actual perpetrator is the whole organism making use of its entire biological toolbox in an effort to achieve the biologically relevant goal of dominance, which itself is the handmaiden of the ultimate goal of all life—reproductive success.

A longitudinal study (van Bokhoven et al. 2006) that assessed multiple measures of antisocial behavior and repeated measures of T among Caucasian males followed from age 12 to 21 found some interesting results. Males who had an official criminal record at age 21 had higher T levels at age 16 than males without a record, and boys who had the highest T levels from ages 13 to 20 were the most highly delinquent. It was also found that boys with the highest T levels were higher in both reactive and proactive aggression. The authors stressed that all males in the study were from lower SES families and were thus vulnerable to antisocial behavior for other reasons. Because all were from lower SES families, SES per se could not account for variance in the T/antisocial behavior relationship.

It is consistently found that while the relationship between T and violent aggression is strong among nonhuman animals, it is much weaker among humans among whom there are layers of culture mediating the relationship. Reviews of the literature (Ellis 2005) report average correlations between T and criminality to be a modest 0.20 to 0.25, although Mazur (2005) indicates that the correlations are higher for behavioral measures than for self-report measures. The interplay between T and social context is illustrated in a longitudinal study of 1,400 boys which found that T levels were unrelated to conduct problems for boys with "non-deviant" or "possibly deviant" friends, but conduct problems were greatly elevated among boys with high T who associated with "definitely deviant" peers (Maughan 2005). Here again we encounter the problem of directionality: do high T boys self-select into deviant peer groups or do the activities of deviant peer groups raise the T levels of boys engaging in them?

The interplay between T and neurological sculpting also appears important. Juveniles who enter puberty significantly earlier than their peers must confront their "raging hormones" with a brain that is no more mature than those of their age peers. Several studies show generally that the earlier the onset of puberty the greater the level of problem behavior for both girls and boys (Felson and Haynie 2002; Haynie 2003). For instance, Cota-Robles, Neiss, and Rowe's (2002) study of 5,550 Anglo-, African-, and Mexican-American boys found that early maturing boys in all three groups reported higher levels of both violent and nonviolent delinquency than other boys. An earlier study found that T level predicted future problem behavior, but only for boys who entered puberty early (Drigotas and Udry 1993). Felson and Haynie (2002) found that boys who experience early onset of puberty were more likely to commit a number of delinquent and other antisocial acts than other boys, but that they were also more autonomous, better psychologically adjusted, and had more friends.

Because of the extensive neuromaturational processes taking place in the adolescent brain, perhaps the worst form of antisocial behavior, from the point of view of the antisocial actor, to engage in during adolescence/early adulthood is substance abuse. As Lubman, Yucel, and Hall (2007: 793) put it: "adolescent substance abuse disrupts neuroendocrine functioning, and can induce greater effects on neural plasticity and cognition than in adults. Substance abuse can also elicit altered sensitivity to later drug exposure." Significantly, early maturers who engage in substance abuse are typically more likely report lifetime abuse and also higher levels of abuse (Patton et al. 2004). As discussed in Chapter 2, experience alters gene expression primarily through the epigenetic processes of methylation (preventing expression) and acetylation (facilitating expression). Addiction appears to be a process by which the substance of abuse modifies gene expression in the brain via histone acetylation (Kumar et al. 2005; Newton and Duman 2006). The greater vulnerability to addiction among juvenile users relative to adult-onset users appears to be the result the greater plasticity of the adolescent brain, although premorbid genetic vulnerability and frequency and length of use also differentiate between experimenters and addicts (Lubman, Yucel, and Hall 2007).

SOCIAL LEARNING, GENE/ENVIRONMENT CORRELATION, AND MOFFITT'S DEVELOPMENTAL THEORY

The social learning tradition tends to treats groups, gangs, behavior patterns, and "definitions" as though they were freestanding entities "out there" rather than abstractions that depend on individuals for their existence. With the exceptions noted earlier, individuals join groups, accept ideas, and engage in activities because these things are attractive to them for one reason or another. Because adolescence is a time of life when rebelliousness becomes an integral part of claiming independence and agency, many youths are attracted to the most rebellious of their peers (Rodkin et al. 2000). Thus peer influences go a long way toward explaining the prevalence of youthful offending, but say nothing about that small proportion of offenders who commit the majority of serious crimes, and do so long after adolescence.

DAT and SL theories are silent about desisting from delinquency, something the great majority of males who offended as juveniles do. In common with most sociological theories, they imply that criminal behavior is self-perpetuating and continuous once initiated (Gove 1985). It is something of a mystery how a theory oblivious to the mechanisms that herald the end of the phenomenon it purports to explain has managed to survive for so long. Of course, one cannot explain the termination of a phenomenon if one does not understand why it begins in the first place (i.e., the neurohormonal changes of puberty). Sutherland's animus towards nonsociological explanations doubtless kept many of his followers from considering what biology and psychology may have to offer. The absence of any explanation for desisting suggests that it is a theory limited to accounting for peer influence on delinquency during the adolescent years.

The concept of evocative and active rGE avers that individuals react to situations according to the subjective meanings the situations have for them, and that they are active shapers of their own environments. Symbolic interactionism shares this view, but the rGE concept goes beyond a mere statement of the obvious to explore the foundations of those subjective meanings. Social learning theories present a passive and static model of delinquency positing a one-way transmission of "definitions favorable to law violation" and an environment functioning as an independent variable that causes the individuals' behavior. There no hint of reciprocal effects or of the possibility that the sociocultural environment could be the dependent variable forged by similar phenotypes selectively aggregating. In a similar vein, Beirne and Messerschmidt (2000:136) complain that Sutherland's image of the social actor is "that of an empty vessel" . . . into which are poured "pro- and anti-criminal tendencies."

In contrast to this static and automaton-like view of social actors, Terrie Moffitt's (1993) biosocial developmental theory is dynamic and developmental. This theory has much to say about adolescence, delinquent peers, and other and environmental factors, but it is far more sophisticated and comprehensive in scope. The theory posits dual pathways to offending: one that begins in early childhood and continues well into adulthood, which Moffitt termed *life course persistent* (LCP) offenders, and the other in which offending is limited to the adolescent years, which she called *adolescent limited offenders* (AL). This does not mean that all offenders fit snugly into one or the other of these categories, or that LCP offenders literally offend across their entire lives. The dual pathway model is simply a convenient typology into which the great majority of offenders fit. Cohort studies in the United States and abroad have consistently shown that although about one-third of adolescents acquire a delinquent record, a small proportion in all cohorts commit a vastly disproportionate percentage of the offenses, particularly the serious violent offenses. It is also consistently found in these studies that high-rate offenders begin their criminal careers in childhood and continue way into adulthood. Data such as these had long suggested that there are two primary trajectories of offending, each with its own developmental history, but Moffitt appears to have been the first to systematically explore this duality.

Moffitt's theory is based on findings from the longitudinal Dunedin Multidisciplinary Health and Development birth cohort study (in its 33rd year as of 2008). It has become perhaps the most empirically supported theory in the literature in terms of the consistency and robustness of its findings, and Charles Tittle (2000:68) has called it "the most innovative approach to age-crime relationships and life-course patterns." The data available to Moffitt and her colleagues come from collaborative efforts by scientists in medicine, genetics, neuroscience, and endocrinology, as well as social scientists. These rich data have enabled the study team to test many biosocial hypotheses about all aspects of social life and individual development, including the MAOA/maltreatment study discussed in Chapter 2 (Caspi et al. 2002).

Studies that differentiate between prepubescent and postpubescent starters consistently find that early starters are the most frequent and serious offenders in all ages categories (Caspi et al. 1995; Farrington 1996). Moffitt proposes that neuropsychological and temperamental impairments initiate a cumulative process of negative person-environment interactions for LCP offenders that result in a life-course trajectory that propels them toward ever hardening antisocial attitudes and behaviors. The temperamental and neuropsychological impairments most often mentioned—low IQ, hyperactivity, inattentiveness, negatively emotionality, and low impulse control—are consistently and robustly found to be correlated with criminality (Agnew 2005; Moffitt and Walsh 2003) and will be further explored in the context of the comorbidity of ADHD and conduct disorder following.

Moffitt proposes that these problems arise from a combination of genetic effects and environmental effects on central nervous system development such as maternal substance abuse during pregnancy, poor nutrition, and birthing difficulties. These initial infant and childhood problems are exacerbated by ineffective socialization because problem children tend to have ineffectual parents. Moffitt describes the antisocial trajectory of LCP offenders as one of "biting and hitting at age 4, shoplifting and truancy at age 10, selling drugs and stealing cars at age 16, robbery and rape at age 22, fraud and child abuse at age 30; the underlying disposition remains the same, but its expression changes form as new social opportunities arise at different points of development" (1993:679). Behavioral consistency across time is matched by cross-situational behavioral consistency: LCP offenders "lie at home, steal from shops, cheat at school, fight in bars, and embezzle at work" (Moffitt 1993:679).

AL offenders have a different developmental history which puts them on a prosocial trajectory that is temporarily derailed at adolescence. They do not have to contend with the temperamental and neuropsychological problems that burden LCP offenders, and for the most part they are adequately socialized. These youths are statistically normal, and we may view their offending as adaptive responses to conditions and transitional events that temporarily divert them from their basically prosocial life-course trajectories. In other words, they are responding to the social and biological changes of adolescence in a totally normative way, including disengaging from parents and bonding with peers. Moffitt (1993) agrees with social learning theories that AL offending is a group social phenomenon and that it does not reflect any kind of stable personal deficiency on the part of offenders. Moffitt characterizes AL offending as motivated by the widening gap between biological and social maturity, learned in association with antisocial peers, and sustained by reinforcement principles.

According to Moffitt (1993), the route by which youths on a basically prosocial life trajectory are temporarily diverted involves biological, social, and economic vectors that are diverging as never before. Health and nutritional advances have continually lowered the age of puberty, and technological advances have continually raised the time needed to prepare for participation in today's complex economy. This divergence has resulted in about a 5- to10-year gap between puberty and the acquisition of socially responsible roles for many of today's youth, making modern adolescence a very wide strain-inducing maturity gap between childhood and adulthood. Because of this wide gap, many more youngsters with no personal history suggesting that they are at risk for antisocial conduct suddenly become antisocial. Thus, "adolescent-limited offending is a product of an interaction between age and historical period" (Moffitt 1993:692). Filled with boundless energy, strength, and confidence, a brain that is almost literally designed at this juncture in life for excitement and conflict, and a strong desire to shed the restrictions of childhood, some youngsters gravitate to the excitement of antisocial peer groups led by LCP youths. Once in these groups, juveniles learn the techniques of offending via mimicry and reinforcement as outlined by social learning theory.

Adolescent antisocial behavior is adaptive, according to Moffitt, because it offers the opportunity to gain valuable resources otherwise temporarily unavailable. The most important of these resources is mature status, yet adolescents are still dependent on parents for almost everything, and adult independence seems to be a distant dream. These youths may turn their envious eyes on LCP offenders, who have already declared their independence (their parents have given up on them) and have obtained a modicum of the resources via both legal and illegal means (cars, nice clothes, access to sex partners) that signal mature status. Because the behavior of LCP offenders seems to bring desired results for them, novice AL delinquents are drawn to them and mimic their behavior. This contention is supported by a study of 59 high school classrooms in rural, suburban, and inner-city schools. This study found that the most popular boys were athletic, cooperative, and sociable, but about a third of the most popular boys were antisocial youths who were frequently disruptive and belligerent and "central members of prominent classroom cliques" (Rodkin et al. 2000:21). Other studies have shown that as youngsters age in adolescence they begin to admire good students less and aggressive antisocial peers more (Bukowski, Sippola, and Newcomb 2000). Popular antisocial youths are thus rewarded with status among their peers while their AL mimics receive vicarious reinforcement by identifying with them. These neophyte delinquents internalize the idea that antisocial behavior and popularity go together, and thus receive validation for their oppositional behavior.

Figure 6.3 compares the static and generalized differential association pathway model to delinquency with Moffitt's dynamic dual pathway model. Particular attention should be paid to the role of association with delinquent peers in the LCP and AL delinquency models. In the case of LCP offenders, *stable* antisocial characteristics precede association with delinquent peers and exemplify active rGE in that like seeks like. By way of contrast, for AL offenders association with delinquent peers precedes the development of *temporary* antisocial characteristics. This suggests that association with delinquent peers may be necessary to initiate delinquency for AL offenders, and that there is little or no genetic influence on delinquency for these temporary offenders at a time in life when peer influence is tremendously important. In other words, teens have a limited ability to choose their environments, so even those with low genetic risk for delinquency may often succumb to it under the influence of their more daring peers, whom they temporarily admire and seek to emulate.

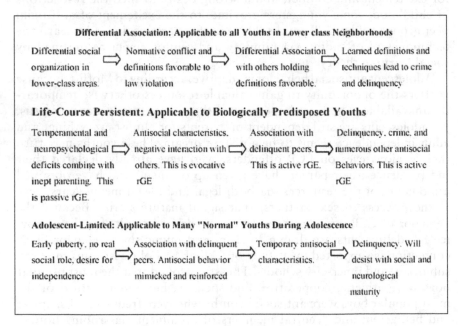

Figure 6.3   Differential association theory and Moffitt's theory.

Desisting

As AL offenders mature neurologically and socially they become freer to structure their environments consistent with their genetic preferences, and they begin to realize that an adult criminal record will severely limit their future options. They begin to knuckle down, take on socially responsible roles, and desist from further antisocial behavior. Unlike LCP offenders, who have essentially burnt their bridges to the prosocial world, AL offenders have accumulated a store of positive attachments (they elicit positive responses from others) and academic skills (they are intelligent and conscientious) which provide them with prosocial opportunities such as a good marriage and a good job. In short, AL offenders desist from antisocial behavior because, in Moffitt's (1993:690) words, they are "psychologically healthy," and "Healthy youths respond adaptively to changing contingencies."

Sampson and Laub (2005) support Moffitt's contention, showing that job stability, commitment, and attachment to a prosocial spouse are important inhibitors of adult criminal behavior among former delinquents. Sampson and Laub did not differentiate between LCP and AL offenders (indeed, they deny such a typology), but they maintain that such positive adult social bonds are "social capital," and the result of prior "social investment." We may expect, therefore, that their findings are more salient for AL than LCP offenders, especially given the data on assortative mating for antisocial behavior and characteristics (Krueger et al. 1998; Quinton et al. 1993). These studies show that the likelihood of persistent offenders securing the support of a nondeviant spouse, while not impossible, is minimal.

Moffitt's theory has been widely tested; a review of these studies conducted 10 years after the theory was first proposed found strong support for it (Moffitt and Walsh 2003). Overall, these studies show that LCP offenders almost always had childhood temperament problems identifiable at age 3 years, and that they were more likely to have weak family bonds, low verbal IQ, and the psychopathic traits of callousness and impulsivity. LCP were much more likely to commit serious violent crimes (robbery, rape, assault) than AL boys (typically offenses such as petty theft and public drunkenness), and although LCP boys constituted only 7% of cohort they were responsible for more than 50% of all delinquent acts committed by it. Note the consistency with other cohort data indicating that serious and frequent offending tends to be concentrated among a very small group, the vast majority of whom began offending prior to puberty.

Why do some Adolescents abstain from Delinquency altogether?

If AL delinquency is normative adaptive behavior, then the existence of teens who abstain from delinquency requires an explanation because their abstinence is at least as interesting as why others offend. The theory predicts that teens abstaining from antisocial behavior would be rare and they must have either structural barriers that prevent them from exercising anti-social behavior, a smaller than average maturity gap, or personal unappealing characteristics that cause them to be excluded from teen social group activities (Moffitt 1993).

Consistent with the prediction, the cohort contained only a small group of males who avoided virtually any antisocial behavior during both childhood and adolescence (Moffitt et al. 1996). Because Moffitt views the age peak to be a function of today's long maturity gap, the most obvious reason for why some adolescents do not offend from her perspective is that they do not experience the maturity gap to the same degree that most young people do. That is, they may experience late puberty and/or early initiation into adult roles and responsibilities. These youths may be considered throwbacks to an age in which puberty arrived later and less preparation was needed to enter the job market (Moffitt 1993). Moffitt also speculates that abstainers may belong to religious and cultural groups in which youths are given early access to adult privileges and adult accountability, or that they are less well adjusted than adolescents who experiment with delinquency. Moffitt cites a study that found boys at environmental risk for delinquency but who did not become delinquent "seemed nervous and withdrawn and had few or no friends" (and thus no antisocial role models) (1993:689).

There are obviously also personality differences that may account for abstaining. These abstainers described themselves at age 18 as extremely self-controlled, fearful, interpersonally timid, socially inept, and latecomers to sexual relationships (i.e., virgins at age 18). Abstainers were the type of good and compliant students who become unpopular with peers (Bukowski et al. 2000) and fit the profile reported for youth who abstained from drug and sexual experimentation in a historical period when it was normative: i.e., they were overcontrolled, not curious, not active, not open to experience, socially isolated, and lacking in social skills (Shedler and Block 1990; Walsh 1995). Follow-up data at age 26 confirmed that abstainers had not become late-onset offenders (Moffitt et al. 2002). Although their teenaged years had been troubling, they became more successful in adulthood. They retained their self-constrained personality as adults, had virtually no crime or mental disorder, were likely to have settled into marriage, were delaying children (a desirable strategy for a generation needing prolonged education to succeed), were likely to be college educated, held high-status jobs, and expressed optimism about their own futures.

The kind of inhibited and introverted personality characteristic of abstainers is consistent with studies of autonomic nervous system (ANS) arousal and criminality. Recall that individuals with a *hyper*arousable ANS are easily conditioned and those with a *hypo*arousable ANS are conditioned with difficulty. Abstainers may be individuals located on the "hyper" tail of the ANS arousal distribution, and thus have excessive guilt feelings and excessive fear of the negative consequences of nonconformity (Moffitt and Walsh 2003).

ADHD, CD, ODD, AND DELINQUENCY

Many of the neurological and temperamental deficits identified as characterizing LCP offenders characterize to various degrees children who have been diagnosed with attention deficit with hyperactivity disorder (ADHD), conduct disorder (CD), and oppositional defiant disorder (ODD). These separate but often linked syndromes are neuropsychological and temperamental deficits that can lead to criminal offending long after adolescence. For those who are afflicted with all three conditions, ADHD symptoms usually appear first, followed by ODD symptoms, and then CD.

ADHD is a chronic neurological condition characterized by impairment in executive functions and manifested as constant restlessness, low self-control, difficulty with peers, frequent disruptive behavior, short attention span, academic underachievement, risk taking, and proneness to extreme boredom. The symptoms vary widely in their severity and frequency of occurrence and most healthy children will sometimes manifest them. However, they cluster together to form a syndrome in ADHD children and are chronic and more severe than simple childish high spirits. Numerous brain-imaging studies find many small differences in brain anatomy and physiology, especially in the right frontal region, between ADHD and non-ADHD children (Raz 2004; Sanjiv and Thaden 2004; Stein et al. 2007).

ADHD affects somewhere between 2% and 9% of the childhood population and is four or five times more prevalent in males than in females (Levy et al. 1997) (males are 10 times more likely to be found among LCP offenders than are females [Moffitt and Walsh 2003]). Heritability coefficients in the .75 to .91 range (Spencer et al. 2002) are consistently found for ADHD regardless of whether it is considered to be a categorical or continuous trait (Levy et al. 1997). Molecular genetic studies show ADHD to be highly polygenic, with at least 50 genes, all with small effect sizes, being involved (Comings et al. 2005). Environmental features identified as playing a role in the etiology of ADHD are fetal exposure to noxious substances, perinatal complications, and head trauma (Durston 2003). Factors such as family and school experiences or peer pressure do not to have any causal impact on ADHD, although they may exacerbate its symptoms (Coolidge, Thede, and Young 2000). ADHD symptoms (particularly impulsivity) usually decline in severity as the PFC becomes more myelinated, although about 90% of ADHD individuals continue to exhibit some symptoms into adulthood (Willoughby 2003).

Neurological deficits associated with ADHD include suboptimal arousal and frontal lobe dysfunction. Some children diagnosed with ADHD show EEG patterns of underarousal (slow brain waves) similar to adult psychopaths (Lynam 1996). Slow brain waves are subjectively experienced as boredom, which motivates the person to seek or create environments containing more excitement. ADHD behavior can be normalized temporarily by administering drugs such as the stimulant methylphenidate (Ritalin) and the nonstimulant norepinephrine reuptake inhibitor atomoxetine. The efficacy of early pharmacotherapeutic approaches in calming ADHD individuals gave researchers their first clues to the underlying neurochemical basis for the disorder (Durston 2003). Stimulant drugs have a calming or normalizing affect on suboptimally aroused individuals by raising the activity of the brain's sensory mechanisms to normal levels. This relieves feelings of boredom because the brain is now able to be more attentive to features of the environment that it could not previously capture. When on medication, ADHD children are less disruptive, become less obnoxious to peers, and can focus more on schoolwork.

The highly polygenic nature of ADHD probably explains why it is clinically heterogeneous and why it is linked to many other problems such as CD, alcohol and drug abuse, pathological gambling, and antisocial behavior in general (Comings et al. 2005). A review of 100 studies conducted prior to 1999 found that 99 of them reported a positive relationship between ADHD and various antisocial behaviors (violent and property crimes, delinquency, drug abuse) while only one was found not to be significant (Ellis and Walsh 2000).

ADHD delinquents are more likely than non-ADHD delinquents to persist in their offending as adults, but this probability rises dramatically for ADHD children also diagnosed with CD. CD is defined as "the persistent display of serious antisocial actions [assaulting, stealing, setting fires, bullying, cruelty to animals, vandalism] that are extreme given the child's developmental level and have a significant impact on the rights of others" (Lynam 1996:211) and is considered one of the most stable diagnoses in psychiatry (Comings et al. 2005). ADHD and CD are found to co-occur in about 50% of cases in most clinical and epidemiological studies (Waschbusch et al. 2002). Conduct disorder has an onset at around 5 years of age. It remains at a steady rate for girls (about 0.8% of all girls) and rises to about 2.8% at age 15, but rises steadily in boys from about 2.1% at age 5 to about 5.5% at age 15 (Maughan et al. 2004). CD is also a neurological disorder with substantial polygenic effects and with heritability estimates ranging between .27 and .78 (Coolidge, Thede, and Young 2000).

Moffitt (1996) has proposed that verbal deficits are what place children at risk for CD. This gains some support from a study of children comorbid for ADHD and reading disorder, which showed that both were caused in part by common genetic effects (Friedman et al. 2003). Neurological evidence suggests that the left frontal lobes contain the mechanisms by which children process their parents' instructions that become their internalized basis of self-control. Children with deficits in these mechanisms fail to profit from their parents' verbal instructions and tend to develop a present-oriented and impulsive cognitive style. Lacking normal ability to connect abstract verbal commands with their own concrete behavior, such children may have to learn lesson through the more painful process of trial and error and may thus experience more frequent punishments for their lack of compliance with instructions. As Moffitt et al. (1994:296) put it: "Children who have difficulty expressing themselves and remembering information are significantly handicapped. Dysfunctional communication between child and his parents, peers, and teachers may be one of the most critical factors for childhood conduct problems that grow into persistent antisocial behavior in young adulthood."

Many of the cognitive and temperamental symptoms of CD and ADHD children are similar. CD children tend to score in the low-normal or borderline range of intelligence, are most likely to be found in impoverished families, and are significantly more likely than children without CD to have parents diagnosed with antisocial personality disorder (ASPD) (Sergeant et al. 2003). One of the psychiatric requirements for a diagnosis of ASPD is a childhood diagnosis of CD. Thus, if a parent is diagnosed with ASPD it means that he (almost always a he) was diagnosed with CD as a child, and if his offspring is also diagnosed with CD, the cross-generation linkage strongly suggests genetic transmission (Comings 2005).

A number of researchers have offered evidence that ADHD is a product of a deficient BIS, and CD is a product of an oversensitive BAS (Levy 2004; Quay 1997). If this is the case, then those afflicted with both ADHD and CD suffer a double disability. First, they are inclined to seek high levels of stimulation because of an oversensitive BAS, and second, they are hampered by a faulty BIS and thus have difficulty putting a stop to their search for pleasurable stimulation once it is initiated. The high level of comorbidity for addictive behaviors in ADHD + CD individuals is certainly indicative of reward dominance induced by BAS/BIS imbalance.

Lynam (1996:22) describes the trajectory from ADHD + CD to criminality in a way that reminds us of evocative rGE, stating that the co-occurrence of ADHD and CD "[M]ay tax the skills of parents and lead to the adoption of coercive child rearing techniques, which in turn may enhance the risk of antisocial behavior. Entry into school may bring academic failure and increase the child's frustration, which may increase his or her level of aggressive behavior. Finally, the peer rejection associated with hyperactivity may lead to increased social isolation and conflict with peers."

Oppositional defiant disorder (ODD) is an earlier appearing and less severe form of a disorder characterized by such behaviors as temper tantrums and lying. Some researchers view ODD as a less severe and earlier form of CD, while others maintain the distinction between the two but acknowledge that ODD frequently develops into CD. Lahey and Loeber (1994:144) describe the distinction between the two: "ODD is an enduring pattern of oppositional, irritable, and defiant behavior, whereas CD is a persistent pattern of more serious violations of the rights of others and social norms." The two syndromes are developmental and hierarchical in that all children who meet the clinical criteria for CD also met the criteria for ODD when they were younger, but not all ODD children will develop CD. Factor analyses of the symptoms and behaviors of the two conditions also produce two or three factors, with a three-factor solution labeled as "ODD," "intermediate CD," and "advanced CD" (Lahey and Loeber 1994).

ADHD is not a form of hopeless pathology that leads to inevitable criminality, particularly when CD is not present. The increasing numbers of children being diagnosed with ADHD most likely reflects a growing intolerance for disruptive classroom behavior than anything else, putting more children at risk for being diagnosed with the disorder. Perhaps it should not even be called a disorder, but rather a natural variant of human diversity. While acknowledging that ADHD has real neurological foundations, and that parents are probably right in the current social context to choose to medicate their ADHD children, Jaak Panksepp (1998) asserts that ADHD-like behaviors are observed in the young of all social species and it is called "rough-and-tumble" play. Some ADHD individuals have above-average IQ's and are creative, so perhaps the symptoms of ADHD are only problematic in the modern context in which children are expected to sit still for long periods learning subjects that they find boring.

ADHD-like behaviors may have even been adaptive in our evolutionary history when restless boldness and curiosity meant exploring beyond known boundaries. If the true rate of ADHD is as high as reported, then genes underlying it have survived natural selection, which means that they must have conferred some benefits in evolutionary environments even if they do not in evolutionarily novel classroom environments (Bjorklund and Pellegrini 2000). Williams and Taylor (2006) offer a plausible evolutionary model of ADHD based on the value of unpredictable and novelty-seeking behavior in changing environments. Their model is based on the 7-repeat version of the DRD4 "novelty-seeking" gene. A study of 2,320 individuals from 39 different racial/ethnic groups (Ding et al. 2002) found that the percentage of 7-repeat alleles ranged from zero in sedentary populations (those that have remained in the same geographic region for the last 30,000 years; i.e., Han Chinese, Yemeni Jews) to an average of 63% in six migratory populations of South American Indians (the worldwide percentage is estimated to be 19.2%). The correlation between percentage of 7-repeats in a population and geographic distance from the original parent population was an impressive 0.85. Williams and Taylor (2006) hypothesize that the restlessness and impulsiveness associated with the 7-repeat DRD4 allele prompted increased exploration of new behaviors and new territories that enhanced the fitness (albeit in frequency-dependent ways) of those carrying it, and of the group as a whole.

The preceding only suggests that *some* endophenotypes (energetic, unpredictable, novelty seeking) of ADHD *may* have been selectively adaptive, not all of them. As previously mentioned, ADHD children often show neurological deficits, but ADHD children who also carry the 7-repeat DRD4 do not show these deficits (Harpending and Cochran 2002). This also indicates the heterogeneity, and hence the polygenic nature, of ADHD.

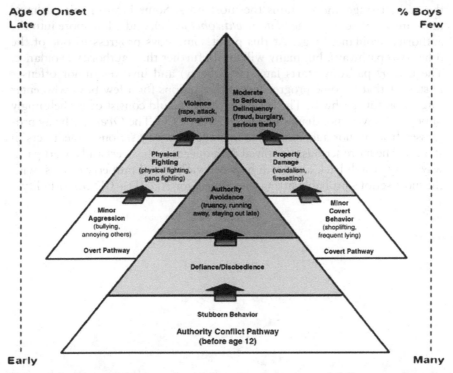

Age of Onset
Late

% Boys
Few

Violence
(rape, attack, strongarm)

Moderate to Serious Delinquency
(fraud, burglary, serious theft)

Physical Fighting
(physical fighting, gang fighting)

Property Damage
(vandalism, firesetting)

Authority Avoidance
(truancy, running away, staying out late)

Minor Aggression
(bullying, annoying others)

Minor Covert Behavior
(shoplifting, frequent lying)

Overt Pathway

Covert Pathway

Defiance/Disobedience

Stubborn Behavior

Authority Conflict Pathway
(before age 12)

Early

Many

Figure 6.4 Three pathways to boys' disruptive behavior and delinquency. Source: Thornberry, Huizinga, & Loeber, 2004. Editor: Note from reference that this is a public domain government document.

PATTERNS OF SERIOUS DELINQUENCY

Just as there are different pathways to offending based on a variety of phenotypic differences, there are different patterns of offending that partially represent these differences and partially represent developmental age. Figure 6.4 presents a developmental pyramid model of offending that focuses on the escalation of seriousness of delinquent acts being committed as boys age (Thornberry, Huizinga, and Loeber 2004). The model is based on three longitudinal studies—the Denver Youth Survey, the Pittsburgh Youth Study, and the Rochester Youth Developmental Study—and includes over 4,000 subjects followed since 1987. The model presents three theoretically distinct offending pathways that represent patterns of behavior; it says nothing about individual characteristics, but rather neatly conforms to Lahey and Loeber's (1994) distinction between ODD, intermediary CD, and advanced CD.

The *Authority Conflict* pathway is the earliest pathway (starting before age 12) and begins with simple stubborn behavior (ODD), followed by defiance and authority avoidance. Note that the base of the triangle represents the earliest stage and contains the most boys. Some boys in this pathway move into the second stage (*defiance/disobedience*), and a few more into the authority avoidance stage. At this point some boys progress to one of the other two pathways, but many will go no further than authority avoidance. The *Covert* pathway starts later (at puberty) and involves minor offenses in stage 1 that become progressively more serious for a few boys who enter stage 3 on this pathway. The covert pathway would consist overwhelmingly of boys who were not diagnosed with ODD or CD. The *Overt* pathway progresses from minor aggressive acts in stage 1 to very serious violent acts in stage 3. The more seriously involved delinquents in the overt and covert pathways may switch back and forth between violent and property crimes, with the most serious probably fitting the criteria for ADHD + CD comorbidity.

7 The Control Tradition and the Family

If humans are to live together in relative peace and harmony, antisocial behavior must be kept to a minimum. Mechanisms that have been devised to accomplish this have been collectively called social control. Social control entails any action on the part of others that facilitates conformity on the part of those toward whom the action is directed. The first agents of social control we meet are our parents, on whose shoulders control theorists place the primary burden of producing good citizens. It is they who must teach their offspring the accepted rules of proper conduct, which is the rational component of our consciences. When we have internalized rules of conduct, we feel guilty, anxious, and ashamed when we misbehave, and happy and self-righteous when we behave well. Those less than adequately socialized feel no such emotions, and thus grant themselves permission to do more or less as they please—to do what comes naturally. Gwynn Nettler put it most colorfully when he wrote: "If we grow up 'naturally,' without cultivation, like weeds, we grow up like weeds—rank" (1984:313).

A class of theories called control theories focus on mechanisms of informal control (socialization, training, education, as opposed to the formal control mechanisms exercised by the criminal-justice system). The two theories presented in this chapter, social control and self-control theories, focus on the family and have "commonsense" appeal as gauged by the results of numerous studies asking laypersons their opinions of the causes of crime. Every such study shows family factors (lax discipline and supervision, broken homes, family discord, and so on) to be uppermost in the public's mind as causes of crime and delinquency (Ellis and Walsh 2000).

SOCIAL CONTROL THEORY

Most criminological theories focus on conditions that are alleged to propel offenders into criminal activity. This view implicitly assumes that people naturally want to "do the right thing" and conform to conventional social rules, and that it takes some sort of negative departure from the "normal" to make them behave badly. This is akin to the largely discredited theory of group selection in biology; i.e., we have evolved to place group interests above our own individual interests, and will do so unless somehow forced to do otherwise. For those who believe that crime and delinquency are behaviors learned by good people in bad environments, it makes sense to ask "What causes crime?"

Control theorists believe that the real question is not why some people behave badly, but why most of us behave well most of the time. They share with evolutionary psychology the assumption that humans are self-interested and will seek benefits to themselves any way they can unless motivated to behave well (to be good reciprocal altruists). We behave well if our ties to prosocial others are strong, but we may revert to predatory self-interest if they are not. After all, human infants express little else other than concern for self-gratification, and it is only through years of socialization that we come to realize that our self-interest is bound to the self-interest of others. Children who are not properly socialized hit, kick, bite, steal, whine, scream, and otherwise behave obnoxiously whenever the mood strikes them. They have to be taught not to do these things, which in the absence of training "come naturally." As Freud reminded us in his *Civilization and Its Discontents* (1961), civilized society is bought at the cost of the repression of our freedom to do as we damn well please. For the control theorist, as it was for Durkheim, it is society that is "good" and human beings, in the absence of prosocial training, who are "bad." Crime is natural, and the criminal is just the unsocialized self-centered child grown strong. Thus instead of asking how natural social conformists are induced to behave badly, control theorists ask how naturally selfish individuals can be induced to behave well.

Both control theories discussed in this chapter are associated with Travis Hirschi, and both begin with what Gottfredson (2006:80) calls "foundational facts." Foundational facts are the prevailing empirical knowledge about crime around which the theories are, or should be, formulated. Social control theory thus begins with the demographic characteristics of typical criminals; i.e., young males who grew up in fatherless homes in an urban slum, and who have a history of difficulty in school and work. Having identified the social situation of the typical criminal, Hirschi (1977) makes series of logical deductions, beginning with the observation that criminal activity is contrary to the wishes and expectations of others. From this he deduces that those most likely to commit crimes are least likely to be concerned with the wishes and expectations of others, have the free time to commit crimes, have little to lose if caught, and are least likely to accept the moral beliefs underlying the law.

The Four Elements of the Social Bond

From these observations Hirschi makes the assumptions that the typical delinquent or criminal lacks social bonds; that is, they lack *attachment* to prosocial others and prosocial institutions, lack *commitment* to a prosocial career and *involvement* in a prosocial lifestyle, and that he does not *believe* in the validity of the moral order. These four elements of the social bond are highly interrelated mechanisms assumed to be present in the lives of noncriminals restraining them from criminal activity and assumed to be largely absent in the lives of criminals.

Attachment refers to the emotional bond existing between the individual and key social institutions like the family and the school, and is the "master" bond that lays the foundation for all other social bonds. Attachment implies strong emotional relationships in which persons feel valued, respected, and admired, and in which the favorable judgments of others in the relationship are valued: "If a person feels no emotional attachment to a person or institution, the rules [of that person or institution] tend to be denied legitimacy" (Hirschi 1969:127).

Parents are the most important behavior-orienting significant others for us all. Children who do not care about parental reactions are those who are most likely to behave in ways contrary to their wishes. Risking the good opinion of another is of minor concern when that good opinion is not valued. Parental opinion may not be valued if parents have not earned the love and respect of their children because of physical and/or emotional neglect and abuse, the lack of intimate communication, erratic and unfair disciplinary practices, emotional coldness, or for any number of other reasons (Hirschi 1977). Lack of attachment to parents and lack of respect for their wishes easily spills over into a lack of attachment and respect for the broader social groups of which the child is a part. Much of the controlling power of others outside the family lies in the threat of reporting misbehavior to parents. If the child has little concern for parental sanctions, the control exercised by others has little influence on the child's behavior. In sum: "The essence of internalization of norms, conscience, or superego, thus lies in the attachment of the individual to others" (Hirschi 1969:18).

If attachment is the emotional component of conformity, commitment is the rational component. *Commitment* refers to a lifestyle in which one has invested considerable time and energy in the pursuit of a lawful career, which provides a valuable stake in conformity. Unlike the negative view of aspirations as strain-inducing contained in anomie/strain theory, Hirschi contends that aspirations, if they energize actual behavior rather than simply being idle dreams of the good life, tie individuals to the social order. The person who has made this considerable investment is not likely to risk it by engaging in crime. Poor students, truants, dropouts, and the chronically unemployed have little investment in conventional behavior and therefore risk little in the cost/benefit comparison. The acquisition of a stake in conformity requires success in school, which requires intelligence and disciplined application to tasks that children do not relish but which they complete in order to gain valued approval. If approval is not forthcoming or is not valued, children will busy themselves in tasks more congenial to their natural inclinations, inclinations that almost certainly do not include struggling with the complexities of algebra or English grammar.

Involvement is a direct consequence of commitment and is part of a conventional pattern of existence. Hirschi's classical view of human nature is evident when he wrote that "many persons undoubtedly owe a life of virtue to lack of opportunity to do otherwise" (1969:21). Essentially, involvement is a matter of time and energy constrictions placed upon us by the demands of our commitments, and involvement in lawful activities reduces exposure to illegal opportunities. Conversely, the lack of involvement in lawful activities increases the possibility of exposure to illegal activities—the devil finds work for idle hands.

Belief refers to the acceptance of the social norms regulating conduct. Individuals who are free of the constraints on their behavior imposed by social bonds evolve a belief system shorn of conventional morality. Beliefs shorn of conventional morality support only narrow self-interest justified by a jungle philosophy. Unlike differential association theory, control theory does not view a criminal belief system as motivating criminal behavior. Rather, criminals act according to their urges and then justify or rationalize their behavior with a set of statements such as "Suckers deserve what they get" and "Do onto others as they would do onto you—only do it first." For control theorists, behavior gives birth to the belief rather than vice versa. It is important to understand that the lack of attachment, commitment, involvement, and conventional beliefs does not constitute a motive for crime; their lack simply represents deficiencies that grant those who do not have them permission to do what comes "naturally."

GOTTFREDSON AND HIRSCHI'S LOW-SELF-CONTROL THEORY

All control theories have one basic idea: the natural inclination to seek self-interest at the expense of others must be controlled. Durkheim addressed the breakdown of controls at the societal level (anomie), Shaw and McKay at the neighborhood level (social disorganization; see Chapter 8), and Hirschi at the family level (attachment and its sequelae). There is one further step to go: control at the individual level. Hirschi took that final step with Michael Gottfredson in developing a "general theory" of crime based on self-control (Gottfredson and Hirschi 1990). Gottfredson and Hirschi claim that their theory "explains all crime, at all times, and, for that matter many forms of behavior that are not sanctioned by the state" (1990:117).

In his social control/social bonding theory, Hirschi repudiated the concept of self-control along with all other individual differences. He notes that "Being true to the assumptions of my discipline (sociology), I did not link social control to self-control. On the contrary, I disavowed the significance of the latter concept" (2004:539). In moving from social- to self-control, Hirschi has violated the assumptions of his discipline and has adopted a psychological approach, with the implication that the relationship between social bonds and crime and delinquency is actually "caused" by the underlying variable of low self-control. That is, self-controlled individuals will easily form social bonds, but people with low self-control will do so only with difficulty. Consistent with this shift, rather than starting with the social location of the typical criminal as their foundational facts, Gottfredson and Hirschi begin with what is known about the typical crime; i.e., an impulsive act designed to provide the offender with immediate short-term gratification. If this is the defining characteristic of the typical criminal act, it follows that the defining trait of the typical criminal is low self-control.

Self-control is defined as the "extent to which [different people] are vulnerable to the temptations of the moment" (Gottfredson and Hirschi 1990:87). Crimes are the result of the natural human impulse to enhance pleasure and avoid pain, and defined as "acts of force or fraud undertaken in pursuit of self-interest" (Gottfredson and Hirschi 1990:15). Low self-control is not a motivator of antisocial behavior; it is a lack of internal constraint that permits it. There is "little variability among people in their ability to see the pleasures of crime," Gottfredson and Hirschi (1990:95), assert, but [due to variability in self-control] "there will be considerable variability to calculate pains." Most crimes are spontaneous acts in response to a tempting opportunity that require little skill and most often earn the criminal only minimal reward or satisfaction. Such criminal opportunities do not tempt people with self-control, but people with low self-control are always strongly tempted to make some quick gain for themselves whenever opportunities present themselves.

According to the theory, people with low self-control possess the following cognitive and temperamental traits that make offending more probable for them than for others (1990:89–90):

- They are oriented to the present rather than to the future, and crime affords them immediate rather than delayed gratification. [They are impulsive.]
- They are risk-taking and physical as opposed to cautious and cognitive, and crime provides them with exciting and risky adventures. [They have low arousal levels, are sensation seekers, and lack cognitive skills.]
- They lack patience, persistence, diligence, and crime provides them with quick and easy ways to obtain money, sex, revenge, and so forth. [They lack conscientiousness.]
- They are self-centered and insensitive, so they can commit crimes without experiencing pangs of guilt for causing the suffering of others. [They lack empathy and conscience.]

The theory asserts that low self-control is established early in childhood, is caused by inept parenting, and tends to persist throughout life. It is important to understand that children do not learn low self-control; rather, it is the "default" outcome that occurs in the absence of adequate socialization. Brannigan (1997:425) puts this concept in evolutionary terms: "In the absence of interaction designed to promote empathy, identification, delayed gratification, a long time horizon, and prosocial values, the infant appears to exhibit older evolutionary scripts of adaptation which favour immediate gratification and egoism."

If children are to be taught self-control their behavior must be consistently monitored, and parents and/or caregivers must recognize deviant behavior when it occurs and punish it. Active parental concern for the well-being of their children is central to the rearing of self-controlled children. Warmth, nurturance, vigilance, and the willingness to practice "tough love" are the necessary links between parental self-control and offspring self-control. Other family-related factors that lead to low self-control include parental criminality (criminals are not very successful in socializing their children not to be criminals), family size (the larger the family the more difficult it is to monitor behavior), single-parent family (the efforts of two parents are generally better than one), and working mothers (which negatively affects the development of self-control if no substitute monitor is provided) (Gottfredson and Hirschi 1990:100–105).

Gottfredson and Hirschi argue that children learn or fail to learn self-control in the first decade of life, after which the attained level of control remains stable across the life course. Subsequent experiences, situations, and circumstances have little independent affect on the probability of offending because these unfolding events are heavily influenced by the level of self-control learned in childhood. Low self-control is thus considered a stable component of a criminal personality, which is why most criminals typically fail in anything that requires long-term commitment and compromise, such as school, employment, and marriage, because such commitments and compromises get in the way of immediate satisfaction of their desires. Note that self-control is not a motivator of any act; it is a brake, not an accelerator. Self-control is the ability to engage the executive functions of the PFC in the service of preventing the transgression of a moral norm in the face of some juicy temptation or angry provocation.

Given the emphasis on parental guidance in the development of self-control, it has been something of a mystery to many criminologists why the four elements of the social bond, particularly attachment, are all but absent in self-control theory. Hirschi now says that the mystery is solved by assuming that, like self-control, "differences in social control are stable [across the life span], that social control and self-control are the same thing" (2004:543). Thus without denying the significance of the psychological construct of self-control, he is moving back to his sociological roots. He welds the two theories together with two simple sentences: "Self-control is the set of inhibitions one carries with one wherever one goes. Their character may be initially described by going to the elements of the bond identified by social control theory: Attachment, commitment, involvement, and belief" (2004:543–544). Hirschi now appears to be thinking along the same lines as Robert Agnew (2005) in his "super traits" theory in that self-control and the social control exercised by the various institutions to which one is attached or committed mutually effect one another constantly across the life course.

Low self-control is not sufficient in itself to account for offending. Because Gottfredson and Hirschi consider it to be stable across the life course, variation in criminal behavior cannot be explained by variation in self-control (they seem to be saying here that low self-control lacks dimensionality; i.e., that all who lack self-control are equally affected). What explains variation in criminal behavior among low-self-control individuals is variation in *opportunities* to commit it. As in rational choice theory, criminal offending is the result of people with low self-control (a motivated offender) meeting a criminal opportunity (something or someone lacking a capable guardian). Gottfredson and Hirschi illustrate what they mean by this by describing typical homicide, rape, robbery, burglary, embezzlement, and auto theft incidents. These descriptions show that most criminal events are not the culmination of foresight and planning; rather, they are environmental events (an open door, an unlocked car with keys in the ignition, a vulnerable person walking down a dark alley) witnessed by someone ready to take advantage of them.

The impulsive nature of the typical crime incident is supported by a study in which 83% of offenders claimed that the crimes for which they were first arrested "just happened." Eleven percent of the offenders said they planned the crime on the day it was committed, and only 6% said they planned their crimes one or more days in advance. The impulsive tendency is, as Gottfredson and Hirschi claim, remarkably stable. When asked about their *last* crime, 83% of those who had not planned their first crime also had not planned their last, while 60% of those who had planned their first crime had not planned their last (Wolfgang, Thornberry, and Figlio 1987:126–127).

Given the assumptions of control theories, their policy recommendations are very much unlike those of other theories that posit that humans have to learn to be selfish and criminal. Because low self-control is the result of the absence of inhibiting forces, Gottfredson and Hirschi are pessimistic about the ability of less powerful inhibiting forces (such as the threat of punishment) in later life to deter crime. They also see little use in seeking to reduce crime by satisfying the wants and needs alleged by other theories to cause crime (reducing poverty, improving neighborhoods, more job opportunities, etc.) because crime's appeal is its provision of immediate gains and minimal cost. People who possess self-control will do fine in the job market as it is. In short, "society" is neither the cause nor the solution to the crime problem. The only way to reduce crime is to strengthen families and improve parenting skills, especially skills involved in dealing with teaching self-control to children.

THE FAMILY: NURSERY OF HUMAN NATURE

Despite the popularity of both social- and self-control theories among contemporary American criminologists, they have not escaped criticism. All control theories agree that the family is central to the developmental mechanisms that affect criminal behavior, but they ignore the social, economic, cultural, and political factors that impede stable and nurturing families (Lilly, Cullen, and Ball 2007). This criticism has merit, but control theory concerns itself with the alleged effects of problematic families, not with the alleged causes of the problems faced by modern families. Control theorist stress that the most important aspect of the environment for the healthy prosocial development of human beings is love and nurturance. Because the love and nurturance of children is *typically* best accomplished (in the context of industrial and postindustrial societies) in a two-parent family, it behooves policymakers to do everything they can to strengthen the family.

A more telling criticism from a biosocial point of view is the lack of attention paid to *why* families are so important. As previously mentioned, when confronted with the challenge of explaining any animal (including the human animal) behavior or practice found to be species wide, the evolutionary biologist's first question is "what purpose did and does it serve in the ultimate evolutionary goals of survival and reproductive success?" The view of the family (for the moment simply defined as a mother/father/child[ren] triad) presented here is that the family is a "specialized and very basic adaptation that greatly extended the investment parents could make in their offspring" (Lancaster and Lancaster 1987:188). This idea of extended parental investment in the form of biparental care exemplified by male provisioning is still pivotal in evolutionary theorizing about the origins of the human family (Emlen 1995; Gross 2005; Rodseth and Novak 2006). As Pillsworth and Haselton (2005:101) put it: "Comparative, physiological, and cross-cultural evidence supports the hypothesis that humans have a suite of adaptations for forming conjugal pairs . . . the motivations to form couples emerges from adaptations with deep evolutionary roots, and thus assumes a central motivational status for most humans." If the family is an adaptation in the fullest sense of the word, it is the "expected environment" (Scarr 1993) of rearing for the human animal, and we can expect a variety of negative outcomes when child rearing does not occur in a family context.

The family is literally the nursery of human nature. The human infant arrives in this world with all the biological equipment it needs to be human, but it is the family that first takes hold of that potential and molds it into a fully social being. Despite the importance placed on the family and on the socialization process by sociologists, criminologists outside of the control tradition have tended to placed more emphasis on factors such as bad neighborhoods, poverty, values, and peer pressure as causes of crime. This may be so because to support the modern nuclear family and the values it embodies is a dangerous thing for anyone wanting to be thought of as a "good liberal" (Popenoe 1993; Hirschi 1995). Judith Stacey is one of those who want to bury what she calls the ideology of "the family" (her sneer quotes), and states that nontraditional arrangements (single-parent, serial monogamy, and same-sex groupings) are preferable because they are "more egalitarian" (1993:547).

In taking her position, Stacey betrays an ignorance of the evolutionary origins of the family, an ignorance that embarrasses biologically informed sociologists such as Alice Rossi. Rossi argues that the kind of "egalitarianism underlying current research on, and advocacy of, 'variant' marriage and family forms is inadequate and misleading because it neglects some fundamental human characteristics rooted in our biological inheritance" (1977:2). Sociologists Brigitte and Peter Berger (1984:188–189) are also cognizant of the evolutionary history of the family:

> We do know enough about the biological constitution of *Homo sapiens* to be able to say that in many areas of behavior it acts as a tendency, rather than a compelling determinant—and there seems little doubt about a tendency toward the centrality of the father-mother-child triad. In other words, the triad may be biologically "natural" even though it is not institutionalized in the child-rearing practices of the Mumbumbu tribe, say in New Guinea, and of some lesbian communes in New Hampshire.

Evolutionary psychologists recognize that there is a variety of reproductive and child-rearing strategies cataloged in the anthropological literature, and that family forms respond to ecological, economic, and cultural contingencies (Daly and Wilson 2005). Natural selection has provided us with "a set of dispositions regarding mating and kin behavior, and these interact with differing social environments to produce a variety of family systems" (Smith 1987:232). In evolutionary terms, the "best" family form is that which optimally nurtures and protects its progeny to reproductive age, and that depends on prevailing conditions, which may favor different family forms under different ecological conditions. Nevertheless, the optimal family-rearing environment is always one in which children are surrounded by many consanguineous individuals, which was the case for most of our species' evolutionary history. Unfortunately, such an arrangement does not fit the economic and social requirements of modern postindustrial societies. Such societies are thus left with the nuclear family (a relatively isolated group composed of mother, father, and child[ren]) as that which "works best . . . to produce offspring who grow up to be both autonomous and socially responsible, while also meeting the adult needs for intimacy and personal adjustment" (Popenoe 1994:94).

Given that the nuclear family is a fairly recent invention, it is a cultural and not a biological adaptation. The term *nuclear*, however, is aptly chosen because the triad of man, woman, and children is the core (nucleus) around which the web of kinship manifested in its various family forms revolves. This nucleus has been characterized as "the prototypical human social organization" (MacDonald 1992:754) that can satisfy human needs, for comfort, security, protection, and love (see also Sesardic 2003). The man/woman/child triad is thus the minimal core of the "expected environment" of rearing for the human species (Scarr 1993), and thus a biological adaptation even if the nuclear family per se is not.

This does not imply that the "natural" reproductive strategy of humans is monogamy; rather, it is likely that we are adapted to serial monogamy (Young and Wang 2004). We cannot have evolved to be pure monogamists anymore than we could have evolved to be pure cooperators, because such a population would be highly vulnerable to invasion by promiscuous cheats. Reproductive strategies involve parenting strategies as well as mating strategies that depend on environmental conditions as well as individual differences. *Homo sapiens* evolved its most human characteristics largely during the Pleistocene epoch in small hunter-gatherer kin groups in conditions characterized by low fertility, low mortality, monogamous mating coupled with high biparental investment, and prolonged, intimate, and intense contact between parents and child (Gross 2005; Rossi 1997; Sesardic 2003). Human infants are the most altricial of all infants, so males would have been considered an important source of parental investment by women in our ancestral environments.

Biparental care is found only in about 10% of mammalian species (Storey et al. 2006), and when it is it is typically found in species in which offspring remain highly dependent for a long time, where food procurement is somewhat problematic, and in which rates of predation are neither too high nor too low (Low 1998; Manica and Johnstone 2004). Pair-bonding will be selected for when the help of a male positively influences the probability of offspring survival by procuring food for gestating and lactating mothers and defending mother and child against predation. In precocial species with ready access to food, and with predation rates so high or so low that male parental investment is unlikely to have any positive effect on offspring survival, pair-bonding is not necessary, and therefore no evolutionary pressures were exerted for its selection (Low 1998; Manica and Johnstone 2004; Quinlan and Quinlan 2007).

THE EVOLUTION OF PROXIMATE MECHANISMS OF ATTACHMENT

The evolution of *Homo sapiens* from a simple social species into a *cultural* species required ever-increasing levels of parental investment because humans were facing selection pressures for intelligence, and hence for larger brains and head size. This conflicted with selection for bipedalism, a conflict that manifested itself in the cephalo-pelvic disproportion problem. As we have seen, the partial solution to this problem was selection for extero-gestation, which meant an extremely long period of dependency for human infants. This dependency allows for the behavioral flexibility of human beings and for the development of culture.

The long period of dependency required selection for strong bonds (attachment) between mother and infant, and this extra demand on females produced selective pressure for male/female bonding. Males and females who bonded to jointly provide parental investment increased the probability of their offspring surviving to reproductive age, and thus improved their own reproductive success. We can thus unromantically view the family as a reciprocity arrangement in which two people agree to cooperate in the propagation of each other's genes.

Evolutionary scenarios relating to the selection for female/infant and male/female attachment extend over millions of years, during which many mechanisms for the intensification of mothering and for male/female emotional involvement were tried, added to, or discarded. With the possible exception of empathy, these mechanisms need not have been explicitly designed by natural selection to serve the particular function of maternal intensification or for the mechanisms underlying male/female bonding. Nature is parsimonious in that it frequently makes one mechanism serve a variety of tasks. For intensification of the nurturant or attachment emotions, any physiological system associated with feelings of joy or satisfaction could have been pressed into service.

Mother/infant and male/female bonds are both similar and different. Because they both involve an active concern for the well-being of another, they share crucial evolutionary goals, and to great extent, a common neurobiology (Esch and Stefano 2005). Functional MRI studies have shown that attachment-mediating neurohormones activate regions in the brain's reward system specific to maternal and romantic love as well as overlapping into regions common to both (Bartels and Zeki 2004). Both types of love also deactivate brain regions associated with the assessment of negative emotions and social judgments, making the lover relatively unconcerned with any negatives associated with the loved one that others may perceive. Love really is blind. Bartels and Zeki (2004:1164) conclude that this body of research brings us closer "to understanding the neural basis of one of the most formidable instruments of evolution, which makes procreation of the species and its maintenance a deeply rewarding and pleasurable experience, and therefore ensures its survival and perpetuation."

However, mother/infant and male/female bonds do involve different motivational systems and goals. Mother/infant bonds had to exist in prehominid times, but prehominid mating was probably largely bereft of emotional involvement. Unlike male/female love, mother/infant love must be unconditional because the infant is incapable of meeting any conditions. Male/female love is very much conditional, and need not *necessarily* be maintained after offspring survival is unproblematic.

MALE/FEMALE BONDING

Romantic love is a biopsychosocial cross-culturally and historically universal feature of human social life (Aron et al. 2005; Hiller 2004). The touching love stories in the Bible, the *Arabian Nights*, and ancient Chinese writings make nonsense of social constructivist claims (e.g., de Rougemont 1973) that it is an invention of the troubadours in the Middle Ages. To the extent that divorce is the result of unrealistic expectations of love, understanding of the neurobiological mechanisms behind it may help to prevent some unnecessary marital fracturing.

Fisher and her colleagues (2002, 2005, 2006) have investigated love anthropologically, chemically, and neurologically, and view it as a three-stage interactive phenomenon—lust, attraction, and attachment. They explain: "The sex drive evolved to motivate individuals to seek a range of mating partners; attraction evolved to motivate individuals to prefer and pursue specific partners; and attachment evolved to motivate individuals to remain together long enough to complete species-specific parenting duties" (Fisher, Aron, and Brown 2006:2173). The sex drive (lust) is driven by the sex hormones, is motivated by the urge for sexual gratification, and is generalized to all desirable opposite-sex conspecifics.

Attraction is motivated by the desire to possess the love and sexual fidelity of one special person. This desire is underlain by a decrease in serotonin activity (foot off the brakes), an increase in dopamine activity (foot on the gas), and in norepinepherine (fifth gear) (Esch and Stefano 2005). Being in love is literally an intoxicating experience which is made so by these and certain other stimulating and slightly hallucinogenic catecholamines observed bombarding the nucleus accumbens in fMRI scans of people in love (Fisher, Aron and Brown 2005).

Natural selection has built neurological reward systems into organisms that are activated whenever they perform actions necessary for survival and reproductive success. Unfortunately, we soon develop tolerance for the effects of these endogenous chemicals the same way that drug abusers develop a tolerance for exogenous drugs. If we believe that this mad helter-skelter phase of bonding is all that there is to love, in our culture of self-absorption we will feel entitled to abandon the former object of our desire and seek a new source of endogenous chemical pleasure. This is perfectly normal; no one is saying that couples who "do not love one another" should stay together, particularly if no children are involved. But what if children are involved? It would be wonderful (but quite unromantic) if all couples were instructed before they get married about the neurochemistry of love. They would learn that what they are feeling in the attraction phase cannot possibly last, and so there had better be other reasons to commit themselves to one another—such as liking, as opposed to merely loving the other person. Liking is a rational *and* emotional motive rather than an entirely emotional motive for moving toward another person.

If one likes as well as loves one's bonded partner, the attachment phase will eventually emerge. This third stage of love is underlain by a different suite of chemicals—the neuropeptides oxytocin and vasopressin. These peptides are only found in mammals and are regulated by estrogen and testosterone (T), respectively (Hiller 2004). If the chemical cocktail associated with attraction is a mixture of endogenous stimulants, the chemicals associated with the attachment phase are endogenous morphine-like substances that soothe and calm rather than excite and stress. These neuropeptides help to cement the male/female bond; oxytocin appears to be necessary for experiencing arousal and orgasm in both sexes, and produces the feeling of calm satisfaction after having sex.

Male/female bonds are fragile, and may not last long beyond the attraction phase. Many bonds do last long enough for the neuropeptides to replace the catecholamines as their basis. We may not have not evolved to be an exclusively monogamous species, but recognizing the importance of the male/female bond for the survival of their offspring, culture has given biology a helping hand by establishing norms of conduct, including the institution of marriage, that favor the maintenance of the bond. There is no other reason for cultural norms to support male/female bonds than to assure that children receive a good start in life.

MOTHER/INFANT BONDING

Mother/infant bonding is the prototypical human attachment. Attachment is similar to MacDonald's concept of *warmth*, which he defines as "a [biological] reward system which evolved to facilitate cohesive family relationships and parental investment in children" (1992:753). Panksepp (1992:559) also views all kinds of human affective experience as emerging "from ancient neurosymbolic systems of the mammalian brain that unconditionally promote survival."

The proximate mechanisms of mother/infant bonds depend on the permutation of several factors, probably organized by the birthing process itself, and consolidated by mother/infant contact and interaction during the immediate postpartum period. Oxytocin—which is synthesized in the hypothalamus and is stimulated by environmental events such as the birthing process, infant distress, and breastfeeding—plays a major role in this process (Curtis and Wang 2003; Nair and Young 2007). A number of studies involving a variety of mammalian species have shown that administering exogenous oxytocin increases maternal behavior, and that administering oxytocin antagonists reduces it (Young and Wang 2004). Breastfeeding combines the panoply of sight, sound, smell, touch, and the tangible evidence in the mother's arms that affirms her womanhood to stimulate the release of oxytocin, which intensifies the nurturing that released it. Oxytocin released in response to breastfeeding is related to mothers' reduced sensitivity to environmental stressors, which allows for greater sensitivity to the infant. Lactating mothers show significantly fewer stress responses to infant stimuli, as determined by skin conductance and cardiac-response measures, than nonlactating mothers, and show significantly greater desire to pick up their infants in response to all infant-presented stimuli (Hiller 2004).

The distribution of oxytocin receptors in the brain is related to the ecology of the species being studied. For instance, prairie voles (small rodents) follow a monogamous mating system in which males and females share a nest and show high levels of parental care, while montane voles, a closely related species, follow a polygynous mating system with no nest sharing and very little maternal investment. Maternal behavior in the montane vole appears only in the postpartum period, when there is a temporary increase in oxytocin receptors. Prairie vole mothers, who are parental throughout their lives, maintain high concentrations of oxytocin receptors at all times (Curtis and Wang 2003; Young and Wang 2004).

Oxytocin has the effect of decreasing care-inhibiting testosterone (T) in new fathers in monogamous rodent species. T is responsible for significant sexual dimorphism in the medial preoptic area (MPOA), an area of the hypothalamus important in the control of maternal behavior. New fathers in one biparental rodent species (*Peromyscus californicus*) undergo changes in the MPOA (males have significantly more T receptors in the MPOA than females) that render it more similar to the female MPOA than to the same area in nonpaternal males (Gubernick, Sengelaub, and Kurz 1993).

These laboratory studies of nonhuman animals do not necessarily mean that analogous processes with the same compelling force occur among humans. The evolution of the human neocortex has the ability to modify or even override subcortical functions. What these findings do mean is mechanisms that facilitate caregiving (whether biparental or not) in different species exert pressure for their selection to the degree that offspring require it to reach reproductive maturity. In no other species do the young require as much parental care for so long a period as our own, leading to the conclusion that the neurobiology of bonding is probably even stronger in humans than in other species. For instance, a number of longitudinal studies have reported that human males in committed relationships have lower circulating T levels (just like *Peromyscus californicus*) than males not in such a relationship, and that in divorced males T levels rise to the level of unmarried males (Gray et al. 2004; Mazur and Michalek 1998). Other human studies have shown that paternal care is associated with lower T levels (Berg and Wynne-Edwards 2001; Fleming et al. 2002). T-modulated reduction of mating effort and increase in parenting effort in marriage may help to explain the decrease in antisocial behavior often noted to accompany marriage (Sampson and Laub 2005).

We may conclude that attachment is a need as normal and natural as is the need for food and drink. Natural selection has provided us with neurohormonal mechanisms that lead us (infants, children, and adults) to want to be, and expect to be, attached to others because of the vital survival role attachment has played over our evolutionary history. If we are able to look at the family through an evolutionary lens, there would be less controversy about its importance.

DISRUPTION OF THE PAIR-BOND

Cultural rules regarding anything touching on survival and reproduction reflect evolutionary logic. Perhaps one of the first cultural rules ever laid out by our species was rules about who has sexual access to whom, and one of the first cultural ceremonies may have been the marriage ceremony. Marriage may have originated with the recognition of the importance of the family in rearing healthy offspring to reproductive age: "Marital alliance and biparental care are part of the human adaptation" (Daly and Wilson 1988:187). Because the family is as much a biological as a cultural phenomenon; i.e., a product of gene/culture coevolution, we should expect significant social problems to arise when it is disrupted.

Broken homes have long been implicated in antisocial behavior. While *broken home* is often a convenient label for a number of problems preceding it, family fracture itself can be a precipitating factor in antisocial behavior via a number of routes such as offspring resentment, decreased supervision of children, or increased parent/child conflict. Children from broken homes are often found to exhibit more traits favorable to offending and fewer favorable to conformity than do children from intact homes. The overall conclusion from a meta-analysis of 92 studies comparing children from broken and intact homes was that children from broken homes had significantly poorer psychological adjustment, lower self-esteem, and poorer academic achievement and social relations (Amato and Keith 1991a). Another meta-analysis of 33 studies of adult children of divorced parents concluded that adult children of divorced parents had significantly more psychological maladjustment and marital instability than adults reared in intact homes (Amato and Keith 1991b).

Children's antisocial behavior is not necessarily attributable to the effects of divorce per se given fairly robust findings that the heritability of divorce is around 0.50 (Booth, Carver, and Granger 2000; Jockin, McGue, and Lykken 1996). Of course, there is no "divorce gene" anymore than there is a crime gene; the coefficient is the result of people with traits which make them difficult to live with that increase the likelihood of unfavorable outcomes in all areas of their lives, including the probability of divorce. For instance, a person may be impulsive or highly sexed, making it difficult to remain faithful, or be susceptible to anger, irritability, or depression. Because parents pass these traits to their children, their children are susceptible to antisocial behavior because of their inherited traits, not because of their parent's divorce. This does not mean that all, or even most, divorced people possess these negative traits. Marriage is a mix of two people that produces a situation-specific "third personality" which is not necessarily predictable a priori from the personalities each brings with them to it.

To the extent that a broken home per se increases children's risk of antisocial behavior, the mechanism typically invoked is the negative effects of father absence. The effects of father absence are not due to its non-normativeness in Western societies because similar effects are found in prestate polygynous cultures where children *typically* grow up in mother–child households separate from their fathers. As Ember and Ember (1998:14) put it: "Societies in which children are reared in mother–child households or the father spends little time in child care tend to have more physical violence by males than do societies in which fathers are mostly around." They go on to speculate that a father-absence upbringing breeds a "supermasculine" male identity. The same hypermasculinity is observed in modern Western societies where father absence is non-normative (Hall 2002; Hayslett-McCall and Bernard 2002), although it has become typical in some subcultures (see Chapter 8).

I do not imply that father absence is always a bad thing. Assortative mating tends to bring antisocial males and females together (Krueger et al. 1998; Rutter 1996), which results in children born to antisocial parents receiving genes for heritable traits predictive of antisocial behavior from both parents as well as an environment modeling it. When an antisocial father resides with mother and offspring there is more risk to the children than when one does not because "children experience a double whammy of risk for antisocial behavior. They are at genetic risk because antisocial behavior is highly heritable. In addition, the same parents who transmit genes also provide the child's environment" (Jaffee et al. 2003:120). However, we should not confuse variation with central tendency. The consequences of not being reared in a two-parent family *range* from positive to negative, but the *central tendency* is negative.

HISTORICAL LESSONS

Notwithstanding the special case of "criminal families," recent history provides insight about what becomes of the young when the family is in disarray. One of the goals of Marxism was to destroy the nuclear family because it was considered a "bourgeois" property-based notion that exploited women. Marx's long-time collaborator, Friedrich Engels, viewed marriage as a kind of class struggle in which the male exploited the female just as the bourgeoisie exploited the proletariat: "The first division of labour is that between man and woman for child breeding . . . The first class antagonism which appears in history coincides with the development of the antagonism between man and woman in monogamian marriage" (1884/1988:720).

Based on their Marxist understanding of marriage, in 1928 the Communist Party of the Soviet Union passed legislation that legitimized unmarried cohabitation and the offspring of such, made divorce available on demand, and encouraged "free love" as the "essence of communist living" (Hazard, Butler, and Maggs 1977:470). As any Darwinian biologist or control theorist would have predicted, the human cost of these "reforms" was staggering. The practice of free love is incompatible with pair-bonding and biparental care throughout the animal kingdom (Clutton-Brock 1991), and this practice, along with easy divorce and the legitimization of out-of-wedlock births, resulted in thousands of fatherless children roaming the streets who "formed into gangs, and who would rob and attack people in the street, or even invade and ransack apartment blocks" (Hosking 1985:213). After assessing the damage, the Soviet government quickly responded with new laws extolling the family, the sanctity of marriage, the evils of divorce, and the joys of parenthood, as well as making divorce more difficult to obtain and restoring the legal concepts of legitimacy and illegitimacy (Hosking 1985).

A similar attack on the nuclear family occurred in China after its revolution, followed by the same rapid about-face when confronted with the consequences (Fletcher 1991). More than all social science studies of family disruption put together, these two "natural experiments" should tell us something about the importance of the family for the development of healthy prosocial human beings. As Alfred Blumstein (1995:12), the "guru of demographics," put it: "teenage mothers, single-parent households, divorced households, unwed mothers" all constitute high risk for criminality, "and certainly don't bode well for the future of the nation."

Crime is only part of the problem. A large-scale economic study of the cost of fractured families (divorce and unwed childbearing) to the American taxpayer put it at $112 billion per year (Scafidi 2008). This figure includes various welfare programs and Medicaid as well as the indirect costs associated with the often antisocial behavior of fatherless children.

ILLEGITIMACY AND CRIME

According to James Q. Wilson (2000:107), to address the problem of illegitimacy "is to risk being called a racist, unless you quickly add that families are not important to children, in which case you may be called either a progressive or a fool." Illegitimacy has been called "the new American dilemma" and "a cultural catastrophe" by a variety of scholars and political figures from across the political and ideological spectra (Hamburg 1993; Hirschi 1995; Wilson 2000). A number of studies have shown the illegitimacy rate to be a strong predictor of many kinds of antisocial behavior (Barber 2003; Demuth and Brown 2004; Walsh 2003). Violent crime in the United States (Lykken 1995) and Britain (Himmelfarb 1994) rose almost directly proportional to illegitimacy rates over the past 30 years in those countries. A study comparing the out-of-wedlock birthrates and crime rates in each of the 50 states found a very strong partial correlation of 0.82 after controlling for unemployment rates (Mackey 1997). A British longitudinal study comparing delinquency among legitimate and illegitimate boys concluded that "the boys born illegitimate were singularly delinquent prone" (West and Farrington 1977:197), and another British study concluded that there were "relatively high levels of behavioral and adaptive problems in the illegitimate group" (Maughan and Pickles 1990:55). An American study comparing delinquents born in and out of wedlock found that the illegitimate boys were almost twice as violent, and that they were significantly more deprived on all indices of emotional, economic, and social well-being (Walsh 1990).

Figures 7.1 and 7.2 show that births to unwed mothers in the United States exploded from 1960 to 1962, as did arrests for violence among young males. The increase among young nonwhite male arrests was particularly striking (just over 100%). Illegitimate births in the black community during this period comprised about 70% of all live births (Lykken 1995). As Matt DeLisi (2003:671) sums it up: "Rampant illegitimacy has catastrophic consequences for the life chances of children and is one of the most important predictors of crime."

Single-parent families increase a child's risk of future offending for a variety of reasons. For instance, it "decreases community networks of informal control," and it increases "the prevalence of unsupervised teenage peer groups" (Messner and Sampson 1991:697). Unwed mothers tend to be younger than women who bear their first child in wedlock, and to come from educationally and economically deprived families (Anderson 2002, Vedder and Gallaway 1993). They also tend to lack the same level of social support enjoyed by married mothers, which socially isolates them and increases the probability of child abuse and neglect. For instance, a nationwide study of children ages 2 through 9 found that children of single parents were 6.7 times more likely to witness family violence, 3.9 times more likely to be maltreated, and 2.7 times more likely to be sexually assaulted than children with both biological parents present. The figures for stepparent families were even worse at 9.2, 4.6, and 4.3, respectively (Turner, Finkelhor, and Ormond 2006).

The relationship between unwed motherhood and criminality cannot be accounted for simply by family structure (single-parent versus intact home) and its demographic consequences (poverty, residing in disorganized neighborhoods, etc.). The personal traits of unmarried mothers and their inseminators must also be considered. According to a study of 1,524 sibling pairs from different family structures taken from the National Longitudinal Survey of Youth, Cleveland and his colleagues (2000) found that heritable traits associated with antisocial behavior may select individuals into different family structures and these traits are then passed on to offspring.

Cleveland et al. found that, on average, unmarried mothers have a tendency to follow an impulsive and risky lifestyle, to be more promiscuous, and to have below-average IQ. They found that the family type that put offspring most at risk for antisocial behavior was one in which a single mother had children fathered by different men. Two-parent families with full siblings were found to be the family type placing offspring at lowest risk for antisocial behavior. Genetic differences accounted for 94% of the difference on an antisocial subscale between the most at-risk group and the least at-risk group. The most at-risk families were single-parent families with half siblings; in evolutionary terms, this is indicative of an emphasis on mating effort; the families least at risk were two-parent families with full siblings, indicative of parenting effort. Similar findings and conclusions from a large-scale British behavior genetic study have been reported (Moffitt and the E-Risk study team 2002).

David Rowe emphasizes the traits of their "feckless boyfriends" who abandon their pregnant girlfriends as another link between illegitimacy and criminality. These traits include: "strong hypermasculinity, early sexuality, absence of pair-bonding capacity, and . . . other hallmarks of 'psychopaths'—are all passed on genetically to offspring" (1997:257). Studies of fathers of illegitimate children have been found to be more than twice as likely to be involved in delinquent and criminal behavior as non-fathers in the same neighborhoods (Stouthhamer-Loeber and Wei 1998; Thornberry et al. 2000). Thus it would seem that many children born out of wedlock have the cards stacked against them from the beginning. The many problems associated with high illegitimacy rates led Gottfredson and Hirschi (1997:33) to focus on illegitimacy as *the* policy recommendation

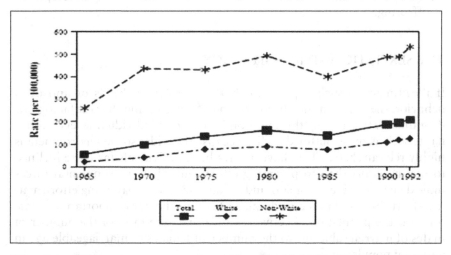

Figure 7.1   Violent crime arrest rates for youths under 18: 1965–1992. Source: FBI Uniform Crime Report, 1993. Age-specific arrest rates and race-specific rates for selected offenses, 1965-1992. Crimes are murder, rape, robbery, and aggravated assault.

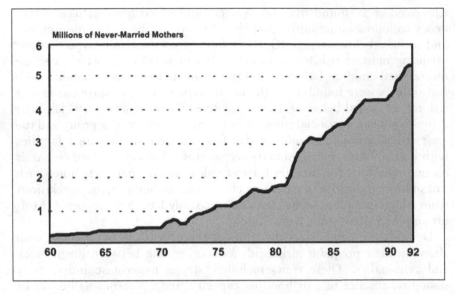

Figure 7.2   Millions of never-married mothers 1960–1992. Source: Bureau of the Census. Current Population Reports-Marital status and Living arrangements. March, 1992.

deducible from self-control theory: "Delaying pregnancy among unmarried girls would probably do more to affect the long-term crime rates than all the criminal justice programs combined." This is not a statement of moral condemnation. It is a statement that implicitly recognizes the evolutionary importance of the reproductive team to the healthy development of offspring.

THE SEX RATIO AND ILLEGITIMACY

If illegitimacy is such a powerful risk factor for all kinds of antisocial behavior, the question then becomes one of determining the risk factors for illegitimacy. In pursuing this question I go beyond individual-level determinants such as those outlined earlier because the wider social environment is highly relevant here. There have always been individuals far more inclined to mating effort than to parenting effort, but what environmental forces moved an increasing number of individuals across the parenting effort/mating effort threshold in recent decades? Perhaps the most important of these forces is the population effective sex ratio; i.e., the ratio of the number of males of marriageable age to the number of females of marriageable age in a mating population.

Across a huge variety of animal species, the sex ratio in a breeding population is the most important environmental factor affecting mating patterns in nominally monogamous species. Field experiments with species in which pair-bonding occurs (at least for the mating season) show that when researchers manipulate the sex ratio such that females are more plentiful than males, male desertion rates rise precipitously as opportunities to re-mate occur (Emlen 1995; Gross 2005). In almost all sexually reproducing species the sex that invests the least in offspring (almost always males in mammals) is facultatively polygynous (Alcock 2005; Krebs and Davies 1993). This essentially means that male mating patterns will shift to mating effort at the expense of parenting effort when a low sex ratio makes it possible because the fitness value of monogamy for males declines with a decline in the sex ratio. In high-sex-ratio conditions (more males than females), males who have gained access to mates jealously guard them and provide them with valuable resources and parenting effort (Alcock 2005; Krebs and Davies 1993).

The laws of supply and demand also help to determine human mating strategies. When there is a greater number of one sex of marriageable age than the other, the less numerous sex is a scarce resource and thus a valued commodity. The scarcer sex holds the dyadic power in dating and mating relationships and can dictate the conditions of the relationship. Because the reproductive strategies of men and women are different (emphasis on mating versus parenting effort), when sex ratios are significantly skewed, mating environments are considerably altered.

The first major work to apply the sex-ratio concept to human beings was Marcia Guttentag and Paul Secord's seminal *Too Many Women? The Sex Ratio Question* (1983). Guttentag and Secord surveyed the historical literature on cultures ranging from ancient Athens and Sparta, Medieval Europe, and orthodox Jewish culture, to the modern United States. Their arguments regarding low-sex-ratio societies are encapsulated by the following predictions, which turn out to be accurate across time and place (1983:20):

> Women in such [low-sex-ratio] societies would have a subjective sense of powerlessness and would feel disvalued by the society. They would be more likely to be valued as mere sex objects. Unlike the high sex ratio situation, women would find it difficult to achieve economic mobility through marriage. More men and women would remain single, or if they married, would be more apt to get divorced. Illegitimate births would rise sharply. The divorce rate would be high, but the remarriage rate would be high for men only.

Guttentag and Secord show that low-sex-ratio societies tend to be unstable, misogynistic, and licentious, and that they breed female responses to these things in the form of feminist movements. On the other hand, in high-sex-ratio societies "women would be valued as romantic love objects" and as mothers, both sexes "would stress sexual morality," women would have a "subjective sense of power and control over their lives," male commitment to marriage would be strong, and society would be stable (1983:19–20). Their argument boils down to the proposition that when males or females are free to choose their mating behavior (because the sex ratio favors them), they will behave in ways compatible with their innate inclinations. Members of the more numerous sex must compromise their natural mating strategies and conform to the strategy of the sex favored by the sex ratio if they are to participate at all in the mating game. As Anne Campbell and he colleagues (2001:487–488) put it: "When males are rare and valuable, women have little option but to engage in short-term relationships despite their preference for a committed partner. This creates a men's 'buyers market,' which in turn makes them less likely to commit themselves to long-term investment."

Guttentag and Secord show that the sex ratio in the United States was consistently high until 1970. This female-favoring ratio helped to shape American mores relating to dating, mating, and marriage and most males found it difficult to adopt a promiscuous strategy under these conditions. High sex ratios meant that men had to compromise with the female strategy; if they did not there were many others who would. Women were prized and respected, marriage was considered an attractive and permanent prospect, sexual intimacy was an expression of love, and adultery was morally unacceptable and legally punishable.

By the 1970 census the sex ratio had become male-favoring with only 78 white men of that age for every 100 white women, and 73 black men for every 100 black women. Now females had to compromise with the male strategy, and predictably, a licentious and misogynistic environment ensued (Walsh 2003). Women did tend to respond to this male-favoring environment by requiring lower levels of commitment before agreeing to have sex with males, although many found it to be distasteful (Campbell, Muncer, and Bibel 2001; Wilson 1983).

Messner and Sampson (1991) attempted to account for the high rate of single-parent families (they did not differentiate between illegitimacy and other possible reasons for the absence of a male head in the home) in high crime areas with reference to the sex ratio. They reasoned that because gender is the major predictor of antisocial behavior, populations with more males than females should have higher rates of crime than populations with more females than males. Populations with a greater number of males than females (a high sex ratio), however, are associated with family stability and lower rates of illegitimacy. Populations with low sex ratios means that there are fewer males available to commit crimes, but it also leads to a licentious environment in which males are reluctant to commit to one women (Barber 2000b; Geary 2000). This situation leads to high rates of illegitimacy and divorce, in short, to family disruption, which affect crime rates in many ways.

Messner and Sampson tested their hypothesis for blacks and whites based on data from 153 cities in the United States with populations greater than

100,000. They first examined the effect of the sex ratio on the percentage of female-headed households in multiple regression models. In the black model, the sex ratio was by far the most powerful of the eight predictors in the model (β =–.60), and in the white model (a population with a less skewed sex ratio) it was the third most powerful predictor (β =–.29), behind per-capita income and welfare availability. In a nutshell, cities with low sex ratios have high rates of female-headed households and crime, and cities with high sex ratios have low rates of female-headed households and crime.

Guttentag and Secord focused on social exchange theory to explain the link between rarity and compromise. Social exchange theory works well as a proximate explanation of dyadic power relationships, but it assumes that all parties in an exchange relationship desire the same things and employ the same strategies to achieve them. Darwinian sexual selection theory is superior to exchange theory in this instance because it informs us that although reproductive success is the fundamental goal of both sexes in all sexually reproducing species, males and females employ quite different strategies (Pedersen 1991).

Sexual selection theory avers that the primary constraints on maximizing reproductive success for males is access to females, and for females it is acquiring resources for herself and offspring (Alcock 2005; Geary 2000). Thus, males can be expected to employ a strategy aimed at maximizing their access to as many females as possible, and females will be attracted to males who can demonstrate their ability to provide resources. For Pedersen (1991:276) the primary criterion for favoring sexual selection theory over social exchange theory to explore the correlates of skewed sex ratios is that "sexual selection theory specifies more directly the criteria for male and female choice on the basis of asymmetries in reproductive constraints." A low sex ratio removes much of the constraint for males, and a high sex ratio removes much of it from females. Pedersen goes on to state that sexual selection theory subsumes everything in social exchange theory and adds substantially to it by making sense of the psychological processes Guttentag and Secord highlight by identifying sex-specific reasons (beyond dyadic power differentials) why the mating environment is radically different when high and low sex ratios obtain, and by pointing to similar processes in other species.

THE EVOLUTIONARY CONTEXT OF CHILD MALTREATMENT

Child maltreatment, in addition to being repugnant in itself, is related to the probability of antisocial behavior in maltreated children through its effects on the child's central and peripheral nervous system functioning. An evolutionary explanation for why child abuse and neglect occurs is almost a contradiction in terms; after all, evolution is all about preserving genetic material. Nevertheless, taking advantage of the theoretical insights provided by evolutionary theory may enhance our understanding of these behaviors by pointing to the conditions under which they occur in other species and then comparing similarities among humans.

The probability of abuse, neglect, and infanticide among nonhuman animals increases when the food supply is low, the litter size is large, when an infant has low reproductive viability, and, in biparental species, when the female lacks the assistance of a mate (Alcock 2005; Allman 1994). These are the same conditions under which most human incidences of abuse, neglect, and infanticide occur; i.e., under conditions of poverty, within large and single-parent families, and against children who are physically or mentally handicapped (Buss 2005; Daly and Wilson 1988a; Gelles 1991). Neglecting or killing offspring under such conditions may have enhanced the killers' inclusive fitness in ancestral environments in several ways. When human ancestral mothers had too many mouths to feed with available resources, lacked a mate, or had children who were unlikely to contribute to the family well-being because of illness or deformity, a "triage" strategy may have been the best one available. A triage strategy would have increased the probability of the survival of the most reproductively viable of offspring, while a strategy of trying to nurture each offspring equally may have resulted in the survival of none. Legalities, morality, and politics aside, these conditions are often the very same conditions that lead women to decide to seek an abortion (Lamanna 1985), which from an evolutionary point of view may be considered the functional equivalent of infanticide (Alcock 2005).

A good proportion of infanticidal behavior is either performed or instigated by males genetically unrelated to the victim. A male in a number of mammalian species claiming a female commences to kill any offspring sired by her previous mate (reviewed in Van Hooff 1990). Killing infants puts an end to breastfeeding and prompts the female's return to estrus, thus providing the new male an opportunity to produce his own offspring. Human males acquiring wives with dependent children may also kill any children from a previous relationship in a number of prestate cultures (Daly and Wilson 1988). Although this increases the genetic fitness of the killers at the expense of the fitness of the fathers of the victims, males in these societies are probably no more aware of this fact than are nonhuman animals. Infanticide may have had some positive fitness consequences in ancestral environments, but as Symons points out, infanticide per se is not an adaptation, but " . . . rather the general mechanisms of emotion [parental solicitude] and cognition [cost/benefit calculations within a stressful context] that are the adaptations, regardless of infanticide's effect on reproductive success" (1987:140). The mental mechanism behind much male-initiated infanticide is probably: "Don't waste precious resources on children for whom I have no warm feelings."

Children in modern state societies are also at greater risk for maltreatment when not raised by both biological parents. The vast majority of stepparents do not abuse their stepchildren, but the risk is greatly elevated in stepfamilies. It was pointed out earlier that a nationwide study found that stepchildren were 9.2 times more likely to witness family violence, 4.6

times more likely to be maltreated, and 4.3 times more likely to be sexually assaulted than children living with two biological parents (Turner, Finkelhor, and Ormond 2006). Another study found that stepchildren were between 9 and 25 times more likely (depending on age, with the risk being greater the younger the child) to be abused than children residing with both biological parents (Daly and Wilson 1985). A child living with a stepfather or live-in boyfriend is approximately 65 times more likely to be fatally abused than a child living with both biological parents (Daly and Wilson 1996).

The dominant social science model of intrafamily homicide is the intimate contact hypothesis whereby intrafamily homicide is explained by the fact that family members are frequently in "striking distance" of one another. Ketalar and Ellis (2000) examine and reject this notion, stating that if intrafamilial homicide simply followed the "striking distance" rule then genetic and nongenetic family members would be equally at risk for homicide. This is never the case, and the fact that it is not provides excellent evidence for the usefulness of Darwinian thinking applied to social science data. In Darwinian terms, stepparenting is a fitness reducer and a chore reluctantly undertaken as a condition for gaining access to the child's mother's reproductive potential or the father's resources.

Stepparenting also significantly increases the risk of sexual abuse of stepchildren (Turner, Finkelhor, and Ormond 2006). Stepfathers or live-in boyfriends may find their stepdaughters as sexually desirable as any other nongenetically related female. Close physical proximity early in life appears to be the evolved mechanism that triggers the incest avoidance mechanism that dulls sexual attraction between individuals, genetically related or not (Ingham and Spain 2005; van den Berghe 1987). The later a stepfather enters the lives of genetically unrelated females, then the greater the likelihood that he will be sexually attracted to them.

The stepparent/stepchild relationship is more tenuous than the biological parent/child relationship because is does not rest on the firm basis of early bonding, and therefore not on the mutual trust, nurturance, and solicitude that such a relationship engenders. In decrying the ever-decreasing number of children who live with both biological parents, Robert Wright (1994:104) insightfully remarks: "[W]henever marital institutions . . . are allowed to dissolve, so that divorce and unwed motherhood are rampant, and many children no longer live with both natural parents, there will ensue a massive waste of the most precious evolutionary resource: love."

The Biology of Low Self-Control

Because of widely varying breed temperaments, some dogs train easily (e.g., border collies) and others require the patience of Job (e.g., pit bulls). Similarly, and for the same reasons, some children learn self-control easily and others only with great difficulty. Learning to appropriately restrain one's behavior, emotions, and desires is an executive function of the PFC and is obviously a good thing in any social group. Self-control theory attributes self-control levels solely to parenting, ignores children's genetic effects, and thus child effects on parents. This is unfortunate, because as Lilly, Cullen, and Ball (2007:110) point out: "research suggests that parents may affect levels of self control less by their parenting styles and more by genetic transmission." Indeed, Naomi Friedman and her colleagues (2008) found that individual differences in executive functions (behavioral inhibition or self-control being one such function) are almost entirely genetic in origin. Harking back to our discussion of how not to read behavior genetics in Chapter 2, this does not mean that executive functioning per se is almost entirely genetic, that is, impervious to socialization or to targeted training. Executive functioning can be quite strongly related to training and thus affect the population *mean*, but *individual differences* (i.e., population *variance*) can still be almost entirely genetic. In other words, the mean of a measure of self-control in a population can move toward higher levels under the influence of parental training, but it will still drag a constant level of variation (the basis of computing heritability) along with it.

Gottfredson and Hirschi chose to ignore the biology of self-control because they believe: "biology connotes fixation, immutability, or even destiny" (1990:135). Yet in the penultimate sentence in their book, they state: "The study of crime is too important to be diverted by arguments about theory ownership or discipline boundaries" (1990:275). Their first assertion is patently untrue, and their second assertion, unless they want to make an exception in the case of biology, indicates that they would, at least in spirit, welcome assistance from outside sociology in their quest for a general theory of crime. In this endeavor they would be well advised to pay attentions to the words of John Wright and his colleagues (2009:83; emphasis in original): "*All* scientific data indicate that self-control is housed in the frontal and prefrontal cortex, that self-control is strongly influenced by genetic factors expressed in the brain, and that self-control involves a complex, dynamic balancing of limbic and cortical functioning." Of course, this "dynamic balancing" requires input from the environment, and this is where parenting interacts with individual propensities; and the ease or difficulty of the parenting task depends on whether the offspring is temperamentally more like a border collie or a pit bull.

Wright and Beaver's (2005) test of the assumption that parents are primarily responsible for their children's level of self-control found that when genes were not taken into account, parenting style had a modest effect, but when genetically informed methods were used, parental effects all but disappeared (see also Wright et al. 2008). Without using genetically informed methods, which is the case in the overwhelming majority of criminological studies, researchers seriously misidentify important causal influences. As Wright and Beaver (2005:1190) concluded: "for self-control to be a valid theory of crime it must incorporate a more sophisticated understanding of the origins of self-control." These researchers, however, cautioned that the results of their study not be naively construed as saying that parents don't matter. Our discussion of this claim in Chapter 2 showed the many ways that parents matter so very much. The Wright and Beaver study examined very young schoolchildren, not criminals and delinquents, matched with a control group, and self-control was assessed via parental and teacher reports of children's self-control. Parents help to establish a certain level of self-control in their children, but parental effects do not account for population *variance* in the self-control construct.

Is Self-Control all that Matters?

Another problem with self-control theory is that it claims too much for its central construct. Calling Gottfredson and Hirschi's assertion that crime proneness can be explained by a single tendency "simplistic psychologically," Caspi and his colleagues (1994:187) contend that crime proneness is defined at a minimum by both low self-control (which they call low constraint) and negative emotionality, which is the tendency to experience many situations as aversive and to react to them with irritation and anger more readily than with positive affective states. Negative emotionality is strongly related to self-reported and officially recorded criminality across countries, genders, races, and methods (Caspi et al. 1994). Constraint is inversely associated with negative emotionality. Individuals high on negative emotionality but also high on self-control are able to hold their anger and irritability at abeyance, but if a person high on negative affect is also low on self-control, the risk for violent behavior is high. As we saw in Chapter 5, Robert Agnew (2005) has developed a promising theory which incorporates both low self-control and negative emotionality (irritability). Impulsiveness, a central component of the concept of low self-control, has been found to be moderately to strongly heritable (Caspi et al. 1994; Lykken 1995) and strongly related to low levels of the serotonin metabolite CSF 5-HIAA (cerebrospinal fluid 5-hydroxyindoleacetic acid) in human and nonhuman animals (Bernhardt 1997). The median estimate of the heritability of serotonin level is between .55 and .66 (Hur and Bouchard 1997). Because serotonin is a modulator of behavior, poor serotonin turnover (weak BIS) is a brain with defective brakes, resulting in a runaway emotional train (a strong BAS) that all too frequently fails to think before acting. Low serotonin, therefore, "may be a heritable diathesis for a personality style involving high levels of negative affect and low levels of constraint (self-control), which generates in turn a vulnerability to criminal behavior" (Caspi et al. 1994:188).

Although some individuals may be at greater genetic risk for higher negative emotionality and lower self-control than others, both traits are influenced by environmental factors. Neurologically, both traits appear to be a function of low serotonin, but we know that serotonin turnover rates reflect, as well as affect, environmental events (Raleigh et al. 1991). Caspi and his colleagues (1994) maintain that both negative affect and low constraint in children are affected by family dynamics that include emotional and physical abuse and neglect. The effects of abuse and neglect on the behavioral activating and inhibiting systems and on the prefrontal cortex were discussed in Chapter 4. At a minimum, Gottfredson and Hirschi's theory would benefit from incorporating negative affect into their theory, by acknowledging that both it and low self-control are heritable, and that serotonin turnover rates underlie both traits.

Age in Self-Control Theory

Gottfredson and Hirschi contend that the age effect on crime and delinquency is basically inexplicable since it is an invariant phenomenon across time and space. They are suspicious of situational explanations of the effect because they contend that, as a stable characteristic of individuals, self-control leads people to differentially involve themselves in situations leading them to engage in antisocial behavior. They believe that the age/ crime curve is a "law of nature," and that it presents a major problem for sociological criminology theories. In Chapter 6 we saw how these theories attempt to explain the adolescent rise in antisocial behavior by emphasizing an increase in peer involvement during this period, and the decline thereafter by the decreasing influence of peers and the acquisition of girlfriends, wives, children, and jobs. The problem, say Gottfredson and Hirschi (1990:141), is that "the offender tends to convert these institutions into sources of satisfaction consistent with his previous criminal behavior . . . there is no drastic reshuffling of the criminal and noncriminal populations based on unpredictable, situational events."

Gottfredson and Hirschi believe that criminological theories are unable to explain the age effect because they conflate the concepts of crime (the prevalence of which changes with age) and criminality, which they say is stable across all ages. Because the age effect is invariant across time and place, they believe that criminologists should simply accept it as a fact on go on about other business. To argue that the age affect is inexplicable because sociological criminology cannot explain it is to ignore a vital component of the crime problem. Biosocial science provides excellent explanations for the changes in the probability of antisocial behavior for youths who start offending at adolescence and then desist, as we have already noted. Gottfredson and Hirschi seem to be saying that only criminality (which they equate with low self-control) needs explaining. In asserting this they appear to concur with Terrie Moffitt (1993) that there are two types of delinquents: the age-dependent (Moffitt's AL offender) and the stable (Moffitt's LCP offender), but they are apparently only interested in the latter. Moffitt also stresses the stability of traits conducive to antisocial conduct across ages and situations, but she does not ignore how age affects the expression of those traits; i.e., how the hormonal and neurological changes occurring in adolescence negatively affect self-control.

A biosocial analysis of control theory leads us to the same conclusion regarding persistent criminality that our analyses of the social learning tradition did: Persistent offenders are "biologically different from the adolescent offender who stops at age 18–21" (Jeffery 1993:494). Whether we concentrate on overall temperament or on specifics such as low self-control, negative emotionality, or cognitive deficits, we arrive at an image of a small set of offenders who are difficult to socialize. They suffer the double liability of being born to parents who tend to have temperaments unsuitable for the task of providing adequate socialization of them, and who have provided them with a set of genes biasing their development in negative directions. If anything is to be done to divert at-risk children from the trajectory they are on deducible from Gottfredson and Hirschi's theory, it is to implement nurturant strategies within families and to do everything we can to strengthen those families.

8 The Human Ecology/Social Disorganization Tradition and Race

The human ecology/social disorganization tradition in criminology was explicitly formulated to exclude "kinds of people" explanations of crime. As such, it is the most sociological of criminological theories. Although human ecologists and evolutionary psychologists differ in their assumptions about human nature, human ecology also drew its theoretical sustenance from the writings of Charles Darwin. The intent of the early human ecology was to "achieve a thorough-going natural science treatment of human behavior" (Hawley 1944:400). According to Amos Hawley (1950:6), at the core of human ecology's conceptual debt to Darwinism lay three main points:

> Scientific [human] ecology, then, is indebted to Darwin for the main outlines of its theory, the essential conceptions being: (1) the web of life in which organisms are adjusted or are seeking adjustment to one another, (2) the adjustment process as a struggle for existence, and (3) the environment comprising a highly complex set of conditions of adjustment.

Ecology is a term used in biology to describe the interrelationships of plants and animals and their environment; how each affects and is affected by the other. Long-term occupation of a particular ecological niche implies a high level of adaptation to it to the point that the niche becomes a "natural area" for the organisms occupying it. Sometimes nonnative species may invade these niches, may come to dominate it, and even drive the former species to extinction. This biological process of invasion, dominance, and succession, describing the course of occupation of ecological niches by species of plant or animal life previously occupied by other species, was borrowed whole by human ecologists.

Human ecology describes the interrelations of human beings and the environments in which they live and views the city as kind of superorganism with areas differentially adaptive for different ethnic groups (e.g., "little Italy," "Chinatown"). In early 20th-century America these "natural areas" (city neighborhoods) saw successive waves of alien ethnic groups invade the areas and human ecologists noted the criminogenic consequences. When alien groups invade natural areas, opportunities are increased for criminal activity because the symbiotic relationship that formerly existed between native organism and environment was lost. Human ecology is thus about how antisocial behavior is amplified or constrained by the social setting in which it is found. Unfortunately, what could have been a true human ecology drifted away from its biological roots to settle within the structural tradition of sociology, which is happiest the further away its analyses are from real flesh-and-blood human beings. In short, human ecologists have tended to focus only on the environmental half of the ecological whole and to ignore the other half—the organism (Savage and Vila 2003).

Ecological criminology was developed in Chicago in the 1920s and 1930s through the works of Clifford Shaw and Henry McKay (1972). Shaw and McKay analyzed Cook County Juvenile Court records spanning the years from 1900 to 1933 and noted that the majority of delinquents came from the same neighborhoods regardless of their ethnic composition. This suggested the existence of areas that facilitate antisocial behavior; that is, there may be certain areas possessing characteristics that persist over time that infect individuals exposed to them in ways that increase the likelihood of antisocial behavior regardless of their race or ethnicity. Shaw and McKay did not assert a neighborhood's milieu was a necessary and sufficient cause of antisocial behavior; rather, they strove to explain why delinquency was so heavily concentrated in certain neighborhoods without having to address kinds of people issues.

Previous work in social ecology characterized the spatial patterns of American cities as radiating outward from central business and industrial areas in a series of concentric circles or *ecological zones* (Park, Burgess, and McKenzie 1925). The Loop area of Chicago (downtown) was surrounded by the factory zone. As business and industry grew, the factory zone invaded the surrounding areas, making them less desirable as residential sites. Housing in these areas rapidly deteriorated, and those who could afford to move to better locations did. Old inner-city neighborhoods in the process of invasion and succession were labeled *transition zones,* and were the homes of the poorest and most recently arrived families, who were typically foreign immigrants and African American migrants from the south. Successive waves of newcomers to the poorest neighborhoods precipitated constant movement from zone to zone as more established groups sought to escape the intrusion of the alien newcomers.

Of greatest interest to the Chicago ecologists was the fact that delinquency and crime rates varied predictably and substantially by zone. For instance, for the years 1927 through 1933, Shaw and McKay (1972) report delinquency rates per 100 males aged 10 to 16 of 9.8 in the poorest zone, decreasing linearly to 1.8 in zone five (the affluent suburbs). Thus, different zones with their distinctive forms of social organization produced widely varying rates of delinquency. The next task for the human ecologists was to identify possible processes that might explain why certain zones are more likely to produce delinquents and criminals than other zones. The answer they came up with is *social disorganization*.

Social Disorganization

Social disorganization is the key concept of ecological theory. Social disorganization is created by the continuous redistribution of the population in and out of neighborhoods disrupting the symbiosis previously established. The mix of transient peoples with limited resources bringing a wide variety of competing cultural traditions to the neighborhood is not conducive to developing or maintaining a sense of community. A system of shared values and rules of conduct transmitted from generation to generation creates the kinds of warm social bonds that sociologists have in mind when they speak of a sense of community or a sense of belonging. The conflict of values, interests, practices, and many other issues that arise when culturally diverse groups are thrown together generates the breakdown, or serious dilution, of the power of social norms to regulate conduct. Social disorganization is thus a sort of Durkheimian anomie operating at the neighborhood level.

Social disorganization facilitates crime and delinquency in two ways. First, the lack of social control facilitates crime by failing to inhibit it. Ecologists viewed the control traditionally exercised by communities as a function of their ability to bring members together to organize strategies to combat community problems, an ability that modern ecologists call collective efficacy (Sampson 2004). Because communities in transition zones were culturally fractured, conventional institutions of control, such as the family, church, school, and neighborhood clubs and voluntary organizations, were unable to exert proper supervision and control over their restless youth. Without these informal social controls, slum youths were freed to follow their natural impulses and inclinations, which, as numerous theorists from Plato on down the centuries have emphasized, do not run in prosocial directions.

Second, social disorganization contributes to crime and delinquency by providing positive incentives to engage in it. In the absence of prosocial values, a set of values supporting antisocial behavior is likely to develop to fill the vacuum. Slum youths thus have both negative and positive inducements to crime and delinquency represented by the absence of social controls and the presence of delinquent values, respectively. These conditions are transmitted across generations until they become, as it where, intrinsic properties of the neighborhood's cultural milieu. This brings us back to the most important finding from social ecology research: Chicago's transitional neighborhoods always had the highest rates of crime regardless of the race or ethnicity of the dominant group.

Findings such as these increased sociological faith in the structural origins of crime. Individualistic or cultural explanations of a neighborhood's high crime rate imply that intrinsic properties of ethnic/racial groups or of individuals are responsible for crime. The essential ecological argument against such approaches was that if they were true they should observe changes in crime rates (either up or down) across different neighborhoods as different ethnic groups move in and out, but they did not. It was reasoned that if members of ethnic group X living in zone 1 have consistently higher delinquency rates than members of the same ethnic group living in zone 5, then it is the area itself, not characteristics of individuals or of their ethnic group, that generates antisocial behavior. Because neighborhoods tend to retain their characteristics regardless of their ethnic composition, according to ecological theorists there must be something about neighborhoods *per se* that either promotes or resists crime.

Few would argue that there are *essential* differences between and among racial or ethnic groups that mark some groups as *intrinsically* antisocial, and others as intrinsically prosocial. But there are large differences among individuals. After all, the majority of the inhabitants of even the worst neighborhoods did not acquire a criminal record (Shaw and McKay 1972), and neighborhoods lose their criminogenic character when "gentrified," meaning of course that they are now inhabited by different kinds of people. Obviously, there must be something about individuals that either promotes or resists crime. Individuals within racial and ethnic groups vary widely on traits that place them differentially at risk for antisocial behavior. Persons from ethnic group X in zone 5 may be quite different from their fellow ethnics in zone 1 in terms of traits associated with antisocial behavior, and individuals from Group X living in zone 1 doubtless resemble their neighbors with respect to these traits, regardless of the ethnicity of their neighbors, more than they resemble their co-ethnics in zone 5.

Related to this is an early criticism that the theory assumed a one-way causal relationship between social disorganization and crime, and ecological analyses require analysis of the *reciprocal* influences of organisms and environments (Kubrin and Weitzer 2003). High crime rates in an area lead to fear, a drop in community activities, and psychological and finally physical withdrawal from the neighborhood. When law-abiding people withdraw from high-crime areas, property values plummet, thus attracting less desirable residents, which generates even more antisocial conduct, further flight from the area, another influx of undesirables, and still more crime. Thus there is a reciprocal rather than a unidirectional relationship between social disorganization and crime (Kubrin and Weitzer 2003).

Modern Ecological Theory: People or Places?

Studies worldwide have supported ecological theory's propositions that socially disorganized slum neighborhoods have higher rates of crime and delinquency than more affluent and stable neighborhoods; this is merely stating the obvious. The assertion that neighborhood effects are more important than individual effects in explaining crime rates, however, is a bone of contention. The "people versus places" argument was perhaps best stated by Ruth Kornhauser's (1978:104) pointed question: "How do we know that area differences in delinquency rates result from the aggregated characteristics of communities rather than the characteristics of individuals selectively aggregated into communities?" Some environments certainly invite crime, but surely we can also acknowledge that people are at least partially responsible for their environments.

The concept of *ecological fallacy* (Robinson 1950) is a related criticism. Even in the heyday of ecological theory, Asian-Americans living in high crime areas had low crime rates (Hayner 1933; Shaw and McKay 1972). Low Asian crime rates in high-crime neighborhoods suggested that the rejection of group and individual differences as explanations for different crime rates may have been premature (Vold and Bernard 1986). That is, while it is true that certain neighborhoods always had high crime rates regardless of their ethnic makeup, perhaps when members of ethnic group X were the most populous inhabitants the rates were significantly higher than when ethnic group Y predominated. Shaw and McKay (1972) did report differences by ethnic group, but failed to incorporate these data into their theory.

Just how well do neighborhood effects stack up against the effects of individuals "selectively aggregating"? Some cross-sectional studies do tend to show fairly robust neighborhood effects on delinquency independent of individual characteristics. For instance, Wikstrom and Loeber (2000) examined 90 Pittsburgh neighborhoods in which the percentage of families living in poverty ranged from zero in the most advantaged neighborhood to 86 in the most disadvantaged. The most advantaged neighborhoods had zero percent African American residents and the most disadvantaged had 99%. Neighborhoods were categorized into advantaged, middle-range, disadvantaged-nonpublic, and disadvantaged-public. The percentage of youths reporting "serious offending" in each of these neighborhoods was 30.9, 43.4, 50.5, and 63.7, respectively, which is indicative of neighborhood effects at the zero-order level. Individual risk and protective factors (e.g., impulsiveness, guilt, parental supervision) of the 1,530 boys in the sample were also measured, and the boys placed into *high risk, balanced,* and *high protective* categories. Not surprisingly, most boys at high risk came from the most disadvantaged areas and most boys in the high protective category were from the advantaged neighborhoods. Most pertinent, however, was the fact that neighborhood effects were only related to delinquency for adolescent-onset boys; early-onset boys were equally delinquent across all neighborhoods.

Longitudinal studies tend to record lower neighborhood effects, however. For instance, a study of 264 rural counties in four states by Osgood and Chambers (2003) traced social disorganization to high rates of population turnover and high levels of ethnic diversity, but the most important factor was high rates of female-headed households. Single-parent households are highly related to the kinds of problems Shaw and McKay identified as being related to social disorganization some 75 years previously (i.e., low economic status, dilapidated housing, and poor supervision of children). The most notable finding of the study was that "a 10 percent increase in female-headed households was associated with a 73- to 100-percent higher rates of arrest for all offenses except homicide [a 10% increase in female-headed households was associated with a 33% increase in homicide]" (Osgood and Chambers 2003:6)

Is "female-headed household" a neighborhood effect or an individual effect? It is the former if we decline to inquire further, but the latter if we do. We have already seen that family structure is strongly conditioned by individual traits; i.e., people create their family environments influenced by their genetic proclivities (Cleveland et al. 2000). Dietrich Oberwittler (2004:228) concluded his multilevel analysis of serious juvenile offending in two German cities and a rural area by similarly stating that "ecological contextual effects are likely to exist, and the neighborhood context appears to play an important role in these effects. However, as has been shown in previous research in U.S. cities, contextual effects on adolescents' behavior are generally very small compared with individual-level influences."

Large-scale studies using experimental data from the federal government's Moving to Opportunity (MTO) program also find minor neighborhood effects (Kling, Liebman, and Katz 2005; Ludwig, Duncan, and Hirschfield 2001). The program involves randomly assigning vouchers to slum residents to move into more advantaged environments and their subsequent behavior compared with similarly situated families left behind in the old neighborhood. Averaged over numerous MTO studies, the general conclusions are that moving to advantaged neighborhoods has no effect on poverty or welfare participation; it does, however, lower female delinquency, but actually increases male delinquency (Kling, Liebman, and Katz 2007). Male delinquency may have increased because of a heightened sense of relative deprivation and envy as newcomers contrasted their resources with those of their new neighbors, or perhaps because of hostility on the part of the inhabitants of the new neighborhood, but neither explanation accounts for the positive effects that moving had on females.

Perhaps more telling were the deadly ripples of crime, particularly violent crime, in cities across the south-central United States following the influx of refugees from crime-infected New Orleans after Hurricane Katrina in 2005. Investigative reporters Grossman, Hylton, and McCulley (2006:61) concluded their piece on the displaced gangs of New Orleans by stating that just like everyone else, displaced gang members will spend time getting their bearings, and "Then they will find one another and start killing one another again. They will go where the housing and drug users are . . . they will carry with them the petty disputes of the past." In other words, criminals tend to take their cultures with them and to infect previously healthy areas.

RACE AND CRIME

No extensive literature review is needed to convince us that inner-city neighborhoods are the most dangerous places in America. When we hear of epidemics of gang violence, teenage pregnancy, crack smoking, and so forth, we know that the reference is to ghetto neighborhoods. Figure 8.1 presents UCR (Uniform Crime Reports) arrest rates for Part I crimes for 2006 broken down by race. The data are presented as rate ratios with the Asian rate set at 1.0 and the white and black rates given as multiples of the Asian rate. For example, the Asian murder rate for 2006 was 1.26, the white rate 3.28, and the black rate was 18.44. Thus the white rate was 2.6 times the Asian rate and the black rate was 14.63 times greater. Note that since the UCR stopped including Hispanics as a separate category in 1986, the vast majority (about 96%) of Hispanics are included in the "white" category and the remainder in the "black" category. The "Asian" category is primarily composed of Chinese, Japanese, and Koreans, but Pacific Islanders are also included.

The rate differences are most clear for violent crimes, and racial differences in violence are further underlined for serial killing. Among all known American serial killers from 1945 to 2004, African Americans are overrepresented in proportion to their numbers in the population by about two times (Walsh 2005), and Hickey (2006:143) claims that blacks have constituted about 44% of the known serial killers operating in the United States between 1995 and 2004, an overrepresentation factor of about 3.4. The only *known* Asian-American serial killer operating in the United States during the 20th century was Charles Ng (Walsh 2005).

Some criminologists argue that racial differences in arrests are a function of police bias; others that they are simply a function of race differences in criminal involvement. To be consistent with the police bias position, one would have to also say that the low Asian rate vis-à-vis the white rate is a function of antiwhite bias, that the greater rate of male versus female arrests was a function of antimale sexism, or the greater rate of young male arrests versus older male arrests a function of antiyouth ageism, which we can all agree are absurdities.

Studies comparing official arrest data from the UCR with National Crime Victimization Survey (NCVS) data find that victimization surveys yield essentially the same racial differentials as do official statistics (LaFree 1996). For example, about 60% of robbery victims describe their assailants as black, and about 60% of the suspects arrested for robbery are black (O'Brien 2001). Two recent large-scale studies using the National Incident-Based Reporting System (NIBRS) report the opposite of the police bias assumption. D'Alessio and Stolzenberg's (2003) data from 17 states and 335,619 arrests found that given that the race of the offender is known, the odds of arrest for robbery, aggravated assault, and simple assault were significantly greater for white offenders than for black offenders, but there

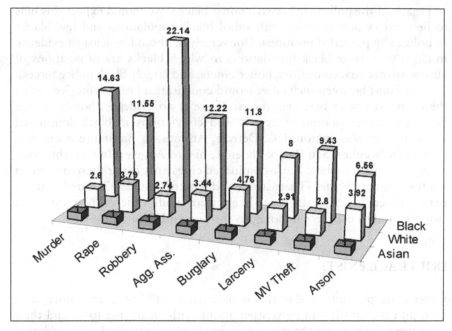

Figure 8.1   Racial differences in arrests for the FBI's Part I Index crimes: 2006. Source: FBI (2007). Crime in the United States: 2006. Black and White rates shown as multiples of the Asian rate with Asian rate set at 1.00.

was no significant racial difference in the probability of arrest for rape. The authors conclude that the disproportionately high black arrest rate is attributable to their disproportionately higher involvement in crime. Pope and Snyder's (2003) analysis of 102,905 incidents of violent crime committed by juveniles found essentially the same thing; i.e., white youths were more likely to be arrested despite the greater overall seriousness of the crimes committed by black youths.

Further arguing against the police bias assumption is the fact that the police have no discretionary arrest power for Part I Index offenses. The offense for which police arguably have the most discretion, and therefore the greatest opportunity to exercise any biases they may have, is driving under the influence. Yet this is the only UCR offense for which blacks are consistently *underrepresented* in the arrest data, which I interpret as less involvement in drunken driving among African Americans. Additionally, the vast proportion of police activity is reactive rather than proactive; i.e., they respond to reports of crimes by citizens rather than initiate arrests proactively, which again restricts them from exercising any biases they may harbor (Goldstein 1990). Indeed, because most crime is intraracial, *not* to arrest in response to citizen reports would be racist and would signify a devaluation of black victims.

Finally, if the police are biased against blacks, we should expect this bias to be most evident in cities with small black populations and few blacks in politically powerful positions. Conversely, it should be least in evidence in cities with large black populations in which blacks are in positions of authority (mayors, councilors, police chiefs, and largely black police forces). Comparisons between such cities would constitute a fairly definitive test of the criminal justice bias hypothesis. There is no evidence, however, that black overrepresentation in arrests and convictions in black-dominated cities such as Washington, DC, Detroit, Atlanta, or Baltimore is any less than in Milwaukee, San Diego, Phoenix, or Los Angeles. In fact, the black arrest rate ratios in these black-dominated cities are greater than they are in white-dominated cities (Thernstrom and Thernstrom 1997). Furthermore, black officers are typically more likely than white officers to arrest black suspects and to use force in doing so (Brown and Frank 2006).

DOES RACE EXIST?

Other concepts addressed in this book such as SES, age, sex/gender, conflict, and the family can be written about without having to defend their objective existence, but the concept of race has been attacked as an arbitrary social construct lacking any defensible empirical basis independent of the social meanings attached to it. The assault on the concept thus necessitates a short digression. Racial classifications are indeed social constructs, but as already acknowledged, so is everything else at one level; nature does not reveal itself to us presorted and labeled; humans must do it for her. Classification systems do have their problems, of course, the biggest of which is how fine a line we wish to draw between categories. Anthropologists have presented us with as few as three and as many as 100 races (Cavalli-Sforza 2000), a fact jumped on to underline the arbitrariness of the race concept. But taxonomical fissioning and fusing occur in every field of human inquiry without the cantankerous debates about the underlying reality of that which is categorized that we experience with racial categorization.

Snyder (1962:19) defined race out of existence a long time ago when he remarked: "The people of the world have become so intermingled biologically that there can be no possibility of an absolutely pure race anywhere." In other words, there is no such thing as a "race gene" that is present in all members of one population and absent in all members of others. If this is what social constructionists mean by race, they won that war years ago, for no one denies that *Homo sapiens* constitutes a single biological unit. Using purity as the defining criterion of race enables one to correctly state that there is no such thing and on that basis to conclude that the concept lacks any scientific merit. This argument destroys a straw man, for it is patently obvious that there are no pure biologically discrete human races; if there were, we would call them species.

Whatever else the "no-races" argument has evolved into, it began as an ideological argument in the American Anthropological Association's (AAA) official *Statement on "Race,"* which stated that human racial groupings differed so little genetically from one another (about 6%, according to the AAA) that it implies the nonexistence of race. As Klein and Takahata (2002:389) point out, however, such a difference can be enormously significant: "Sewell Wright, who can hardly be taken for a dilettante in the question of population genetics, has stated emphatically that if differences of this magnitude were observed in any other species, the groups they distinguish would be called subspecies." Nevertheless, the AAA argued that even if race exists it is a socially dangerous concept, and should be dropped and replaced with *ethnic group* (1997:4). Substituting *ethnic* for *racial* does nothing but replace one term with another; it does not shift the reality underlying them, nor do "official" professional fiats. The idea that changing names makes the world less dangerous is rather disingenuous, as Jonahan Marks points out: "If biologically diverse peoples had no biological differences but were marked simply on the basis of language, religion, or behavior, the same problems would still exist" (1996:131).

Ernst Mayr, the doyen of population genetics, defines race as "the descendants of a once-isolated geographical population primarily adapted for the environmental conditions of their original country" (2002:91). He further states that those who say that there are no human races "are obviously ignorant of modern biology" (2002:89). He does not deny that racial boundaries are fuzzy, relative, and dynamic rather than absolute and static, but the precision of our classifications depends on the degree of "lumping" or "splitting" required by the research issue. For instance, geneticist Peter Smouse states that it would be difficult to distinguish a Scot from a Welshman (very fine ethnic splitting) by their DNA: "But ask me if someone is from Norway or Taiwan [gross racial lumping], sure, I could do that" (in Shreeve 1994:57).

It is estimated that about 90% of skeletal remains can be racially identified by skull morphology alone (Sauer 1992), yet almost one-half of physical anthropologists claim that they do not believe that races exist (Shreeve 1994). Thus a significant minority of physical anthropologists function within a classificatory system they allegedly do not believe in, although this should be put in the context of the official position of the AAA on race and of the professional consequences of voicing contrary opinions. However, note that if they define race in terms of purity, they can satisfy both their political correctness and professional needs. That is, they can rightly assert that *pure* race (pairs of individuals without or with reduced capacity to mate) does not exist, and at the same time use race in their work to identify human remains.

The Human Genome Project has reframed the race issue such that we have "become increasingly focused on genetic diversity identified along the classificatory lines of race" (Abu El-Haj 2007:284). In other words, geneticists are looking for evidence at the molecular level rather than the level of morphological traits to identify race. A worldwide study of genetic racial variation using autosomal, mitochondrial, and Y-chromosome polymorphisms concluded that its findings support "the current practice of grouping reference populations into broad ethnic categories" (Jorde et al. 2000:985). Between-racial (African, Asian, and European) groupings were significantly different at < .0001 for all DNA analyses. In another worldwide study, researchers were able to correctly assign 99 to 100% of individuals to their continent of origin using markers from only 100 genetic loci (Bamshad et al. 2003).

A number of other studies have also concluded that race can be identified genetically, but perhaps the most convincing of them is that of Tang and his colleagues (2005). Using data from 326 microsatellites and blind to the phenotypes of their subjects, using cluster analysis, only 5 out of a total of 3,636 (0.14%) subjects were not correctly classified according to their self-identified race/ethnicity. Given the large number of studies coming to the same conclusion with such consistency, robustness, and extraordinary accuracy, it is difficult to maintain the argument that race is a social fiction akin to witches and unicorns. We can call the genetic clusters anything we want (populations, ancestry indicators, clines, ethnicities, and so on), but relabeling will not change the reality underlying whatever label we chose to apply to those genetic clusters.

IS POVERTY THE REASON?

If differential black involvement in crime accounts for differential black arrest rates, the question becomes why blacks are differentially involved. I attempt an explanation within the human ecology tradition and want to make it clear that I am concentrating on the ecology of the inner city. There is no such thing as a homogeneous "black community" any more than there is a homogeneous white community. Even within the inner city the majority of families are "decent" as opposed to "street" families, to use Elijah Anderson's (1999) terms.

We tend to talk of "black poverty" as if every black family in America is poor, but the great majority of them are not. Higher rates of poverty among blacks are largely accounted for by their higher rate of single-parent families. The U.S. Census Bureau's (McKinnon and Humes 2000) breakdown of family types by race and income showed white single-parent households were more than twice as likely as black two-parent households to have an annual income of less than $25,000 (46% versus 20.8%). To state it in reverse, *a black two-parent family is less than half as likely to be poor as a white single-parent family.* These figures constitute powerful evidence against the thesis that black poverty is the result of white racism, as well as powerful evidence that high rates of illegitimacy are a major cause of family poverty for all racial/ethnic groups. The prevalence of single-parent families is so high in the black community that "[A] majority of black children are now virtually assured of growing up in poverty, in large part because of their family status" (Ellwood and Crane 1990:81).

All of this evidence tends to be ignored by criminologists who have taken it as an article of faith that poverty causes crime, and that once poverty is controlled for racial differences, crime will disappear (Young and Sulton 1996), but it doesn't work out that way. Bryne's (1986) sample of 910 U.S. cities found that the best predictor of robbery rates, controlling for a host of city characteristics, such as density and housing type, and population compositional variables, such as average income and education levels, was percentage of black residents. Sampson's (1985) study of homicide rates in the 55 largest U.S. cities found that "percentage black" had more than twice the explanatory power of other variables in the regression such as population size, poverty, income inequality, and unemployment, and Chilton's (1986) study of the 125 largest SMSAs controlled for a wide variety of variables in a number of different regression models found percentage of black residents emerged as the best predictor of murder and assault rates, although population size was a better predictor of rape and robbery.

Thus the problem is not simply one of poverty, for even when compared to similarly situated whites, blacks still commit more crime (Sampson 1995). But the response to that is being poor and black does not equate with being poor and white: "even given the same objective socioeconomic status, blacks and whites face vastly different environments in which to live, work, and raise their children" (Sampson and Wilson 2000:152). White poverty is considerably more dispersed across different neighborhoods than is black poverty, where it is highly concentrated in single neighborhoods—hyper-ghettoization, as it is termed by Sampson and Wilson (2000). Hyper-ghettoization breeds even higher concentrations of poverty and racial segregation as those who are able to do so flee the neighborhood, thus making room for more undesirables to move in to create even higher rates of antisocial behavior.

Yet as Patterson points out (1998:25), millions of people the world over live in disadvantaged racially, ethnically, or socioeconomically segregated areas without resorting to the kinds of behavior prevalent in our inner cities. Moreover, hyper-ghettoization has been invoked as an explanation for *low* crime rates among Jews, Chinese, and Japanese: "what is striking is that the argument used . . . to explain *low* crime rates among Orientals— namely, being separate from the larger society—has been the same argument used to explain *high* rates among blacks" (Wilson and Herrnstein 1985:474; emphasis in original).

Amidst all this talk of poverty we often forget that there is a thriving black middle class which African American economist Glen Loury describes as "the most privileged, empowered people of African descent anywhere on the globe," and about which Walter Williams writes: "Blacks spend enough money each year to make us, if we were a nation, the 14th richest" (cited in Walsh and Hemmens 2008:307). Two of the contemporary Americans I admire most, ex-Secretary of Defense General Colin Powell and philosopher/economist Dr. Thomas Sowell, were both born in the 1930s at the back end of the Jim Crow era but rose to extraordinary heights of achievement. Powell was the son of Jamaican immigrants and raised in poverty conditions; Sowell's father died before he was born, and he also was raised in poverty conditions. Both men, however, were imbued with the values of hard work and personal responsibility and did not succumb to the fatal disease of claiming victimhood. These men show that there is nothing in either American society or in the black character that dooms blacks to poverty and crime.

EDUCATION AND POVERTY

To escape poverty one needs to acquire and retain decent employment, and in order to do those things one has to be prepared. Notwithstanding the growing black middle class, many African Americans are disadvantaged in the workplace by educational credentials that do not permit them to compete successfully in the workplace. This is not the result of denial of opportunity, for as James Clarke (1998:255) has put it: "Despite unprecedented opportunities for 'equal' employment, and the best intentions of liberal politicians, millions of blacks simply lack the basic skills required for employment in a technology-driven economy." This lack of preparation is underlined in racial gaps in SAT scores. As the *Journal of Blacks in Higher Education* (TJBHE 2008:3) pointed out, in 2005: "Whites from families with incomes below $10,000 had a mean SAT score that was 61 points higher than blacks whose families had incomes of between $80,000 and $100,000." Additionally, blacks with parents with college degrees score lower than whites or Asians whose parents went no further than high school (McWhorter 2000).

A widespread belief is that the poor educational performance of blacks is the result of low funding for schools in predominantly black districts. The assumption underlying this opinion is that more resources equal better pupil performance. The data on this are extremely mixed, with some studies showing a positive relationship between per-pupil expenditure and student achievement, some showing no relationship, and some showing a negative relationship (Al-Samarrai 2006). Perhaps the experience of the so-called magnet schools, particularly those in Kansas City, Missouri, provide the best evidence that school spending is not the answer to black educational woes. To improve the education of black students and enforce desegregation, in 1982 a federal judge ordered the Kansas City Missouri School District (KCMSD) to implement methods to draw white students from their suburban schools into the mostly black inner-city schools and to provide the funds to do so by doubling property taxes. Flush with cash, the school district built 15 new schools and renovated 54 others, increased teacher salaries by 40%, reduced class sizes, and built educational and recreational facilities that rivaled those existing in the most elite universities in the country (Ciotti 1998; Gewertz 2000). At one point, 44% of the K–12 education budget was being spent on the 9% of Missouri's students enrolled in Kansas City and St. Louis magnet schools (Ciotti 1998).

The results of this enormously expensive experiment were disappointing. White students were still from three to five years ahead of black students on standardized tests, which is about the same gap seen in nonmagnet schools (Ciotti 1998). The magnet schools failed in all of 11 indicators of performance—e.g., test scores, dropout rates, truancy rates, vocational and college preparation (Gewertz 2000). The 17-year KCMSD experiment finally ended in 1999. Similar programs in other parts of the country have proved equally disappointing, thus highlighting once again the error of thinking that throwing enough money at a problem will solve it.

Much of the blame for the low level of achievement among blacks, and thus the high level of poverty, is the devaluation of education. Anderson (1999) points out that although there are many "decent" families that value education in inner-city neighborhoods, the cultural ambience is set by "street" families. "Decent" individuals often have to adopt the oppositional attitudes and behavior of "street" individuals to survive. Valuing education and striving for upward mobility is viewed as "dissing" the neighborhood and street people often "mount a policing effort to keep their decent counterparts from 'selling out' or 'acting white' " (Anderson, 1999:65). Patterson (1998:278) notes that "there is now such chronic anti-intellectualism among Afro-American youth that those few, who by some sociological miracle, become involved in their studies must find ways to camouflage their interests." Another black commentator blames black leadership and white guilt for black anti-intellectualism: "Black leadership (think Jesse Jackson and Al Sharpton) exploits white guilt, which instead of liberating blacks locks them in an attitudinal gulag of victimhood, a self-imposed Orwellian blackthink, a psychological woundedness that perpetuates a faith in past oppression as the root of present failure" (Fields 2002:2). Finally, James Clarke views the devaluation of education as just one symptom of the general rejection of white culture and its values: "In black inner-city schools . . . academic success has become an invitation to ridicule or worse" (1998:287).

African American educator John McWhorter also blames black educational failure on the "cult of victimology" (2000:31). McWhorter argues that a cult of victimology leads to a "cult of separatism," and as a natural consequence to a cult of "anti-intellectualism." Separatism does not automatically lead to anti-intellectualism, for it was arguably separatism that spawned a robust intellectualism among the Jews, history's perennial victims. Jews are well aware of their ancestor's historical victimization, but they are also aware that historical victimization is just that—historical. They get on with their lives in whatever country they find themselves, realizing that to dwell on the past is to sabotage the future. For McWhorter, the answer to black educational failure is to follow the Jewish example and shed their mantle of victimhood.

STRUCTURE AND CULTURE

There is a great deal of debate among sociologists regarding the relative power of structure versus culture to account for human behavior. Ecological effects are most often seen as structural effects, but according to Matsueda, Drakulich, and Kubrin (2006:335), among the conditions said to qualify as "structural barriers to success" are concentrated poverty, joblessness, and drug use. But surely variables such as these can only be considered structural if we ignore human choice. Aggregate data are shaped by processes forged by numerous individual choices, and once aggregate phenomena are formed from those processes they serve as a context in which further choices are made. The poverty, joblessness, or drug-abusing rates in a neighborhood are determined by aggregating the number of individuals in it who are in poverty, unemployed, or abusing drugs, which is a measurement strategy that confounds the processes that have resulted in these outcomes (Sobel 2006).

"Neighborhood" is a macromirror reflecting the combined micro-images of all the individuals who live in it. As John Wright (2009:148) explains: "It should be expected that individuals with similar traits and abilities, who have made many of the same choices over their life-course, should tend to cluster together within economic and social spheres. In other words, a degree of homogeneity should exist within neighborhoods, within networks within those neighborhoods, and within families within those neighborhoods." Obviously, choices people make are often constrained by factors beyond their power to control. We cannot do anything about the genes or rearing environment our parents bequeathed to us, the fact that the factories have moved out of town, or that the people around us sell drugs and are not very nice. We can, however, decide to respond to the inevitable travails of life constructively or destructively within the limits of our abilities.

As understood by ecologists, a neighborhood is a context in which culture is expressed. Culture is a sense of reality or an "information system" shared by a particular group of people and transmitted across generations. As a worldview ordering cognitive reality, culture guides individual choices in ways that are presumably adaptive. Culture is thus an adaptation to geographic, political, economic, and social environments. Like biological adaptations, cultural adaptations fit organisms into their environments as comfortably as possible given existing conditions. Also like biological adaptations, they can become maladaptive when environmental conditions change and the carriers of the culture do not change with them.

The structural conditions blacks have historically had to adapt to in the United States were *structural* in the real sense of the word, not conditions impeding individuals that are largely self-created and then defined ex post facto as structural. These structural conditions were objectively real barriers to participation in mainstream (white) society such as slavery, the Black Codes, disenfranchisement, racial stigma, Jim Crow laws, and any number of other things supported by the full force of law (Walsh and Hemmens 2008). Thus structure ("the material conditions of one's existence," as a Marxist would put it) necessarily precedes cultural adaptation to it. Black inner-city culture emerged as a strategy for survival in the context of these conditions. What were adaptively functional behaviors in earlier settings, however, have become maladaptive and dysfunctional in modern times when these structures no longer exist.

The problem with cultural explanations is that some see them as "thinly veiled racist" (Almgren 2005:220). The unfortunate tendency of some to pin the racist label on any attempted explanation other than structural "blame society" explanations puts a damper on efforts to unravel the conundrum of black crime. As African American sociologist William Julius Wilson has pointed out, social scientists have tended to ignore issues attending "the tangle of pathology in the inner city," or to address them in "circumspect ways" (1987:22). Many shy away from dealing forthrightly with matters of race because of fear of the racist label, which has led to "an unproductive mix of controversy and silence" (Sampson and Wilson 2000:149). The racist label has been stapled to such a range of imagined sins that all meaning has been washed out of it. Whatever the reason for avoiding the race/crime issue, it is both ethically and scientifically the wrong thing to do. LaFree and Russell (1993:279) argue that the crime/race connection should be studied honestly and courageously because "no group has suffered more than African-Americans by our failure to understand and control street crime." The corollary of this is that no other group can benefit more from a candid examination of race and crime.

Cultural arguments are considered racist by those who descend to ad hominem attacks because the carriers and perpetrators of culture are people, and thus to invoke culture is to "blame the victim." Yet Orlando Patterson (2000:6) points out that critics of cultural explanations often use cultural explanations themselves when countering argument such as the genetic origins of racial differences in IQ. Likewise, Richard Felson (2001:231) writes that cultural arguments are acceptable when celebrating "positive aspects of the protected groups [Felson defines protected groups as any groups other than straight white males] . . . for explaining why certain groups such as Jews and Chinese-Americans are economically well-off [and are] accepted when offending groups [read "whites"] engage in discrimination."

Robert Sampson, the leading luminary of modern human ecology and certainly no racist, along with coauthor Lydia Bean (2006:29), goes further than the culture-as-adaptation argument to insist that interacting individuals create the conditions in which they live in real time, and that the structure versus culture argument produces a false dichotomy because they are actually mutual creations of each other:

> Individuals are part of creating violent neighborhoods; put differently, without the cultural agency of neighborhood residents expressed in on-going engagements in violent altercations, the neighborhood context would not be violent. The relational approach understands culture not as a simplistic adaptation to structure in a one-way causal flow, but an intrasubjective organizing mechanism that shapes unfolding social processes and that is constituent of social structure. From this perspective culture is simultaneously and emergent product and a producer of social organization, interaction, and hence structure.

There may be too much faith in free agency expressed here given that agency is constrained by external conditions and the nature of one's internal intellectual and characterological resources. Nevertheless, the point is that there are no fixed boundaries to culture or structure in a modern society; they are always in flux as interacting individuals create their worldviews and live with the consequences of their creations.

SLAVERY AND THE ORIGINS OF THE OPPOSITIONAL CULTURE

According to African American sociologist Elijah Anderson (1999:107), many black neighborhoods have spawned a hostile and violent oppositional subculture that spurns everything mainstream America values, as in "the rap music that encourages its young listeners to kill cops, to rape, and the like." The concentration of numerous angry and alienated individuals lacking adequate institutional and personal coping skills to counteract daily stress creates a constant feedback loop of angry aggression.

Continuing the line of thought from such black scholars as W. E. B. Du Bois, E. F. Frazier, and Kenneth Clark, I trace the origins of the oppositional culture to the odious institution of slavery. Slavery constituted a total institution for blacks, i.e., one in which large groups of people live together under tightly restricted and coercive circumstances under the control of others. In total institutions (e.g., prisons), controllers and controlled are socially and psychologically isolated from one another and are mutually hostile. In order to adjust to life in prison, inmates develop an inmate culture that guides their interactions with each other and with their controllers which is contrary to the code of the controllers. The inmate code includes injunctions against cooperating with "the man" at levels beyond that which is necessary to avoid trouble, against showing subservience, against being friendly with controllers unless you can use them for your own ends, and against ratting on other inmates. Acceptance of and compliance with this code is necessary to become a "good convict" and to be accepted by other inmates.

Just as prison inmates develop oppositional subculture in response to their imprisonment, blacks developed a subculture of their own in response to their predicament. Slavery and Jim Crow laws are in the past, but the subculture born of them remains, just as the inmate code remains an integral part of the psychology of the long-term convict long after release. As Richard Wright argued long ago: "The Negro's conduct, his personality, his culture, his entire life flows naturally and inevitably out of the conditions imposed upon him by white America" (in Thernstrom and Thernstrom 1997:51). Clarke (1996:50) also traces the high rates of crime, domestic violence, illegitimacy, and child neglect in the black community today to an evolved cultural system born out of slavery and other grave injustices, and which now "accounts for the social and sexual chaos that reigns in America's inner cities."

Clarke (1998) points out that slaves did everything in their power to deceive and get back at their masters, and even considered it their duty to do so (shades of the inmate code). They did not consider stealing from whites to be morally wrong; it was considered so "smart" that black folklore is replete with stories and songs about slaves who outwitted their masters or white folks in general. Cultural norms developed that lauded thievery and deception and warned about the perfidious nature of whites. Such behaviors and attitudes were understandable and even healthy and admirable forms of resistance during slavery. After emancipation, black lawbreakers came to be widely viewed as heroic figures among blacks, especially if they had been to prison. The outlaw tradition of the "bad nigger" was born from this form of hero worship (Milner and Milner 1972). Even though the majority of victims of black crime were other blacks, black criminals still tended to be excused. The failure of blacks to condemn other blacks for their criminal behavior was viewed by the great African American scholar, W. E. B. Du Bois (1903/1969) as a major factor in the high rates of black-on-black crime in the late 19th and early 20th centuries.

The lawless ways of the black community were reinforced by white indifference to black-on-black crime. Characterizing the relationship between white employer and black laborer as "feudal," John Dollard (1988:418) states that powerful white employers often actively protected their black employees from arrest and prosecution "under threat of electoral retaliation." Blacks who were valuable and submissive to whites secured immunity from punishments for a variety of other offenses for which whites would be prosecuted. This state of affairs, writes Dollard, led to many blacks having "extraordinary liberty to do violent things to other Negroes" (1988:201).

This double standard in the legal response to crime is open incitement to violence. An investigator in 1930 wrote that since blacks had no faith in the police to protect them, they often resorted to "the ready use of firearms in trivial matters." He further commented: "Negroes are often allowed without interference from the police, to commit crimes on one another. Many Southern leaders, white and Negro feel strongly that the inadequate police protection provided within Negro communities virtually breeds crime" (in Clarke, 1998:212).

Being left to their own devices, and building on a tradition of rule challenging, blacks did not look to white standards to determine their worth. Like convicts, blacks were expected by their cultural code to settle matters "like a man," and to take care of their own beefs. "Taking care of business" often involves violence in a subculture where it is not viewed as illicit but rather the successful application of aggression—both as a manifestation of subcultural values and as a disavowal of mainstream cultural values—is a source of "juice." Anderson points out that in the inner city "there are always people looking around for a fight in order to increase their share of respect—or 'juice' " (1999:73). The trouble with this is that juice earned through violence can only be cashed in for resources in the neighborhoods in which it is earned. Outside of those environments it is considered counterfeit coinage and attempts to cash it will likely land one in prison. Nevertheless, Kenneth Stampp's classical work on slavery concluded that in the black community "[S]uccess, respectability, and morality were measured by other standards, and prestige was won in other ways" (1956:334).

THE SEX RATIO, MISOGYNY, AND ILLEGITIMACY

One alternative standard of success and prestige is the number of females a man can impregnate (Anderson 1999), and the black sex ratio is an aid in this endeavor. The sex ratio is a structural variable entirely beyond individuals' ability to change. We saw previously that one effect of low-sex-ratio mating environments is the devaluation of women as little more than sex objects. The lower the sex ratio, the lower the value of women will tend to be, and it is especially low in the black community: "American blacks present us with the most persistent and severest shortage of men in a coherent subcultural group that we have been able to discover during the era of modern censuses" (Guttentag and Secord 1983:199). The black sex ratio was 0.87 in 2000 (U.S. Census Bureau, 2001); contrast this with the near parity of the European and Asian American sex ratios of 1.04 and 0.99, respectively (Humes and McKinnon 2000). Black sex ratios of between 0.85 and 0.87 have been noted since the days of slavery (Wilson 2001).

One of the reasons for racial differences in the sex ratio is differing levels of female gonadotropins (a class of hormones that regulate the functions of the gonads). High levels favor more female births, and black females apparently have the highest levels, Asians the lowest, and whites have intermediary levels (James 1986, 1987). Black females are also more likely to abuse drugs, alcohol, and tobacco and less likely (or less able) to seek medical care when pregnant than whites or Asians (Lubinsky and Humphries 1997), resulting in more *in utero* deaths. Because female fetuses are more robust than male fetuses, a greater proportion of female fetuses will survive to parturition (Davis et al. 2007). Additionally, black males of all ages die at higher rates than white or Asian males from homicides, alcoholism, drug overdoses, and accidents.

The effective sex ratio (ratio of males to females of marriageable age) is lower than a simple ratio of males to females in the population. A greater proportion of black males are incarcerated in prisons and mental institutions than males of any other racial or ethnic group, and if current trends continue, according to Mauer and King (2007), one in three black males born today can expect to be imprisoned at some point in their lives. Although many African American women may want marriage, few are willing to consider such men as reliable sources of parental investment (King 1999; Wilson 1987).

A number of theorists have commented on the low-level gender war existing in the inner cities (D'Sousa 1994; Hutchinson 2001; King 1999). Millner and Chiles (1999) extensively document the negative attitudes that inner-city black males and females have toward one another, with females desperately seeking commitment and males doing their best to avoid it. Weeks and her colleagues (1996:348) state that "Several studies of African American communities describe black women's distrust of black men, and their assumption that most black men are 'naturally' or inherently bad, sinful, and untrustworthy—particularly in their relationships with black women." The attitudes of these women have a solid foundation in fact because about one-quarter of black males between the ages of 16 and 24 and one-half between the ages of 25 to 34 are noncustodial fathers (Holzer and Offner 2004), which essentially means that they have abandoned both their female partners and their offspring. Holzer and Offner (2004) further indicate that increasing state pressure on deadbeat dads to pay child support has led many such men to abandon legitimate employment for fear of wage garnishment and to disappear into the underground economy.

Misogynistic themes permeate inner-city African American street culture. Generic terms for females such as "bitch" and "ho" and for sexual intercourse such as "killing the pussy" speak volumes in themselves. Machismo and misogyny have been a major theme in black music from the days of rhythm and blues (Ward 1999) to the truly obnoxious "gangsta rap" of modern times (Johnson 1996). Because such "artistic" expression is typically defended as a reflection of cultural reality, or at least "hyper-reality" (Boyd 1996), we are at liberty to assume that the attitudes toward women contained in them are fairly accurate assessments of the value placed on women by a significant proportion of males in the inner cities of the United States.

Further indicative of the devaluation of women is the rate of violent wife/lover abuse, which is at least four times more common in the black than in the white community (reviewed in D'Sousa 1995). Because of this, black *females* have had a higher homicide rate than white *males* in the

United States since at least the 1930s (Barak 1998). The vast majority of the homicides committed by black females are in self-defense situations in which they are being attacked by their male partners (Mann 1995; Serran and Firestone 2004).

How do black women adjust their reproductive strategy under such conditions? We know that a woman's reproductive success does not depend on the number of copulations she experiences, but on keeping such offspring as she does produce alive to reach reproductive age themselves. In light of this, Sarah Hrdy (1999:246) states that it may benefit the fitness of a mateless female with offspring to set up "networks of well-disposed men to help protect and provision her offspring [in exchange for sexual favors]". Hrdy further states that this is "the emotional calculus behind the decisions that inner-city mothers make every day" (1999:251). According to the 2005 *World Almanac* (in Elder 2005), these decisions have led to almost 70% of black children in the United States born out of wedlock (in 2005 the corresponding percentages for whites was 30% and for Asians it was 15%).

Kenneth Clark (who grew up in ghetto poverty) saw illegitimacy as the root of the "institutionalized pathology of the ghetto. . . . Not only is the pathology of the ghetto self-perpetuating, but one kind of pathology breeds another. The child born in the ghetto is more likely to come into a world of broken homes and illegitimacy; and this family and social instability is conducive to delinquency, drug addiction, and criminal violence" (1965:81). Charles Murray agrees, writing that illegitimacy is "the most important social problem of our time—more important than crime, drugs, poverty, welfare, or homelessness, because it drives everything else" (in Lykken, 1995:195). Jarred Taylor (1992:305) also agrees in stating that "If there is a single statistic that underlies the crime, poverty, and failure that beset blacks in America today, it is an illegitimacy rate of 66 percent."

Democratic Senator Daniel Moynihan wrote in his famous report: "At the heart of the deterioration of the fabric of Negro society is the deterioration of the Negro family" (1965:5). Moynihan viewed this deterioration primarily in terms of the black illegitimacy rate, and although he was viciously attacked for his observation, he received support from none other than Martin Luther King, Jr., who also noted the "alarming statistics on Negro illegitimacy," and wrote that the black family had become "fragile, deprived, and often psychopathic" (in Norton 1987:53). Such statements have never sat well with those who refuse to recognize the role of individual responsibility in human affairs, and who see those who do as "blaming the victim."

Moynihan traced the state of the black family to the legacy of slavery. When we talk of the slave family in the Americas, we are actually talking about a *reproductive unit* formed by slave owners to breed offspring. Although the reproductive unit was not a legitimate *social* unit (slaves were not allowed to marry), it provided the prototype for the postslavery black gender relationships (Patterson 1998). The male role in a bonded relationship is to supply his mate and their children with resources, protection, security, and a family name/identity, but the slave could offer none of these things.

This situation led to a carefree and irresponsible attitude toward women on the part of male slaves and to a pattern of independence and mistrust of men on the part of female slaves (Wilson 2000). In response to these circumstances: "Afro-American men and women developed a distinctive set of reproductive strategies in their struggle to survive. Tragically, the strategies that were most efficient for survival under the extreme environment of slavery were often the least adaptive to survival in a free, competitive social order" (Patterson 1998:41). Patterson (1998:51) continues: "Significantly, the sexual aggression against women did not stop at mere compulsive sexuality; rather, we find throughout the decades of the rural South, and throughout the underclass today, the vicious desire to impregnate and abandon women, as if Afro-American men were unable to shake off the one gender role of value (to the master) thrust upon them during slavery, that of progenitors."

Patterson's observations are not unique to the late 20th century. Commenting on the habits of lower-class blacks in Philadelphia toward the end of the 19th century, W. E. B. Du Bois wrote: "The lax moral habits of the slave regime still show themselves in a large amount of cohabitation without marriage," and a "lack of respect for the marriage bond" (1899/1967:69). Elijah Anderson also supports this position when he states that in the inner city, access to females is "taken quite seriously as a measure of the boy's worth"; a young male's "primary goal is to find as many willing females as possible. The more 'pussy' he gets, the more esteem accrues to him" (1999:150). Sampson and Wilson make much the same point when they write of "Ghetto-specific practices such as an overt emphasis on sexuality and macho values" (2000:156).

Commentaries such as these would be angrily dismissed as racist if made by whites, but W. E. B. Du Bois, Kenneth Clark, Orlando Patterson, Martin Luther King, Elijah Anderson, and William J. Wilson are all respected black scholars with impeccable liberal credentials and are thus free to be brutally candid in areas where only the most courageous whites are willing to go. If we accept their analyses it is difficult to deny that there is continuity in sexual practices among underclass blacks running from their African origins and continuing through slavery to the present day which has resulted in rampant illegitimacy and all the problems that accompany it.

DEVELOPMENTAL FACTORS

We may now examine the developmental impact of these structural and cultural conditions on children experiencing them in light of earlier discussions about the impact of early environmental experiences on central and peripheral nervous system development and functioning. Not surprisingly, children raised in poverty and subjected to abuse and neglect are the most likely to evidence dysregulation of the stress mechanisms and cortisone-induced neuronal loss (Evans et al. 2007). Because the inner cities are violent places in general, it is no surprise that we find child abuse/neglect to be more prevalent there than elsewhere. The Child Trends Data Bank (CTDB 2002a) reported that of the 879,000 known maltreatment cases in 2000, the black rate was about three times the white rate and about five times the Asian rate. More alarmingly, infant homicide rates per 100,000 were 25.6 for African Americans and 6.0 for whites (Asian rates not reported) (CTDB 2002b).

The discussion of the adaptive significance of violence in Chapter 3 emphasized that violent actions in neighborhoods where social control has broken down are rational when we or our property are threatened (Gaulin and McBurney 2001; Kelly 2005). The evolutionary point of view avers that the major long-term factor in violence instigation is how much violence a person has been exposed to in the past. As Caulin and Burney (2001:83) explain, when many acts of violence are observed, "there is a feedback effect; each violent act observed makes observers feel more at risk and therefore more likely to resort to preemptive violence themselves." The neurological literature is consistent in suggesting that impulsive aggression is the proximate behavioral expression of a brain wired by consistent exposure to violence (Niehoff 2003). If our brains are wired in abusive, neglectful, and violent environments we naturally come to expect hostility from others and behave accordingly. But by doing so we invite the very hostility we are on guard for, thus confirming our belief that the world is a dangerous and hostile place and setting in motion a vicious circle of negative expectations and confirmations.

In addition to heightened threat vigilance, chronic exposure to violence desensitizes those exposed to it to the suffering of others and makes them callous and indifferent (Cooley-Quille et al. 2001). Desensitization has been assessed by self-reports and by measures such as heart rate and galvanic skin responses (Carnagy, Anderson, and Bushman 2007) and fMRI studies. These fMRI studies show decreased activity in brain structures that regulate aggression (e.g., anterior cingulate gyrus and prefrontal cortex areas) and increased activity in structures associated with increased aggression such as the amygdala (Murray et al. 2006; Sterzer et al. 2003).

All such studies are conducted with subjects exposed to chronic *vicarious* violence; how much worse it must be for those chronically exposed to real violence. One survey of young children attending an inner-city pediatric clinic found that 10% of them had witnessed a shooting or stabbing, and almost all of them had witnessed violence many times in the home or in the streets (Taylor et al. 1994). In another study in Chicago, 33% of schoolchildren had witnessed a homicide and 66% had witnessed a serious assault; and another found that 32% of Washington, DC children and 51% of New Orleans children had been victims of violence, and 72% of Washington, DC and 91% of New Orleans children had witnessed violence (reviewed in Osofsky 1995). Witnessing and experiencing so much violence cannot help but stamp on the neural circuitry of these children that the world is a dangerous place in which one must be prepared to protect one's interests by violent means if necessary.

Fetal Alcohol Syndrome, Breastfeeding, and Lead Exposure

Other developmental factors not previously discussed disproportionately affecting African Americans include fetal alcohol syndrome, lead exposure, and a lower prevalence of breastfeeding. Each of these factors is related to variation in IQ, and given that the low average IQ of blacks relative to Asians and whites, and that low IQ has been invoked to explain black crime, they warrant mentioning.

From a single precursor cell, embryonic neurons called neuroblasts form at a rate of about 250,000 every minute and then sprout axons that reach out to make connections with other neurons to begin the process of building a brain (Toga, Thompson, and Sowell 2006). Neurons must migrate from their birthplace to their eventual home guided by molecular messengers. It is during the migratory phase of maturation that the brain is most vulnerable to insults because these chemical guides are susceptible to being confused by teratogenic chemicals. If the messenger molecules become contaminated by alien compounds they may send neurons to the wrong area or even direct them to self-destruct (Prayer et al. 2008). The most common teratogens are those associated with alcohol. When mothers drink while pregnant they introduce their fetuses to neurotoxins that produce a number of neurological disorders, the most serious of which is fetal alcohol syndrome (FAS). The physical symptoms of FAS are readily recognized by physicians, but for criminologists the primary concern is how FAS effects behavior via its effects on the frontal lobes, amygdala, hippocampus, hypothalamus, the serotonergic system, and on the myelination process (Goodlett, Horn, and Zhou 2005; Noble, Mayer-Proschel, and Miller 2005).

The prevalence of fetal alcohol disorders (not all are full-blown FAS) in the United States is around 1% of live births (Manning and Hoyme 2007). FAS is found in all racial groups, but African American children are five times more likely than white children to exhibit it (Sokol, Delaney-Black, and Nordstrom 2003). The behavioral symptoms of FAS include low IQ, hyperactivity, impulsiveness, poor social, emotional, and moral development, and a highly elevated probability of alcoholism (Jacobson and Jacobson 2002; Kelly, Day, and Streissguth 2000). Each of these deficits is linked to high levels of antisocial and criminal behavior independent of FAS.

All children exposed to fetal alcohol do not develop FAS; much depends on when they were exposed, how much and how often they were exposed, and on their mothers' constitutions. Mothers who drink during pregnancy but who possess the ADH2*3 allele that codes for a more efficient alcohol dehydrogenase enzyme are less likely to have an FAS child (Goodlett, Horn, and Zhou 2005). Drinking during pregnancy is more common among low-income and unwed mothers (Sokol, Delaney-Black, and Nordstrom 2003), both of which are risk factors that are more common in African American than in white or Asian American populations.

While drinking during pregnancy is a sign of maternal irresponsibility, breastfeeding indexes a desire to care for and be close to one's child.

Breastfeeding has many positive developmental effects, one of which is superior cognitive development (Li et al. 2003). In a large study of 13,889 Belarusian breastfeeding mothers, half the mothers were randomly given incentives that encouraged prolonged and exclusive breastfeeding while the remaining half continued their usual maternity hospital and outpatient care. The researchers assessed the children of these mothers six years later and found that the experimental group children had a mean IQ almost six points higher than the control group children and received higher academic ratings from teachers (Kramer et al. 2008). The randomized design allowed researchers to measure breastfeeding effects on cognitive development without biasing confounds such as the known positive relationship between mothers' IQ and the probability of prolonged breastfeeding.

Although confident that the breastfeeding/IQ relationship was causal, Kramer and his colleagues could not determine if it was due to the constituents of breast milk or to mother/child interactions and the warm kin-to-skin contact experienced during the process of breastfeeding. The authors mentioned the epigenetic effects of licking and grooming of rat pups we discussed in Chapter 3 and that it may have been this tactile closeness rather than the nutritional benefits of the milk that accelerates positive neurocognitive development the same way in humans that it does in experimental animals. Tactile stimulation of human infants confers enormous benefits on the infant, and is recommended by physicians for the optimal brain development of low-birth-weight infants and infants who have suffered some kind of head trauma (Elbert and Rockstroh 2004; Weis, Wilson, and Morrison 2004).

Unfortunately, black infants are the least likely to receive the benefits of breastfeeding, and are thus more likely to be deprived of important experience-expected input. According to the Department of Health and Human Services' National Immunization Program survey (2004), in 2001 only 29.3% of black infants were breastfeeding at 6 months versus 43.2% of whites and 53.7% of Asians. Thus significantly more African American infants than white or Asian infants are deprived of a valuable contributor to positive neurocognitive development.

If breastfeeding increases IQ, exposure to noxious substances in the environment outside the womb such as lead decreases it. The federal guidelines for "acceptable" levels of lead in children is less than 10 micrograms per deciliter of blood (< 10 µg/dl), although exposure to lower levels before 2 years of age, when the brain is most plastic, have been shown to have some neurotoxic effects (Needleman 2004). With the banning of leaded gasoline and paint, the median lead level of 1- to 5-years-olds in the United States was 1.5 µg/dl in 2006 (Bellinger 2008). The IQ decrement per 1 unit increase in µg/dl is an average of 0.50 points (Koller et al. 2004). The main culprit in terms of lead exposure today is lead dust from paint in older houses. Toxic levels of lead distort enzymes, interfere with the development of the endogenous opiate system, disrupt the dopamine system, and reduce serotonin and MAOA levels (Wright et al. 2008).

An fMRI study found that gray matter was inversely and significantly correlated with mean childhood lead concentrations in a sample (n = 157) of mostly black young adults taken from the longitudinal Cincinnati Lead Study (Cecil et al. 2008). The mean childhood blood lead concentration was 13.3 µg/dl, far from 40 µg/dl considered poisonous, but significantly greater than the 2006 average of 1.5 µg/dl for the general population. Although the gray matter lost to lead exposure was considered small (about 1.2%), it was concentrated in the frontal lobes and the anterior cingulate cortex, vital behavior-moderating areas responsible for executive functioning and mood regulation.

A larger sample (n = 250; 90% black; mean age 22.5) from the same Cincinnati Lead Study examined the relationship between childhood blood lead and verified criminal arrests. Among the males ever arrested (n = 136), the mean number of arrests was 5.2 (1.1 for females). Mean childhood blood lead levels were almost identical (13.3 versus13.5 µg/dl) for never-arrested and ever-arrested subjects. The primary finding was that after adjusting for covariates for every 5 µg/dl increase there was an increase in the probability of arrest for a violent crime of about 50% (Wright et al. 2008). Although the independent effect of lead in this study was small, it is dependably there, and it is another risk factor found more often in black neighborhoods than in nonblack neighborhoods. There are a number of other studies that have found these weak to moderate effects of lead on crime, particularly violent crime (reviewed in Needleman 2004).

The associations between blood lead levels and the various negative outcomes are likely modified by genetic polymorphisms, particularly those associated with the vitamin D receptor gene (VDR) (Chakraborty et al. 2008). Lead competes with minerals such as calcium, and when children do not get enough calcium in their diets lead tricks the body into believing it is calcium and is absorbed as if it were. African Americans have lower average calcium intake than whites and are therefore more likely to absorb more lead given equal levels of lead in their environments (Heaney 2006). To make matters even worse, certain polymorphisms of the VDR (an enzyme known as Fok1) make the absorption of calcium more efficient for their carriers, and these polymorphisms are more prevalent among blacks than among whites or Asians (Chakraborty et al. 2008; Haynes et al. 2003). If these polymorphisms render calcium absorption easier, given lead's ability to mimic calcium, lead is also more easily absorbed. This creates a gene x environment interaction in which blacks absorb more lead than similarly exposed members of other racial groups.

Testosterone and Serotonin

The roles of testosterone (T) and serotonin in status and dominance struggles were addressed in Chapter 5. It has been proposed that T might be the primary biochemical mechanism underlying racial differences in both sexual and criminal activity because T levels appear to follow the same black > white > Asian gradient (Lynn 1990). A study of over 4,000 males found blacks to have 3.3% more T than whites (Ellis and Nyborg 1992). This racial difference, however, may reflect the greater status challenges black males face in their honor subcultures rather than true racial differences in T base levels since no racial differences in T levels tend to be found among prepubescent males, older males, males who have attended college, and males raised outside honor subcultures (Mazur and Booth 1998).

However, levels of circulating free T only inform us of the amounts available, not the efficiency with which it is used. As with any other hormone, T functions according to the efficiency of its receptors. Responsiveness to T is governed by the androgen receptor (AR) gene, which has different trinucleotide (CAG) polymorphisms (Schildkraut et al. 2007). Males with a shorter repeat version (< 22 repeats) of the AR gene have a greater binding affinity for T, thus making them more receptive to its effects. All studies done thus far indicate that African American males have a greater frequency (about 0.76) of the short version of the AR gene than whites (about 0.62) or Asians (about 0.55) (reviewed in Nelson & White 2002). If blacks are more receptive to the same level of T than whites or Asians, identical levels of the hormone will have stronger activating effects for blacks than for other races.

As previously noted, elevated T is most likely to result in violent aggressiveness when it is present in conjunction with low serotonin (Birger et al. 2003). There are average racial differences in the polymorphisms in the serotonin transporter and receptor genes (Galernter et al. 1997; Lin 2001), but there is no consensus as to whether these "genetic polymorphisms might lead to differential vulnerability of ethnic differences in psychopathology" (Lin 2001:17). However, experiments with rhesus monkeys have shown that peer-raised monkeys (read "fatherless, gang-raised children" for humans) have lower concentrations of the serotonin metabolite 5-HIAA than parentally raised monkeys (Bennett et al. 2002; Kreamer et al. 1998). Of course, this indexes an environmental rather than a molecular cause of low serotonin, and again points to important environmental effects on the functioning of biological systems, particularly the deleterious effects of parental deprivation to which inner-city African Americans are especially vulnerable.

Figure 8.2 presents a simplified model of the progression from structural conditions imposed on blacks in the past, the psychological responses of those exposed to them, the oppositional culture forged by the sum of those responses, the allostatic (see Chapter 4) impact of living in such a culture, and finally to manifested behavior. The diagram is in the spirit of Sampson and Bean's (2006) contention that structure and culture as they exist today create one another via individual actions in a constant feedback loop.

It is interesting to see what Douglas Massey, ex-president of the American Sociological Association, has to say about the kind of biosocial analysis presented here (2004:22):

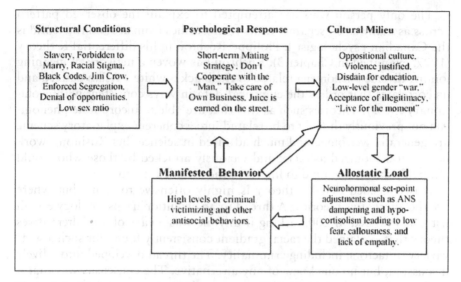

Figure 8.2 From structural conditions to manifested behavior.

. . . by understanding and modeling the interaction between social structure and allostasis, social scientists should be able to discredit explanations of racial differences in terms of pure heredity [of course, no scientists that I am aware of see things in terms of "pure heredity"]. In an era when scientific understanding is advancing rapidly through interdisciplinary efforts, social scientists in general—and sociologists in particular—must abandon the hostility to biological science and incorporate its knowledge and understanding into their work.

The Black > White > Asian Pattern

Having attempted to explain the crime rate of Americans of African descent, one wonders if such explanations can be generalized to black crime rates outside of the United States. It was noted in Chapter 3 that one of the most interesting patterns in the human sciences is the consistent black > white > Asian or Asian > white > black pattern in almost all areas of inquiry. This pattern has been seen in the present chapter from crime rates to breastfeeding rates to genetic polymorphisms, and there are many others ranging from age at first intercourse to dizygotic twinning rates to life expectancy (see Chapter 3). For instance, the black > white > Asian pattern of crime rates has been found everywhere in the world in which the three races coexist (Eysenck and Gudjonsson 1989; Rushton and Whitney 2002; Walsh 2004), which makes it difficult to generalize the structural and cultural arguments I have advanced here beyond the United States.

The only person who has attempted to explain the observed pattern across as many as 60 separate traits from studies from around the world is the Canadian psychologist J. Philippe Rushton in his differential K theory (1997; examined in Chapter 3). Rushton has woven a network of meaning joining diverse phenomena related to the black > white > Asian pattern and its correlates that orders the empirical data in ways consistent with evolutionary theory. Theories such as this that are able to incorporate phenomena not previously thought to be related into a coherent explanatory scheme are generally welcomed and much admired in science, but Rushton's work has been demonized as racist and viciously attacked by those who would put padlocks on the mind to imprison inconvenient truths.

It is conceded that his theory is highly offensive to some, but where are the alternative theories? A thread in an evolutionary psychology e-mail list was initiated in 2003 asking if anyone was aware of any alternatives theory that explained the racial gradient consistently found for such a wide variety of factors, including criminality. The thread developed into a lively discussion, but no one knew of any alternative. The consensus was that it is almost impossible to imagine what strictly environmental factors could account for the systematic alignment of the three racial groups on such a wide variety of traits consistently documented across cultures. There was considerable agreement that if ideological opponents of Rushton believed their arguments they would be in the forefront in collecting comparative racial data to support them. Rather than do this, they either offer ad hoc explanations for a single phenomenon or avoid collecting such data at all and to intimidate others into doing the same. Such tactics can only be described as scientific malfeasance.

Writing honestly about race no more makes Rushton or anyone else a racist anymore than writing about sex makes one a sexist or writing about class makes one a classist. Doing one's damnedest with one's mind, going where the data lead us, being open to all alternatives, and being willing to change course when the data demand it are the hallmarks of a no-holds-barred scientist. Censorship, even if self-imposed, only advances causes dear to bigots and retards understanding, and thus any ameliorative action that might be possible. The suffering African American community deserves a forthright evaluation of black crime at all levels from neurons to neighborhoods, for as LaFree and Russell (1993:273) point out: "All roads in American criminology eventually lead to issues of race."

9 The Critical Tradition and Conflict

Unlike the theories discussed in previous chapters, those discussed in this chapter emphasize social conflict rather than consensus. Although most mainstream sociological theories of criminal behavior necessarily contain elements of underlying conflict, they are best described as consensus theories. They are broadly construed as consensus theories because they tend to judge alternative normative systems from the point of view of mainstream values, and because any policy recommendations flowing from them do no require a major restructuring of the social system. Although consensus theorists may criticize certain factors in society as criminogenic, they do not attack society in its totality nor seek the radical changes advocated by many critical theorists. Critical theorists view the law, criminal behavior, and differential sanctions for those who break it as all originating in class, race, or gender conflict within an inequitable social system.

The term *critical* is used here as a generic name for a variety of left-wing criminologies united only by the assumption that the nature of society is best characterized by conflict and by power relations. These criminologies sail under labels such as *conflict, radical, critical, neo-Marxist,* or *left realists,* and it is fair to say there is far more agreement among them about what they are against than what they are for. Probably the major differences within these various schools is the degree to which they accept the ideas of Karl Marx (1818–1883) versus Max Weber (1864–1920) on the nature of conflict and of society into their thinking, and whether they are reformist or revolutionary in their recommendations.

The more radical theorists favor Marx and revolution, and the less radical theorists favor Weber and reform. Marx and Weber represent two different ontologies that philosophers have argued about for centuries: materialism and idealism. The structural (materialism)/cultural (idealism) debate discussed in the last chapter is a specific example of these two positions. Materialism is a philosophical position that seeks the causes of human action in material things such as economic factors and social structure, and that only material things are real. These material things are said to lead humans to embrace the ideas, attitudes, and values that they do, not the other way around. Idealism posits the opposite: ideas create reality and cause human action, the structure of society, and how individuals and groups behave. Marx was a materialist and Weber was an idealist, and this difference in philosophical positions (although both adopted the position of the other when necessity intervened) led them to view, dissect, and evaluate a number of nominally identical concepts in different ways.

Critical theorists concentrate almost exclusively on crime (they ask questions such as what acts are criminalized, why they are criminalized, and what factors lead to differential rates of crime) and only rarely and very lightly do they touch on criminality. Because they advance no notions of why individuals are differentially susceptible to criminal behavior, it is difficult to see how these theories could benefit from the insights of modern genetics and neuroscience. However, critical theories do share a deep interest in many things with evolutionary biology, particularly conflict and status striving. Evolutionary theory can assist critical theories in developing a better understanding of human nature and to ground many of their policy recommendation in reality rather than "foolishly attempt[ing] to cut against the grain of human nature" (Wilkinson 2005:15). I thus limit discussion in this chapter to the features of human nature revealed to us by Darwinian biology relevant to critical theories.

Given the collapse of communism almost everywhere and the trend toward abandoning of the communist ideal of state ownership (nationalization) of the means of production in social democracies one may wonder why a discussion of Marx in any context at all is relevant today. It is relevant to us because much of the content of critical/conflict criminology can be traced to Marx's ideas, and his thoughts on equality and alienation still resonate strongly among radical and some liberal criminologists. One may loathe or love Marx, but one cannot ignore him. Indeed, a 1999 British Broadcasting Company (BBC) poll declared Marx to be the greatest thinker of the millennium (above Einstein, Newton, and Darwin!). This may well reflect the socialist gift for ballot stuffing or the isolation from reality of the limousine liberals that constitute much of the BBC's audience, as anti-Marxists have claimed. None the less, Marx *was* a great thinker of stupendous intellect whose penetrating critique of 19th-century capitalism was incisive. I am agnostic on the question of whether the state evils and inefficiencies we witnessed in 20th-century communist regimes are inherent in Marxism regardless of how attractive its principles sound in the abstract, but given the general worldwide retreat from those ideas the burden of proof lies with Marxists who claim that they are not.

ARE CRITICAL THEORIES INCOMPATIBLE WITH BIOSOCIAL CRIMINOLOGY?

Critical theorists are the most biophobic of all criminologists. Wright and Miller's (1998) analysis of textbooks discussed in Chapter 1 found that writers embracing critical/radical perspectives were by far the most likely to espouse the view that biosocial explanations of crime and delinquency are racist, sexist, and classist and to engage in ad hominem attacks on those who advance them. Perhaps this is so because they are incorrigibly unconstrained visionaries, the most committed to changing the status quo, to the infinite malleability of human nature, and to the dream of perfectibility. Apparently, the belief is that if they admit that human behavioral differences are even partly "biological," they are admitting that the differences are immutable and predetermined. This position is derived from their understanding of biology, of which they are, to quote van den Berghe once more, "militantly and proudly ignorant" (1990:177).

Critical theories are not *intrinsically* incompatible with biology—after all, humans *are* biological beings, but it is certainly incompatible with the proposition that humans are infinitely malleable, or that humans and their societies can be perfected (whatever this is taken to mean). Degler (1991) reminds us that the French political left enthusiastically embraced sociobiology, the forerunner of evolutionary psychology, as a sort of neo-Rousseauism even as the American political left just as enthusiastically rejected it. Leading figures in evolutionary theory and supporters of the idea of an innate and universal human nature such as J. B. S. Haldane, John Maynard Smith, Noam Chomsky, and Robert Trivers have all been active supporters of radical left-wing causes (Singer 2000), and prominent social scientists such as Melvin Konner and Alan Mazur are well known for both their liberal politics and their commitment to biosocial science (Degler 1991).

More enlightening are the attempts of Chinese Marxist scholars who saw the terrible heritage of Marxism/Maoism in their country and who "thought that sociobiology might correct the extreme ideas of Marxism and Maoism" (Jianhui and Fan 2003:575). Boshu Zhang, a scholar steeped in Marxism and biology, tried to synthesize what he saw as the positive aspects of Marxism and sociobiology because "The fundamental defect in Marx's historical-materialist logic is the neglect of human nature . . . . In contrast, sociobiology is useful in its capacity to help us understand our biological inheritance and its influence upon our future" (1994:113–114).

Karl Marx himself appeared to look favorably on the integration of the social and biological sciences when he claimed that "Darwin's book [*The Origin of Species*] is very important and serves me as a natural-scientific basis for the class struggle in history" (in Singer 2000:20). Marx also wrote that "Natural science will in time subsume under itself the science of man, just as the science of man will subsume under itself natural science: there will be one science" (1978b:91). Marx did not mean the kind of integration that biosocial criminologists aim for because the genomic and neurosciences did not exist in his time. If they had existed, however, I cannot see Marx pulling the wool over his eyes and ignoring them as his contemporary followers are wont to do. Marx viewed Darwin's theory of evolution as a way of conceptualizing historical change using the scientific method from which predictions could be made. Marx saw transitional movement from his posited historic periods (Asiatic, ancient, feudal, and bourgeois) as an evolutionary process which would *inevitably* lead to communism through the workings of the "natural" laws of history. Like so many of his contemporaries, he saw Darwin's evolutionary theory as a theory of progress leading to better and better results (optimal design and adaptation of organisms) in the biological world. Of course, evolutionary theory has no sympathy for claims of inevitability; evolutionary trajectories can and do change at any time, and so can historical trends.

Jerome Barkow (1989:310) has written that "Like Marxism, a sociobiological view of society yields conflict theory." It is indeed true that many concepts central to Marxism such as conflict, self-interest, dominance, false consciousness (self-deception), and exploitation are also central to evolutionary psychology, suggesting that the two theories may be proximate- and ultimate-level explanations of essentially the same set of phenomena. At least three books of fairly recent vintage have been written claiming this to be so, that the left has nothing to fear from Darwin, and that there can be a happy marriage between Darwin and Marx (Lopreato and Crippen 1999; Sanderson 2001; Singer 2000). For Marxists the social processes and individual traits of concern to them are "social constructs," the fruits of a particular mode of production of a particular historical period. Yet these social processes and individual traits have been in evidence throughout recorded history, as Marx and Engels (1948:9) themselves point out: "Freeman and slave, patrician and plebian, lord and serf, guildmaster and journeyman, in a word, oppressor and oppressed, stood in constant opposition to one another." The ubiquity across time and culture of these processes and traits strongly suggest that they are integral to human psychology and sociality rather than temporary aberrations.

The mode of production Marx was concerned with was capitalism because it alienated individuals from themselves and from others. Capitalism created the self-interest and dominance that led to exploitation, and socialism will supposedly eliminate them. By way of contrast, evolutionary psychology views these traits as having evolved to promote individual survival and reproductive success, morally repugnant though some expressions of them may be. They are universal species characteristics that cannot be eliminated in any absolute sense, although they are certainly expressed facultatively rather than stereotypically. Marx and Engels appeared to have agreed with this when, true to their materialist ontology, they wrote that right-thinking communists "do not preach *morality* at all . . . They do not put the moral demand: love one another, do not be egoists, etc; on the contrary, they are well aware that egoism, just as much as selfishness, *is* in definite circumstances a necessary form of self-assertion for individuals" (in Replogle 1990:695, emphasis in original). There is much other evidence throughout Marx's writings that, like a good evolutionary biologist, he saw humans as primarily motivated by self-interest (Jost and Jost 2007), which if true leaves one wondering how Marx could possibly have thought that communism could be viable in a modern context.

MARX'S CONCEPT OF HUMAN NATURE

With all this similarity of conceptual interest and with Marx's own apparent enthusiasm for Darwin, why is the mainstream radical left so miffed when Darwinism is applied to human behavior? The short answer is that early Darwinism was captured by the radical right, and thus the friend of an enemy became an enemy. The radical right transmuted Darwinism into the notorious aberration called social Darwinism. This philosophy, with which Darwin had no part (the phrase was coined and popularized by British sociologist Herbert Spencer), justified and even sometimes exalted exploitation, the class system, "dog-eat-dog" competition, poverty, and the entire status quo as "natural," and therefore good (the naturalistic fallacy again). The insidious philosophy of social Darwinism had to be opposed, and the leaders of the opposition were predominantly left-leaning humanitarians. However, rather than striving to show how Darwinism was being misunderstood and misused, they apparently determined that the point of least resistance was to deny the role of evolution in human affairs and the concept of a universal human nature altogether. As Dawson (2002:43) described the estrangement of the radical left from naturalism: "It is as though Darwin and Marx had put up signposts pointing to two radically divergent paths to the understanding of cultural evolution, and the two have never met."

This line of reasoning also applies to the denial of the role of genes in human behavior. Stalin even had geneticists sent to the infamous gulag camps to protect the crackpot ideas of the Lamarckian agronomist Trofim Lysenko from "anti-progressive" Darwinism and Mendelism. Lamarckism was considered more "progressive" and optimistic than Darwinian natural selection or Mendelian genetics because it implied that even plants can be "trained" by progressively transplanting them in less and less suitable environments so that they may acquire and pass on the characteristics desired by the agronomist (Black Sea oranges grown in Siberia!). Just as Soviet politics and economics had disastrous effects on the polity and the economy, Lysenko's geneless "people's" biology had disastrous effects on Soviet agronomy.

Genetics also received a bad name from the left in the aftermath of the eugenics movement, which some interpret as a sort of proactive social Darwinism. However, those who decry eugenics most fiercely today forget that it was most enthusiastically embraced by the left as progressive and humanitarian before it was turned into a nightmare by the Nazis (Abu El-Haj 2007). Eugenics was an integral part of the radical left's program for human perfectibility, as an early Marxist explained: "unless the socialist is

234 Biology and Criminology

a eugenicist as well, the socialist state will perish from racial degradation" (in Paul 1984:568).

Given this and numerous other such statements from Marxists writing at the time when eugenics was in vogue in the Western world, it becomes clear how Matt Ridley (1996:253) could write that "Hitler was merely carrying out a genocidal policy against 'inferior', incurable or reactionary tribes that Karl Marx and Friedrich Engels had advocated in 1849 . . . . It is even possible that Hitler got his eugenics not from Darwin or Spencer but from Marx, whom he read carefully . . . and echoed closely on the topic." Marx and Engels did write at length about the "Jewish question" and even that "the task of abolishing the essence of Jewry is in truth the task of abolishing Jewry in civil society" (1845/1956:148). Marx was not advocating the destruction of Jewish people, but rather of the destruction of avarice, which he saw Jews as exemplifying. It is nothing but politicking to trace Hitler's monstrous actions to Darwin *or* to Marx, and a perversion of eugenics as Margaret Sanger, Sydney and Beatrice Webb, George Bernard Shaw, Havelock Ellis, or any of a number of other left-leaning supporters saw it, to equate it with the Nazi holocaust as many modern writers are inclined to do.

The contemporary left's denial of a universal innate human nature was not simply forced on them in reaction to the radical right. Left-wing ideology was already hostile to the idea of a constraining human nature; the evils of social Darwinism just helped to cement that position (Pinker 2002). The general position held by the left vis-à-vis human nature is characterized in Marx's classic statement that "The mode of production of material life conditions the social, political and intellectual life processes in general. It is not the consciousness of men that determines their being, but, on the contrary, their social being that determines their consciousness" (1977:389). This is the quintessential statement of Marx's materialistic ontology: systems of ideas and relationships among individuals are rooted in the material conditions existing in historical periods and each mode of production shapes society and the nature of its members in its own image.

No one would deny that the expressions of human nature in its specifics and that the values, attitudes, and behaviors of people are shaped by historical and cultural circumstances, but this fact does not constitute evidence for denying an unchanging human nature underneath its elaboration by time and place. If there is no evolved human nature shared by all humans everywhere, and if human nature is nothing but a veneer constructed by culture, then it follows that it changes from skin to core each time the mode of production changes. Such an unconstrained view of human nature is required to support optimism in the perfectibility of humanity and the possibility of a socialist utopia because to change it all that has to be done is to change the mode of production. This vision of humans as completely malleable blank slates affects to varying degrees the thinking of most criminologists of radical and leftist persuasions.

It is interesting to read what Mikhail Bakunin, Marx's principal rival for leadership of the *First International* (ostensibly a workingman's association, but consisting mainly of a grab bag of anarchists, expatriates, and refugees from all over Europe), thought about Marx's vision of a conflict-free utopia. Bakunin wrote that if the proletariat were to gain control of the state, some number of these ex-workers would become rulers, and "From that time on they represent not the people but themselves and their own claims to govern the people. Those who can doubt this know nothing of human nature" (in Singer 2000:4).

Marx did recognize, however, that the "dictatorship of the proletariat" would be only the first stage of communism ("raw" or "crude" communism"), which would be no better than what it was to replace because people would still be contaminated by capitalism. It was the second stage ("ultimate communism") that would lead to the purification and perfection of humans in a workers' paradise. It is this quasi-religious prophecy of "salvation" that is in serious need of a reality check, because as long as there are those who believe that to get to the heaven of the ultimate we have to suffer the purgatory of the crude, we will repeat the tragic course of the 20th century in the 21st. This is precisely why the left must ground its social-political-economic agenda in a more realistic understanding of human nature, for as the great biologist E. O. Wilson famously remarked about Marxism: "Wonderful theory; wrong species."

Marx's work is so voluminous that a hermeneutic tradition has been erected around it. No one can claim definitive knowledge of exactly what Marx's position was on human nature since he seems to have vacillated so much or was not clear enough in his writings that, as has been said of the Holy Bible, even the devil can quote it to suit his own ends. Marxists hostile to the idea of a human nature quote the sixth of Marx's 1844 *Theses on Feuerbach* as his basis for denying a human nature: "Feuerbach resolves the essence of religion into the essence of man. But the essence of man is no abstraction inherent in each single individual. In its reality it is the ensemble of the social relations" (1978a:145). While it is self-evident that "man" only becomes fully human in the "ensemble of social relations," to assert that he or she is nothing but these relationships extinguishes human individuality and dignity and provides totalitarians with the philosophical underpinnings for pounding a blank-slated humanity into anything they see fit. It is by no means necessarily Marx's position that humans are "nothing but" this social ensemble, although it is certainly understandable how future Marxists could interpret the sixth thesis in this way, especially if buttressed by Marx's materialist view that human consciousness is determined by the mode of production. It is not for nothing that Marx frequently found it necessary to disavow the "Marxism" of his disciples (Jost and Jost 2007).

On several occasion throughout his writings Marx rather strongly implies the existence of a general human nature independent of the elaborations attached to it by a particular mode of production. For instance, in his

1844 *Economic and Philosophical Manuscripts* he wrote: "Man is directly a natural being. As a natural being and as a living natural being he is . . . furnished with natural powers of life—he is an active natural being. These forces exist in him as tendencies and abilities—as impulses" (1978b:115). Such a statement is dismissed as the immature thinking of the young idealist Marx by many contemporary Marists. However, I can imagine an evolutionary psychologist writing the following passage, although it was Karl Marx (not the idealist "young" Marx, but the "mature" materialist Marx of *Capital*) who did so: "To know what is useful for a dog, one must study dog nature . . . Applying this to man, he that would criticize all human acts, movements, relations, etc., by the principle of utility, must first deal with human nature in *general*, and then with human nature as *modified* in each historic epoch" (Marx 1967:609, emphasis added). In this passage Marx appears to be saying:

1. A social theory (critical or otherwise) must necessarily start with a theory of human nature.
2. There is a discoverable universal human nature, as well as one that is modified by a particular mode of production.
3. The explication of human action must be predicated on an understanding of that nature.
4. Aspects of human nature are developed or constrained contingently; e.g., we must not take human nature as we observe it instantiated in a particular mode of production as exhausting all that human nature is.

In another place Marx anticipates Durkheim's contention that human wants are insatiable in that each satisfied need "leads to new needs—and this production of new needs is the first historical act" (1978c:156). This perpetual dissatisfaction leads humans to "antagonistic cooperation" with one another, a strategy evolutionary psychologists call tit-for-tat cooperation or reciprocal altruism in which each person adjusts his or her strategy according to what others have done (cooperate/defect) in previous interactions. Social classes emerge when the division of labor emerges, and individual antagonism then becomes group or class antagonism (conflict between alliances, in evolutionary terms).

Yet the view that collective interests drive class structure and class conflict is only true insofar as collective interests subsume individual interests. That is, like group selectionism in biology, class conflict is only *apparently* driven by group interests. Analyzing group processes is interesting and valuable in its own right, but it has led to the reification of class and away from the understanding that "the undercurrent of alliances is the quest for individual power" (Lopreato and Crippen 1999:211). This may be more the sin of Marx's followers than of Marx himself, for he warned us "to avoid above all . . . the re-establishing of 'Society' as an abstraction vis-à-vis the individual" (1978b:86). Marx also stated: "The separate individuals form a class only in so far as they have to carry on a common battle against another class; otherwise they are on hostile terms with each other as competitors" (in Coser 1971:48). Except for the overemphasis on the hostility of competition, this sentiment is entirely consistent with evolutionary psychology. For instance, Stephen Sanderson (2001:149) states: "Cooperative social relations exist because they are the relations that will best promote each individual's selfish interests, not because they promote the well-being of the group or society as a whole."

WORK, ALIENATION, AND HUMAN NATURE

Work and alienation are central to the understanding of Marxism's view of both human nature and criminal behavior. For Marx the essential being of any species is the activity that distinguishes it from every other species. Nonhuman animals instinctively act on the environment *as given* to satisfy their immediate needs, but humans distinguish themselves from other animals by consciously *creating* their environment instead of merely submitting to it. Free, creative activity, then, is for Marx the distinguishing human *species being*, "man's spiritual essence, his human essence" (in Sayers 2005:611). Furthermore, man *qua* man "only truly produces when he is free of physical need" (Nasser 1975:487). Marx thus divorces "truly" productive work performed when workers are their own masters owing their existence only to themselves from work done out of brute necessity.

In this idealized image of what work should be, if workers are not passionately involved with their work as an expression of their species essence, such as a creative scientist or artist might be, they are estranged from their species being and reduced to working like animals; i.e., as a means to an end (survival) rather than as a creative end in itself. Alienation is thus a state of incongruity between one's human nature (one's species being) and one's behavior (e.g., mindlessly stamping out an endless chain of widgets in a dark, dank factory in order obtain the necessities of survival and thus becoming merely an appendage of the machine). Wage-labor thus literally dehumanizes human beings by taking from them their creative advantage over other animals—robbing them of the species being and reducing them to the level of animals. As Tucker (2002:98) puts it: "Capitalism crushes our particularly human experience. It destroys the pleasure with labor, the distinctively human capacity to make and remake the world, the major distinguishing characteristic of humans from animals." This is the essence of Marx's concept of alienation.

When individuals become alienated from themselves they also become alienated from others and from their society in general. Alienated individuals may then treat others as mere objects to be exploited and victimized as they themselves are exploited and victimized by the "system." Since the great majority of workers do not experience their work as creative activity, they are all dehumanized ritualists or conformists (to borrow from Merton's modes of adaptation). If one accepts this Marxist notion, then perhaps one can view criminals as heroic rebels struggling to re-humanize themselves, as indeed some Marxist criminologists have done. It has been claimed that the concept of alienation is absolutely central to critical criminology and that it can subsume other causal concepts of mainstream criminology (Smith and Bohm 2008).

MARXISM AND INDIVIDUAL DIFFERENCES AND EQUALITY

If the radical left can be said to stand for any one thing in general, it is the moral principle of equality. This blind adherence to egalitarianism at all costs, to rights without obligations, permeates the extreme left wing of social science. Psychologists strive to neutralize talk of abnormality and caution against judgmentalism, anthropologists want us to believe that all cultures are equal, and postmodernist criminologists urge us to "appreciate" the point of view of criminals. If there is anything the radical left cannot acknowledge, lest doing so hinder the quest for equality as they define it, it is that individuals differ widely in their talents and abilities for both genetic and environmental reasons. In criticizing sociology's lack of interest in biology in a special edition of the journal *Social Forces* devoted to an appeal to link the two disciplines, guest editor and sociologist Guang Guo (2006:145) writes: "One of the implicit assumptions of sociological inquiry is that individuals are the same at birth; the differences among them are then attributed to the position each occupies in the social hierarchy."

The assumption of human equipotentiality is biological fiction and is an insult to human dignity, but it is central to some contemporary versions of Marxism. Notwithstanding Guo's criticism of sociology in general, it is probably only the radical left that still holds fast to the belief that human nature is a blank slate and that we all have equal potential. It is doubtful that many social scientists really believe it is so, although most may still carry out their research as if it were. It matters not that the radical left's position flies in the face of mountains of hard genetic evidence to the contrary; the quest for equality (which tends to be defined by the radical left as equality of outcome rather than simply equality of opportunity) trumps any evidence. Much of the problem lies in ignorance of genetics, thus allowing for the erroneous belief that anything "biological" is fixed and unalterably deterministic.

Marx's analysis of human nature only differentiates between species, not between individuals. Marx never concerned himself with a systematic study of individual differences, but he clearly recognized them, albeit as "social constructs." In the *Economic and Philosophical Manuscripts* (1988:132), Marx wrote: "The diversity of human talents is more the effect than the cause of the division of labour." This is, or course, true if one is speaking about acquired skills; one cannot be a rocket scientist or a plumber until rockets and plumbing are part of the cultural landscape and one is trained in matters pertaining to them. Because the goal of Marxism was to eliminate alienation by eliminating the division of labor, if we follow this line of thinking, doing so would also eliminate the diversity of talents it allegedly produces. As one scholar of Marxist thought sums up the implications of Marx's position: "the alienation inherent in human relations is eliminated through the disappearance of the differences between people. This means that individuality is annihilated in the utopian society; if the suppression of alienation means doing away with the difference between the self, and others and nature, then it also means the end of human liberty" (Tralau 2005:393)

Not all Marxists deny the reality of human differences as the modern biosocial scientist understands them; most presumably would readily recognize that rocket science and plumbing would require immensely different intellectual capacities to acquire their respective skills. Long before the structure of DNA was discovered, and at a time when the computation of heritability coefficients was considered pretty exotic stuff, J. B. S. Haldane expressed a position widely held by his fellow Marxists of the day when he wrote: "The dogma of human equality is no part of Communism . . . the formula of Communism: 'from each according to his ability, to each according to his needs,' would be nonsense, if abilities were equal. I think the world would be a much duller place if there were no differences in innate powers between the different individuals and groups of individuals" (in Paul 1984:567).

Although many early 20th-century Marxists believed that individuals possessed innate natural abilities (particularly intelligence), they insisted that such talents and abilities could only be cultivated and realized in societies that provide equal opportunities for all (Paul 1984). As discussed in Chapter 2, the full force of genetic differences among individuals in all kinds of traits, talents, and abilities only become apparent in direct proportion to the equalization of environments conducive to their development. That is, as environmental variance decreases (as the environment becomes more equal) genes associated with these traits, talents, and abilities assert themselves more strongly as indexed in the increasing values of heritability coefficients for those traits. One way of equalizing the environment is to provide education equal to the individual's ability to absorb it. It was for this reason that the British left heartily embraced IQ testing in the belief that they would turn schools into capacity-catching institutions that would lift able working-class children into the middle and upper classes and thus subvert a class system built on ascription rather than achievement. Indeed, it was the British right that initially opposed IQ testing (Pinker 2002:301).

For all the apparent concern about equality among contemporary Marxists, it is not at all clear that this principle was a concern to Marx himself. Ron Replogle (1990:675) writes: "in the *Critique of the Gotha Program*"

Marx "dismisses the notions of 'equal right' and 'fair distribution' as 'obsolete verbal rubbish.' There Marx straightforwardly denies the rationality of assessing basic social arrangements according to ethic criteria." Much of Marx's *Critique* had a more realistic ring to it than the socialist *Gotha Program*, which asserted the equality of human intellectual abilities, and showed Marx to be more meritocratic in his thinking than many contemporary leftist sociologists (Shapiro 1991).

However, while Marx recognized that people had different abilities, he did not seem to approve of the idea that merit should guide the distribution of rewards in society. Given that the *Gotha Program* was a formula for socialism ("crude communism") rather than "real" stage II communism, what Marx was evidently critiquing was the classical idea of distributive justice as exemplified by the Aristotelian notion of treating equals equally and unequals unequally according to relevant differences. In his critique, Marx argues that "one man is superior physically or mentally and so supplies more labour in the same time . . . This *equal* right is an unequal right for unequal labour. It is, *therefore, a right of inequality in its content, like every right*"(in Green 1983:439). Marx considered this unjust since under the principles of distributive justice the superior person will receive more benefits as a result of being blessed with a natural superiority he or she did nothing to earn, and thus is reaping undeserved rewards. Rewards must be based on equal effort within the constraints of one's natural abilities. Thus while Marx plainly recognized that people have different innate abilities, in his perfect society they would not be granted license to benefit from them.

How one might operationalize "equal effort" is anyone's guess, for it would have to mean equal in terms of this or that person's physical and mental condition relative to some other person's. The great 18th-century French philosopher Voltaire, who may have been called a conflict theorist had such labels been in vogue at the time, realized what Marx apparently did not when he wrote: "equality is at once the most natural and the most chimerical thing in the world: natural when it is limited to rights, unnatural when it attempts to level goods and power" (in Durant 1952:245). I leave it to the Marxist hermeneutists to decide what Marx's "real" position on equality was, but whatever it was, it is clear that there is no hegemonic orthodox Marxist position on individual differences or equality.

MARX AND ENGELS ON CRIME

Another central concept of Marxist philosophy relevant to understanding the critical view of crime causation is the concept of *class struggle*. This struggle takes place between the wealthy owners of the means of production (the *bourgeoisie*) and the working class (the *proletariat*), with the former striving to keep the cost of labor at a minimum, and the latter striving to sell their labor at the highest possible price. These opposing goals are the

major source of conflict in a capitalist society. According to Marx, the bourgeoisie enjoy the upper hand because capitalist societies have large armies
of unemployed workers anxious to secure work at any price, thus driving
down the cost of labor. Worker exploitation led to a period of abject misery
in the industrial cities of Europe and the United States in the 19th century.
Herein lies one of the most important inherent contradictions of capitalism
identified by Marx. On the one hand, the unemployed and unemployable
form the core of the "criminal classes" that threatened the capitalist system, but on the other, they were functional for capitalism. According to
Marx, these economic and social arrangements—the material conditions
of people's lives—determine what they will know, believe, and value, and
how they will behave. It follows from this that if people act and think in
antisocial and criminal ways, they do so because their economic and social
positions are not conducive to acting and thinking prosocially.

The ruling class always develops ideologies to justify and legitimize their
exploitation of the masses: "The ideas of the ruling class are in every epoch
the ruling ideas; i.e., the class, which is the ruling material force of society, is
at the same time its ruling intellectual force" (Marx and Engels 1978c:172).
These ideas (which include the law) are typically accepted as valid by the subordinate classes, an acceptance that Marx called *false consciousness* because
those ideas ran counter to their objective class interests. The ruling classes are
able to generate false consciousness because they control the cultural superstructure—law, religion, education, the arts, philosophy, and so forth—that
distort the truth of social reality for the proletariat. False consciousness functions to deflect anger and criticism away from the ruling class.

Marx believed that capitalist competition would drive the less able capi-
talists into the ranks of the proletariat, as indeed it often did. These men
would provide the intellectual leadership that would overthrow capitalism,
and help the workers to understand their false consciousness (a touch of
Weberian idealism here in his recognition of the power of ideas). In time,
false consciousness would be replaced by *class consciousness*; that is, the
recognition of a common class condition and the development of a common
unity in opposition to capitalist exploitation (to be a class *for* itself rather
than simply a class *in* itself). This would set the stage for revolution and
the "dictatorship of the proletariat," marking the end of exploitation and
social conflict. Because exploitation and conflict are considered the causes
of crime, the revolution would end most forms of crime as individuals discard their deformed natures and rediscover their true species being.

Marx only wrote about crime to illustrate the bitter fruits of capitalism, and did not develop any systematic criminological theory. As Alvin
Gouldner (1973:xii) has pointed out: "Viewing criminals and deviants as
a Lumpenproletariat that would play no decisive role in the class struggle,
and indeed, as susceptible to use by reactionary forces, Marxists were not
usually motivated to develop a systematic theory of crime and deviance."
Unlike thinkers such as Durkheim, who viewed crime as a natural part of

social life, and even functional in some ways, Marx and his collaborator, Friedrich Engels, saw it as a palpable indicator of social sickness that had to be cut out, and both made plain their profound disdain for criminals. In the *Communist Manifesto*, Marx and Engels referred to them with the moral outrage that would have done a bourgeois banker or New York cop proud: "The dangerous class, the social scum, that rotting mass thrown off by the lowest layers of the old society" (1948:11).

Although Marx and Engels viewed crime and criminals as the products of an alienating social structure that denied productive labor to masses of unemployed—"the struggle of the isolated individual against the prevailing conditions"—these conditions were not construed as a justification of criminal behavior. After all, these "scum" were victimizing the honest laboring class, or "providing demoralizing services such as prostitution and gambling" (Bernard 1981:365). For Marx and Engels (1965:367) crime was simply the product of alienating social conditions—"the struggle of the isolated individual against the prevailing conditions." This became known as the *primitive rebellion* hypothesis, one of the best modern statements of which is Bohm's (2001:115): "Crime in capitalist societies is often a rational response to the circumstances in which people find themselves."

In a manuscript on surplus value that was to have been part of volume 4 of *Capital* and published after his death, Marx made some little-known comments about the functional value of crime and criminals with a Durkheimian ring to them. The switch from viewing criminals as "scum" to individuals having some social use (of course, the two views are not necessarily mutually exclusive) in capitalist societies once again underlines the notion that there is no one Marxism. Marx mentioned that criminals give birth to the criminal law, provide work for police, jailers, and judges, and lead to new inventions to protect against them. He goes on to say:

> The criminal interrupts the monotony and security of bourgeois life. Thus he protects it from stagnation and brings forth that restless tension, that mobility of spirit without which the stimulus of competition would itself become blunted. . . . Crime takes off the labour market a portion of the excess population, diminishes competition among workers, and to a certain extent stops wages from falling below the minimum, while the war against crime absorbs another part of the same population. The criminal therefore appears as one of those natural "equilibrating forces" which establish a just balance and opens up a whole perspective of "useful" occupation (in Bottomore 1956:159).

The First Marxist Criminologist

Dutch criminologist Willem Bonger has been credited with the first work completely devoted to a Marxist analysis of crime. His book, *Criminality and Economic Conditions* (1914/1969), is widely recognized as a classic.

Like any good Marxist, Bonger saw the roots of crime in the exploitive and alienating conditions of capitalism. Unlike many modern Marxists, Bonger incorporated individual differences into his work. He believed that some individuals are at greater risk for criminality than others due to biological differences, and that people varied in their "moral qualities . . . according to the intensity of their innate social sentiments" (1969:88). The social sentiments that concerned him were altruism—an active concern for the well-being of others—and its opposite, *egoism*—a concern only for one's own selfish interests. As noted in Chapter 3, various studies have placed the heritability of altruism, and hence also of egoism, its polar opposite, at around 0.5, which corroborates Bonger's view that individuals differ in their "innate social sentiments." He was adamant, however, that these individual factors are secondary to environmental ones, for it is only by the transformation of society from capitalism to socialism, from oppression to equality, that it is possible to enhance the altruistic sentiment, and thus reduce crime.

Regardless of the individual's innate propensity for altruism or egoism, then, whether one or the other is the dominant sentiment in society depends on the way society is organized, or more specifically, how it produces its material life (its mode of production). Altruism, Bonger (1969:33) wrote, is the predominant sentiment among "primitive" peoples among whom the ideas of surplus, profit, and social classes did not exist (Marx's "primitive communism"). In short, the uniformity of existence, and thus of interests, is conducive to a general culture-wide altruistic sentiment, or a Durkheimian mechanical solidarity, if you will, regardless of any innate individual variability in this sentiment. Capitalism, on the other hand, by its very nature as an alienating forced division of labor, strengthens the egoism sentiment and blunts the altruism sentiment. The intense competition for wealth, profits, status, jobs, and so forth, leaves many of the less talented and less fortunate in its wake. Such a moral climate generates frustration, envy, and greed, which may lead to crime. All individuals in capitalist societies are infected by egoism because they are alienated from authentic social relationships with their fellow human beings, and all are thus prone to crime, the poor out of economic necessity and the rich from pure greed.

Bonger (1969:108) provides an example of how greed and envy are generated by capitalism that has a very Mertonian "American Dream" ring to it:

> Modern industry manufactures enormous quantities of goods without the outlet for them being known. The desire to buy, then, must be excited in the public. Beautiful displays, dazzling illuminations, and many other means are used to attain the desired end. [In the modern department store] the public is drawn as a moth to a flame. The result of these tactics is that the cupidity of the crowd is highly excited.

Although Bonger never wavered from the position that the "root cause" of crime is the capitalist mode of production, he was careful to delineate the mechanisms by which capitalist-generated egoism translated into criminal behavior. For instance, he traces the effects of poverty on family structure (broken homes, illegitimacy) and on parental inability to properly supervise their children. He also writes about "the lack of civilization and education among the poorer classes" (1969:195). This emphasis on family structure, the moral deficits of the poor, and on interdisciplinary positivism show an affinity for the concepts of control theory, an affinity for which he has been criticized by other Marxists as non-Marxist (Taylor, Walton, and Young 1973). As we saw in Chapter 5, the third author of the work criticizing Bonger, Jock Young (2003), who is perhaps the most influential of contemporary Marxist criminologists, is now favorably disposed to individual differences, at least in the form of differential central and peripheral nervous system arousal levels being predictive of differential involvement in crime.

Modern Marxist Criminology

Because Marx wrote so little about crime, it is better to characterize modern Marxist criminologists as left-wing radicals for whom Marxism serves as a philosophical underpinning. Marxist criminologists typically explain crime and criminality in terms of Marx and Engels' primitive rebellion thesis, i.e., the struggle of alienated individuals against the prevailing social order. These criminologists, especially those writing in the latter part of the 20th century, tended to defend, excuse, and romanticize the acts of street criminals (Marx's "social scum") as rational responses to the brutality of their existence, or even as revolutionary heroes. For instance, William Chambliss (1976:6) viewed some criminal behavior to be "no more than the 'rightful' behavior of persons exploited by the extant economic relationships," and Ian Taylor (1999:151) sees the convict as "an additional victim of the routine operations of a capitalist system—a victim, that is of 'processes of reproduction' of social and racial inequality." David Greenberg (1981:28) even elevated Marx's despised lumpenproletariat to the status of revolutionary leaders: "criminals, rather than the working class, might be the vanguard of the revolution." Many Marxist criminologists also appear to view the class struggle as the *only* source of *all* crime and to view "real" crime as violations of human rights, such as racism, sexism, imperialism, and capitalism, and accuse non-Marxist criminologists of being parties to class oppression. Tony Platt even wrote that ". . . it is not too far-fetched to characterize many criminologists as domestic war criminals" (in Siegel 1986:276).

Contrast all this maudlin sentimentality with the views of statistician Karl Pearson, a Marxist eugenicist and a person more in tune with Marx both in time and in sentiment, in his essay entitled *The Moral Basis of Socialism* published in 1887: "The legislation or measures of police, to be taken against the immoral and antisocial minority, will form the political realization of Socialism. Socialists have to inculcate that spirit which would give offenders against the State short shrift and the nearest lamp-post. Every citizen must learn to say with Louis XIV, *L'etat c'est moi!*" (in Paul 1984:573). This was written well before the first socialist revolution in 1917, but all such revolutions have produced societies that have indeed viewed their criminal offenders in "true" Marxist terms and given them "short-shrift and the nearest lamp-post."

The origin of such illiberal attitudes is that "archaic" concepts of individual rights and procedural limitations on state power only had purchase as long as the state and the individual were distinct entities and at odds with one another. In a socialist society the state and the laws that support it would "wither away" and whatever would be construed as the "state" thereafter would become one and inseparable with the individual (*L'etat c'est moi!*). Thus individuals would not be in need of bourgeois procedural protections (Lovell 2004). Indeed, procedural rights are conspicuously absent in contemporary socialist legal systems under the principle of the oneness of the state and the individual (Walsh and Hemmens 2008). It is partly because of this that Gerhard Lenski, dean of American social stratification and social change theorists, over 30 years ago wrote that the empirical evils of communism are indeed inherent in the abstractions of Marxist theory. Lenski further elaborates: "Marxism, like sociology, rests on much too naive, much too innocent, much too optimistic an assumption about human nature. Marxism, like sociology, still employs an eighteenth-century view of man. Environment is everything, genetics nothing" (1978:380).

Some modern Marxists are critical of any romanticized view of common street criminals because crime prevents the formation of proletarian class consciousness, and thus diverts attention away from true revolutionary activity. Still others find the primitive rebellion thesis simplistic and strive to formulate a specific Marxist criminology relating crime rates to the political economy of a given society. These are the so-called Marxist *structural* theorists (Greenberg 1980). However, these theories differ very little from non-Marxist theories except they attempt to link the concepts of these theories (e.g., strain, social control, peer group influences) to the broader context of the political economy of the particular society they are examining (Vold, Bernard, and Snipes 1998). For instance, Young (1994) stated that criminal behavior is simply the operation of capitalist principles, albeit in illegitimate ways, in that criminals are simply investing their labor for a return. Only a slight turn of phrase differentiates this from Merton's innovators who are seeking by illegitimate means what their capitalist culture has taught them to want.

Class and Crime

Another important way in which Marxist and non-Marxist theories of crime differ is the links between social class, values, and crime. There can be little argument that social class is an important concept in criminology;

it plays some role in many sociological theories of crime and delinquency. But there is a fundamental difference between the way theorists such as Merton and Sutherland and Marxist theorists view the class/crime nexus. For subcultural theorists, crime and delinquency are motivated by conformity to lower-class values and beliefs, meaning that people commit crimes because they have learned that it is something almost demanded by their class heritage. This is an idealist position; i.e., how a person *thinks* causes his or her conduct. It is crucial for Marxists to counter this view, for if crime is really a valued social activity in some settings it would mean that it is not an indicator of alienated social relationships, a concept central to their explanation of crime.

Marxists accuse cultural and subcultural theorists of never identifying the material origins of the values and beliefs thought to generate crime, and are skeptical of values and beliefs as behavioral motivators, especially if one neglects to point out their origins. All behavior, as well as the values and beliefs said to motivate it, must be viewed as generated by the concrete, material conditions of social life; i.e., people have the values they have because they occupy a particular place in the socioeconomic structure; they don't occupy that place because of the values they have (idealism). Values are simply a reflection of self-interest; therefore we cannot stop crime by changing values. It is the *material* source of criminal values that must change, and this will only occur, as leftists used to be fond of saying, with the collapse of capitalism and the birth of socialist society.

Vold, Bernard, and Snipes (1998:267) agree with the preceding when they write: "Marxist theories describe criminal behaviors as the rational response of individuals confronted with a situation structured by the social relations of capitalism. This view is consistent with the general view found in Marxist theory that, in general and in the long run, individuals act and think in ways that are consistent with their economic interests." Vold and his colleagues' statement is a reiteration of the proposition advanced by Marx that people support class interests only as long as those interests correspond with their own. Darwinians have no quarrel with this position. As emphasized in Chapter 5, the struggle for status is ubiquitous in primate species and in almost all mammalian species. Animals, including humans, form and break alliances (classes) with changes in status threats and opportunities. The pursuit of status—the coinage of reproductive success—is the primary reason that conflict among individuals and coalitions will always exist except in conditions of brutal state repression.

LEFT REALISM

Left realism is a branch of Marxist criminology disenchanted and somewhat embarrassed by left idealists who overlook the real pain and suffering of crime victims in their obsessive "critiques" of capitalism and capitalist criminal justice. Left realists may agree with the argument that the "real" crime is committed by politicians and the wealthy upper classes, but insist that it is not an argument for taking street crime any less seriously or for romanticizing it. Crime reduces the quality of life for everyone in all classes, but it is overwhelmingly directed at working-class victims and hardly ever touches the wealthy (Catalano 2006). It was this point more than any other that led to the movement among radical criminologists that became known as left realism, although they also wanted to make sure that right-wing realists did not monopolize the crime-control agenda.

Left realist criminologists believe that the path of least resistance is to work within the system. They had to acknowledge that predatory street crime is a *real* source of concern among the working class, who are the primary victims of it, and they have to translate their concern for the poor into practical, *realistic* social policies. In a statement that could have been made by a rational choice criminologist, left realist Steven Box (1987:29) asserted that people *choose* to act criminally, and "their choice makes them responsible, but the conditions make the choice comprehensible." Box is saying that people make choices for which they must be held accountable and not excused or romanticized. He is also saying that a variety of conditions make some choices more probable and understandable than others; i.e., people respond to their environments contingently.

This shift in left-wing thinking signals a move away from the left idealists' singular emphasis on the political economy to embrace the interrelatedness of the offender, the victim, the community, and the state in the causes of crime (the so-called "square of crime"). It also signals a return to a more orthodox Marxist view of criminals as people whose activities are against the interests of the working class, as well as against the interests of the ruling class. Although unashamedly Marxist in ideological orientation, left realists have been criticized by more traditional Marxists, who see left realism's advocacy of solutions to the crime problem within the context of capitalism as a sellout (Bohm 2001).

MAX WEBER ON HUMAN NATURE, ALIENATION, CONFLICT, AND CRIME

German lawyer, economist, and sociologist Max Weber is the other less revolutionary half of the view that society is best characterized by conflict. As is the case with Marx, social theorists argue about and struggle to understand the complexities of Weber's work. According to George Ritzer (1992:110), Weber is generally considered to be "the best known and most influential figure in sociological theory." I would seriously dispute the greater notoriety and influence of Weber vis-à-vis Marx, but Weber was more nuanced and realistic in his thinking and certainly less radical, but then Weber had 40 more years of social and scientific progress than Marx on to which he could base his work. Before I go on to discuss Weber's ideas about conflict and how they have come to influence modern conflict criminologists, a short point-by-point comparison of Marx and Weber on key concepts is in order.

We have already seen that while Marx saw cultural ideas as molded by its mode of production, particularly its economic system (materialism), and that for Weber a culture's economic system is molded by its ideas (idealism). Darwinians would side with Marx on this, but with Weber on his concept of society. Weber was less likely to fall into the trap of reifying collective phenomenon such as "society" or "class," seeing such things as "only a certain kind of development of actual or possible social actions of individual person" (1978:14). Darwinists would also agree with Weber that the tit-for-tat truck and barter of private enterprise is more in tune with evolved human nature than the command economies characteristic of socialist economies.

Human Nature

Weber had a more naturalistic view of human nature than Marx apparently did. Runciman (2001:14) points out that although Weber explicitly rejected Darwinian principles of selection and adaptation, he was nevertheless a selectionist in spite of himself in that "He still allows that in the unending competition for social advantage some individuals are better fitted than others to succeed in occupying coveted roles because of their inherited abilities and personal characteristics." Sanderson (2001:85) also points out that Weber frequently used selectionist logic in the sense that certain types of individuals are favored over others in the struggle for wealth and status in different types of societies, and that competition for these things (at least for status) will always be a constant under all social conditions.

Walter Wallace (1990:209) notes that "Weber quite unmistakably regards personal self-interest as innate and universal among humans." This anchoring of self-interest in biology is consistent with Emile Durkheim, and is somewhat consistent with Willem Bonger's view of egoism, but inconsistent with Marx's social constructionism because, as Wallace continues to explain, for Weber "personal self-interest is already fixed by genetic inheritance in all human individuals and needs no further fixing there by external imposition." Despite his overall faith in idealism, Weber asserts that self-interest trumps ideas in the governance of human conduct, an assertion that contradicts Weber's idealism and turns him into a materialist. In Weber's own words: "Yet very frequently the 'world images' that have been created by 'ideas' have, like switchmen, determined the tracks along which action has been pushed by the dynamics of interest" (in Wallace 1990:209).

A concern for self-interest is the essence of Weber's concept of *zweckrationalitat*, i.e., "instrumental rationality" or logical behavior that it is assumed will result in some positive outcome for the actor. Instrumental rationality is also at the heart of the classical criminologists' notions of human behavior—the desire to maximize pleasure and minimize pain. Weber augments this form of rationality with another form that is learned rather than innate called *wertrationalitat*. Wallace (1990:210) tells us that in *wertrationalitat* Weber "regards that end or value (which may well be duty-bound altruism and thus antithetical to personal self-interest) as learned and variable from one culture, subculture, or person to the next."

Wallace (and presumably Weber) clearly separates nature (*zweckrationalitat*) from nurture (*wertrationalitat*), but as we saw in Chapter 3, biologists and psychologists have shown this dichotomy to be false—altruism is far from being antithetical to self-interest. Altruism is learned and variable, but it is something for which we are genetically well-prepared to learn. Preparedness means that certain associations are learned more readily than others because of their relevance in evolutionary environments. Human beings act purposively in accordance with an evolved hierarchy of preferences and utilize evolved traits such as cooperative altruism in those endeavors.

Accordingly, some psychologists prefer the word *acquired* to describe the unfolding of evolutionarily relevant traits such as altruism in the sense that native language is acquired with ease rather than "learned" with some difficulty in the sense that, say, calculus is. Learning is easy when it comports with built-in circuitry as in language, but much more difficult when what we are learning is evolutionarily novel, like calculus. In such cases we have to co-opt circuitry that evolved for different purposes, rather like using pliers to do the job of a wrench; the pliers may get the job done, but certainly not with the wrench's ease. We also saw in Chapter 3 that while behaving altruistically (and in conformity with all the other "value traits" such as duty and honor that Weber was concerned with) confers large benefits on others, is also benefits the actor enormously by conferring status, respect, admiration, and affection upon him or her, all of which is extremely gratifying and draws others to him or her. In other words, even the noblest of ideas are at bottom, if one looks closely through an evolutionary lens, expressions of self-interest. When we get away from seeing self-interest as crabbed egoism (which *is* ultimately antithetical to self-interest) we see how it is vital to human sociality.

Weber perhaps understood the role of emotion in human affairs better than Marx. His concern with types of authority (traditional, charismatic, rational-legal) led him to view charismatic authority as dominated by emotional faith and the opposite pole of the dour rationalism of traditional and rational-legal authority. According to Turner (2007:370), "Weber hinted that charisma could not be fully understood without reference to biology." This is another false dichotomy in which "biological" emotion is opposed to "social" rationality (we can forgive this, however, since the opposition was commonly believed before modern neuroscience showed their complementarity). We have seen how emotion and rationality are intimately related and inseparable guides to our behavior and counterbalance to one another. Without the emotion/rationality partnership we would be ill-prepared to participate as decent human beings in social life. Recall that the defining neurophysiological trait of psychopaths is precisely the lack of this link and the stunted development of the social emotions (Wiebe 2009).

Alienation

Weber shared with Marx the view that the capitalist mode of production generated alienation, but he traced it to bureaucracy rather than to property relations (Sanderson 2001). He also differed from Marx's belief that alienation was specific to capitalism; rather, he simply viewed alienation under capitalism as a special case of a universal trend toward the rationalization and bureaucratization of the world. Closely akin to Marx's view, Weber characterized alienation as an "iron cage" that walled people off from each other and distorted their nature, but the cage "also provides openings for incursive awareness [in which] people could experience the stirrings of awareness" (Kalekin-Fishman 2006:526). Thus, alienation is seen by Weber more as a subjective appraisal of an objective condition, while for Marx it is an objective appraisal of one's objective class position, and can only be perceived as such.

The most striking difference between Marx and Weber relative to alienation is that while Marx saw capitalism as the root cause of alienation, Weber perceived capitalism as a *remedy* for alienation. For Weber: "an involved worker is a type of man, bred by free association in which the individual have [sic] to prove himself before his equals, where no authoritative commands, but autonomous decisions, good sense, and responsible conduct train for citizenship" (Sarfraz 1997:48). Weber reasoned this way because he saw growth of capitalism and state bureaucracies marching in lockstep with democratization, which implies greater liberty to make one's own decisions (Giddens 1977). Weber's stance vis-à-vis alienation is probably the reason why there is rarely any mention of the concept in the work of Weberian conflict criminologists. Crime for them is about conflict and little else, which makes most of them about as mono-causal in their thinking as most Marxist criminologists.

Conflict

Lopreato and Crippen (1999:213) view Weber's contribution to conflict theory as being "in some respects richer and also more kindred to an evolutionary perspective [because he] grants competition and conflict a broader and deeper role in social relationships and social development." For Marx, the only conflict that mattered was between two monolithic and homogeneous classes defined in solely economic terms. These classes were headed toward an inevitable Armageddon which was to result in the victory of the proletariat, after which all conflict would slowly wither away. Weber's ideas of social stratification were quite different, and therefore so were his ideas about conflict.

For Weber, social life is at bottom all about the pursuit of social power, which is in turn a way to acquire social honor (an evolutionary biologist would take it a step further and say that "social honor" is universally valued because it has always been the coinage of reproductive success). Weber demarcates three spheres in which this power is sought: class, status, and party. *Class* denotes economic power and is for Weber nonsocial or "barely social" because it is instrumental in orientation and lacks any sense of shared social belonging (a situation that Marx wanted to rectify: "A class *for* itself"). *Status,* on the other hand, is fully social because it is communal; i.e., status groups hold common values, express a common lifestyle, and possess a sense of belonging. *Party* represents political power and is based on associative social relationships to which people belong through free recruitment. Party associative relationships are rational-legal in outlook rather than communal and traditional, so are less "social" than status groups (Gane 2005).

Given this tripartite basis of stratification, it is no surprise that Weber viewed society as composed of multiple temporary as well as permanent alliances of self-interested individuals that coalesce and dissolve as their interests wax and wane. These coalitions exist within more permanent alliances such as economic classes, religious groups, political parties, and racial/ethnic groups. Many of these conflicts are considered just as important, and often more so, by their participants than Marx's economic class struggle (Sanderson 2001).

Weber's demarcation of power spheres is too radical in my estimation, for there is much overlap (although of course the class/status/party trichotomy is Weberian "ideal types" that serve as analytical constructs useful in comparing and contrasting concrete cases). We can be Marxists here and recognize that lifestyle (status) and political power (party) are determined to a great extent by economic power (class). It is also true that status and party power can lead to economic power (Gane 2005). Nevertheless, these ideal types, especially status, will serve us well in our discussion of conflict as a source of crime.

Marx's optimism that socialism would eliminate conflict (just as it would alienation) by ending social inequalities was seen as a biological fantasy by Weber. Weber (1978:39) wrote that "even on the utopian assumption that all competition were completely eliminated, conditions would still lead to a latent process of selection, biological or social, which would favor the types best adapted to the conditions, whether their relevant qualities were mainly determined by hereditary or environment." Thus Weber contended that conflict will always exist, regardless of the social, economic, or political nature of society. Furthermore, even though individuals and groups enjoying wealth, status, and power have the resources necessary to impose their values and vision for society on others with fewer resources, Weber viewed the various class divisions in society as normal, inevitable, and acceptable, as do many contemporary conflict theorists (Gane 2005).

Weber wrote even less about crime than Marx, although like Marx he saw the law is a resource by which the powerful are able to impose their will on others by criminalizing acts that are contrary to their class interests. Because of this, wrote Weber, "criminality exists in all societies and is the result of the political struggle among different groups attempting to promote or enhance their life chances" (in Bartollas 2005:179). With this one sentence Weber reduces criminal behavior to a single cause—conflict—just as Marx did with respect to capitalism.

MODERN CONFLICT THEORY

American criminologist George Vold was to Weber what Willem Bonger was to Marx, i.e., the first criminologist to apply Weber's ideas about conflict to the study of crime. In his book *Theoretical Criminology*, first published in 1958, Vold produced a version of conflict theory that moved conflict away from an emphasis on value and normative conflicts (as in the Chicago ecological tradition) to include *conflicts of interest*. Vold saw social life as a continual struggle to maintain or improve one's own group's interests, which ultimately meant one's own interests—workers against management, race against race, ecologists against land developers, and the young against adult authority—with new interest groups continually forming and disbanded as conflicts arise and are resolved rather than crystallizing into two colossal homogeneous and irrevocably antagonistic classes.

Conflicts between youth gangs and adult authorities were of particular concern to Vold, who saw gangs in conflict with the values and interests of just about every other interest group, including those of other gangs. Gangs are examples of *minority power groups*, or groups whose interests are sufficiently on the margins of mainstream society that just about all their activities are criminalized. Vold's theory concentrates entirely on the clash of individuals loyally upholding their differing group interests, and is not concerned with crimes unrelated to group conflict (Vold and Bernard 1986:276).

Minority power groups are excellent examples of Weber's status groups in which status depends almost solely on adherence to a particular lifestyle. In Weber's own words: "Status honour is normally expressed by the fact that above all else a specific *style of life* is expected from all those who wish to belong to the circle" (1978:1028). We have already discussed this kind of status group in terms of young males in so-called honor subcultures who literally risk life and limb in the pursuit of status as it is defined in those subcultures. The acquisition and defense of a "badass" reputation means so much more to these young men than any concerns related to economic status (class) (Anderson 1999). This latter point is perhaps why Marx would abhor the criminal activity in the present day inner cities of America as siphoning off the energies needed to fuel the revolution, and why he would abhor status group power as false consciousness blinding those who seek it to the realities of their class (economic) position.

Vold's thinking is in the Weberian tradition in that he viewed conflict as normal and socially desirable. Conflict is a way of assuring social change, and in the long run, a way of assuring social stability. A society that stifles conflict in the name of order stagnates and has no mechanisms for change short of revolution. Since social change is inevitable, conflict theorists maintain that it is preferable that it occur peacefully and incrementally (evolutionarily) rather than violently (revolutionarily). Even the 18th-century archconservative British philosopher Edmund Burk saw that conflict is functional in this regard, writing that "A state without the means of some change is without means of its conservation" (in Walsh and Hemmens 2008:246).

Conflict Theory and the Process of Lawmaking

Another way in which conflict criminology differs from Marxist criminology is that it concentrates on the *processes* of value conflict and lawmaking rather than on the social structural elements said to underlie those things. It is also relatively silent about how the powerful got to be powerful, makes no value judgments about crime (is it the activities of "social scum" or of "revolutionaries"?); conflict theorists simply analyze the power relationships underlying the act of criminalization. Conflict theorists tend to holds the idea that crime is a social construct with no intrinsic meaning. For them, criminal behavior is normal behavior subject to criminalization and decriminalization depending on the power relationship existing between those who commit certain acts and those whose interests are harmed by those acts. It is also fair to say that conflict theory really does not attempt to explain crime, but rather identifies social conflict as a fundamental fact of life that leads to activities that are contrary to the wishes of those with the power to criminalize them.

It has even been said that "Conflict theory does not attempt to explain crime; it simply identifies social conflict as a basic fact of life and a source of discriminatory treatment" (Adler, Mueller, and Laufer 2001:223). Crime is thus some act that is arbitrarily defined as such by those with the power to do so. Conflict theory's assumption that crime is just a "social construct" without any intrinsic properties minimizes the suffering of those who have been assaulted, raped, robbed, and victimized in any number of other ways. These acts *are* intrinsically bad (mala in se) and are not arbitrarily criminalized because they threaten the privileged world of the powerful few. Walsh and Ellis (2007) point to a large number of international studies showing that there is wide and robust agreement among people across classes and nations about what acts should be criminalized—laws exist to protect everyone, not just "the elite."

Conflict criminologists have a history of taking their "social construction of crime" argument too far by using mala prohibita crimes to make a point about all crimes—a classic "bait-and-switch" tactic. For instance, William Chambliss showed how the vagrancy laws in 14th-century England were implemented by the elite for the purpose of providing cheap labor for the elite landowners following the decimation of the population due to the Black Death, and concluded, "What is true of the vagrancy laws is also true of the criminal law in general" (in Nettler 1984:197). Thus, in common with Marxist theorists, many conflict theorists prefer case studies to illustrate some point to the empirical methods of mainstream criminologists. Such cases studies can be useful as long as the conclusions drawn from them are not stretched to the breaking point, which unfortunately they often are.

Saving the Left

When all is said and done, theorists in the critical camp are social structuralists working from a model of society that grants primacy to conflict rather than consensus. As structuralist theorists, Marxists believe that any meaningful changes in human behavior are unlikely to occur unless social structure changes first. The Marxist goal is to structure our societies so as to maximize equality and in such away that exploitation and alienation can no longer occur and conflict held to a minimum. Criminologist Richard Quinney (1975:1999) exemplified this position when he wrote: "Only with the collapse of capitalist society and the creation of a new society, based on socialist principles, will there be a solution to the crime problem."

If recent history is anything to go by, the collapse of capitalism in favor of socialism or any other economic alternative is not likely to occur in any advanced society. There are certainly many small anticapitalist movements that revolve around single issues such as opposition to globalization, but they have no credible or constructive alternatives to offer and thus go nowhere. People from all over the world risk everything they have and are in an effort to be part of the "dehumanizing" and "alienating" life of Western capitalist societies. It is in the capitalist societies of the modern world where "bourgeois" human rights are most respected and human wants and needs are most readily accessible. Richard Quinney himself has recognized this and has abandoned his call for social revolution in favor of a religious inner transformation of each individual as a "solution" to the crime problem (Lanier and Henry 1998). What Marxists such as Quinney did in the past (and some die-hard Marxists still do) was to compare the problems of existing capitalist societies (appropriately magnified, of course) with some utopian future communist society, not with the actual atrocities that owe their impetus to Marxist principles, whether we consider those principles to have been distorted or not.

One reason why so many of Marx's predictions never came to pass is that he vastly underestimated capitalism's ability to reform itself. Ironically, much of this reform was forced on capitalist regimes by the threat posed to them by the specter of communism, and for that we can all be thankful to Marx. For instance, Milton and Rose Friedman (1980) tell us that almost every item on the 1928 economic platform of the American Socialist Party has been adopted in the United States. These items include a 40-hour workweek, unemployment benefits, social security, public works, legal trade unions, child labor laws, and the establishment of government unemployment offices; these were co-opted by the Democratic and Republican parties because they were demanded by the American people as just. Other reforms predating 1928, such as a graduated income tax, free education for all children, and the abolition of child labor, while not original with Marx, were part of his social agenda. These things have been so integrated into American life that few today would call them "socialist" or "un-American," or even imagine the whiff of Marx on them.

One does not have to be a radical, or even a liberal, to acknowledge that there are faults in capitalism that still remain, such as the subordination of almost everything to the god of the bottom line, and that they can and should be rectified to the extent that they can be without stifling the entrepreneurial spirit that benefits us all and while still retaining democratic freedoms. Capitalism has shown many times over the past 200 years that it can reform itself. As left realists understand, all such reforms have come from within capitalism, not by movements that stand in opposition to it and outside of it, although fear of these movements has often provided the necessary impetus. If needed changes are to be made, the left (and changes will probably only come at the instigation of the left) will need a more realistic view of what changes are possible and desirable, and thus a more realistic view of human nature.

Peter Singer wants to rescue his beloved intellectual left by presenting to his leftist colleagues what he acknowledges is "a sharply deflated vision of the left, its utopian ideals replaced by a coolly realistic view of what can be achieved" (2000:62). Singer then offers his fellow leftists a number of propositions that a "Darwinian left" would and would not accept (2000:60–62). I concur with Singer that to the extent that a Darwinian left accepts these propositions they will enhance both their reputations as scientists and their chances of having their policy recommendation taken seriously. According to Singer, a Darwinian left *would not*:

1. Deny the existence of human nature, nor insist that it is inherently good and infinitely malleable.
2. Expect to end all conflicts between humans by revolution, social change, or better education.
3. Assume that all inequalities are due to such things as prejudice, discrimination, oppression, and social conditioning.

On the other hand, a Darwinian left *would*:

1. Accept the reality of human nature and seek to understand it so that leftist policies can be grounded in it.
2. Reject any inference from what is "natural" to what is "right."
3. Expect that under any social, political, or social conditions, many people will act competitively to enhance their own status and power and those of their kin.
4. Expect that under any social, political, or social system, most people will respond positively to opportunities for beneficial cooperation.
5. Promote structures that foster cooperation and attempt to channel competition in prosocial directions.
6. Stand by the traditional values of the left by supporting the weak, poor, and oppressed, but think carefully about whether proposed changes will really benefit them.

In summary, Marxism and Darwinism are not natural enemies; they have simply become estranged due to a number of unfortunate misunderstandings. Learning more about biosocial approaches can only have a beneficial affect on the theory and practice of the left. Marxists may well be more interested in changing the world than in interpreting it, as Marx asserted in the eleventh of his theses on Feuerbach, but surely change can be more easily accomplished and be more acceptable to all if predicated on a strong biosocial knowledge of human nature as revealed to us by the modern natural sciences. The left has nothing to fear from such a human nature. In fact, Peter Grosvenor (2002:446) opines that "the intellectual left is likely to be the prime beneficiary [of a Darwinian worldview] if the social sciences and the humanities can be rescued from residual Marxism and obscurantist postmodernism." As Robert Wright (1995) pointed out to us in Chapter 3, it is surprising how far to the left a modern Darwinian view of the human mind can take us.

10 Feminist Criminology and Gender

Broadly defined, feminism is a set of theories and strategies for social change that take gender as their central focus. In common with Marxism, feminism is both a social movement and a worldview of which criminology is but a small part. A prominent feminist theme is that women suffer oppression and discrimination in a society run for men by men who have passed laws and created customs to perpetuate their privileged position. Feminist criminologists take these core elements and apply them to criminology.

There are many varieties of feminist criminology just as there are of "male" critical criminology, and they are just as hard to capture with a single stroke of the pen. These varieties (liberal, radical, socialist, Marxist) and various positions in between have at least one thing in common with critical criminology: they are largely opposed to mainstream ("malestream"?) culture. The extent of the opposition varies according to the faction, but all are conflict-oriented and shift from class and power to emphasize gender and power. Feminists see women as being doubly oppressed by gender inequality (their social position in a sexist culture) and by class inequality (their economic position in a capitalist society). Class-based conflict in the context of capitalism gives way to gender-based conflict within the context of a patriarchal society, and the bourgeoisie and powerful interest groups of male critical theorists become gendered, and defined as "rich white *men*" (Sokoloff and Price 1995:14).

As expected, because most feminists are clearly at one with critical criminologists, they tend to discount biological explanations for gender differences. However, just as there is nothing in Marxist and critical/conflict theories that is *intrinsically* incompatible with biology, there is nothing incompatible with it in feminist theories. Indeed, feminist criminologists would benefit from biosocial explanations of their major issues and concerns more than criminologists from any other theoretical tradition. Many feminists such as Anne Campbell, Diane Fishbein, Sara Hrdy, Terrie Moffitt, Alice Rossi, and Barbara Smuts are leading figures in biosocial science.

Feminist criminologists complain that female crime has been virtually ignored by mainstream criminology. They want to put women on the criminological agenda and to interpret female crime from a feminist perspective. The two major issues in feminist criminology are thus: (1) Do traditional male-centered theories of crime apply to women? And (2) what explains the universal fact that women are far less likely than men to involve themselves in criminal activity (Price & Sokoloff, 1995). The first issue does not concern us; the second does.

GENDER DIFFERENCES IN CRIMINAL BEHAVIOR

At all times and in all places males commit the overwhelming proportion of criminal offenses (Campbell 2009). Figure 10.1 shows percentages of males and females arrested for seven of the eight FBI index crimes (FBI 2007) in 2006 (rape is omitted because the 77-fold difference cannot be sensibly graphed without distorting the remaining comparisons). Note that there are about nine males for every one female arrested for murder and robbery, but less than two male for every female arrested for larceny/theft. Thus the more serious, brutal, and violent the offense, the more males dominate in its commission. For instance, a study of 1,072 gang-related homicides in three cities found that female gang members committed only eight (0.7%) of them (Shelden, Tracy, and Brown 2001), and a four-year study from Chicago found that only one gang-related homicide out of 345 was committed by a female (Spergel 1995).

It would be difficult to propose a viable structural explanation for gender differences in crime rates because female offenders are overwhelmingly found in the same places as their male counterparts, that is, among single-parent families located in poor, socially disorganized neighborhoods: "Males and females are not raised apart and exposed to an entirely different set of developmental conditions" (Bennett, Farrington, and Huesmann 2005:280). Campbell, Muncer, and Bibel (2001:484) report Pearson correlations between male and female violent and property crime rates in the United States of .95 and .99, respectively. They report almost identical correlations from British data (.98 and .99), and that the average correlations across a number of countries for a variety of crimes are all in the mid to upper .90s. These correlations tell us that no matter how wide the gender gap is across and within nations, male and female crime rates march in lockstep; that is, large gender differences in crime exist over a wide range of social and cultural conditions even though these conditions affect the crime rates of both sexes similarly. Furthermore, the individual-level correlates of male offending (e.g., low self-control, conduct disorder, ADHD) are the same for both sexes, although more males than females have these things and are more affected by them (Moffitt et al. 2001). These data suggest very different sex-based threshold levels for criminal behavior; i.e., while the same factors influence offending for both sexes, they apparently need to be more severe and protracted for females to cross the law abiding/criminal offending threshold. Before we venture into that territory, however, we should examine the typical feminist alternative explanations for gender differences in all kinds of attitudes, aptitudes, and behavior.

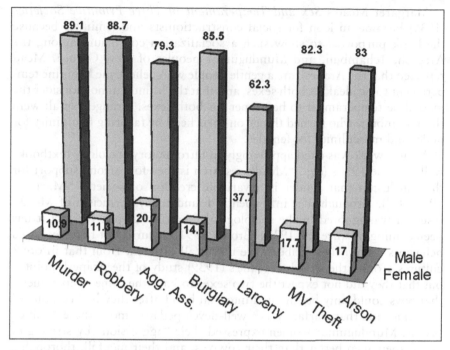

Figure 10.1 Arrests for index crimes in 2006: male/female percentages. Source: FBI (2007). *Uniform Crime Reports: Crime in the United States, 2006.*

Gender and Socialization

Psychologist Sandra Bem (1987) once famously claimed that sex-differenti-ated socialization is an entirely arbitrary historical accident, and softer versions of this position still pervade feminist sociology (Kennelly, Mertz, and Lorber 2001). From this perspective, sex-differentiated evolutionary pressures leading to sex-differentiated neurohormonal structure and functioning have no purchase, and so must be discounted. To illustrate the lengths to which the social construction of gender position sometimes goes, Lopreato and Crippen (1999:143) quote fellow sociologist Judith Lorber (1994:46) as writing that "a purely biological substrate [of gender] cannot be isolated because human physiology is socially constructed and gendered." They then comment on this gem of postmodernist wisdom by writing: "One wonders whether members of the medical profession are aware of this extraordinary discovery." If it were the case that gender differences were arbitrary social constructs decoupled from biology, then simple probability tells us that about one-third of cultures would have decided to socialize the sexes in traditional ways, one-third would have decided to reverse this process and socialize males in female-typical way and females in male-typical ways, and in about one-third of the cultures socialization would be sex-neutral.

Margaret Mead's *Sex and Temperament in Three Primitive Societies* (1935) became an icon for social constructionists and feminists because the book purported to show such a socialization continuum among the Arapesh, Tchambuli, and Mundugumor peoples of New Guinea. Mead reported that the Arapesh are a gentle people who believe the feminine temperament to be ideal for both sexes, and that the Mundugumor consider the masculine temperament to be proper for both sexes. Strangest of all were the Tchambuli, who turned things on their head by favoring femininity for males and masculinity for females.

Mead's work was cited approvingly in introductory sociology textbooks well into the 1990s (e.g., "Mead's research is therefore strong support for the conclusion that gender is a variable creation of society" [Macionis 1989:317]), although it is infrequently found today. Criticism of Mead's research was ignored in the sociological literature despite the fact it has been coming in since the 1930s (Brown 1991). It turns out that sex-based behaviors within these tribes were not at all different from that in other cultures around the world. Fortune's (1939) study of the Arapesh pointed out that they did not expect the two sexes to have the same temperament, that boys could only be initiated into manhood after they had committed homicide, and that warfare was a well-developed art among these "gentle" people. Mundugumor women expressed their "aggression" by striving to please their men better than their cowives, and their menfolk thoroughly dominated them. Deborah Gewertz's (1981; Gerwertz and Errington, 1991) fieldwork among the Tchambuli showed it to be a thoroughly male-dominant society where "aggressive" behavior on the part of women earned them a beating from their "feminine" husbands.

Perhaps the most devastating blows to the social construction of gender argument to come from the social sciences were Medford Spiro's (1975, 1980) studies of Israel's kibbutzim. The kibbutzim movement provides us with a natural cultural experiment, the scope and size of which could never be duplicated by scientists, with which to explore the gender-as-product-of-socialization hypothesis. Begun in 1910 and heavily influenced by the Marxism of Russian immigrants, the purpose of the communal movement was to strip its members of all vestiges of "bourgeois culture" and to emancipate women from their socioeconomic and sexual shackles. Boys and girls were raised collectively, taught the same lessons, given equal responsibilities, shared the same toys, games, living quarters, and even the same toilets, dressing rooms, and showers. All this sex-neutral socialization was supposed to result in androgynous beings devoid of observable differences in nurturance, role preferences, empathy, aggression, and so forth.

Spiro believed in the social construction of human nature and thus thought that he was setting out to discover and document the dimension of the changes in human nature brought about by the movement. What he actually found forced on him what he describes as "a kind of Copernican revolution on my own thinking" (1980:106). "As a cultural determinist," he wrote, "my aim in studying personality development in 1951 was to observe the influence of culture on human nature or, more accurately, to discover how a new culture produces a new human nature. In 1975 I found (against my own intentions) that I was observing the influence of human nature on culture" (1980:106).

What Spiro found was a counterrevolutionary feminization of the sabra (kibbutzim-born and -reared) women. Despite the decades of sex-neutral socialization and the exhortations of their ideologically committed fore-mothers, sabra women fought for formal marriage vows, for greater contact with their children, for separate toilets, showers, and living arrangements prior to marriage, for modesty of dress in the company of men, and for pos-session of "bourgeois" means of enhancing female charms. Even in an ear-lier work, Spiro (1975) found that the activities and fantasy lives of young children varied significantly between the sexes despite adult role models striving assiduously to eliminate those differences. As a reluctant apostate, Spiro could not fully bring himself to impose a biosocial interpretation on his data, preferring to write that his data supported the notion that "sexually appropriate role modeling is a function of precultural differences between the sexes" (1980:107). Yet what else could "precultural" differ-ences mean other than biological differences?

This is not to say that gender is not variable; culture can and does mold masculine and feminine characteristics in diverse ways, but the biasing framework supporting these characteristics is plainly biological. In other words, biological sex is a constraint on just how malleable gender can be (Udry 2000). Mead's own later work supported this when she wrote: "If any human society—larger or small, simple or complex, based on the most rudimentary hunting and fishing, or on the whole elaborate interchange of manufactured products—is to survive, it must have a pattern of social life that comes to terms with the differences between the sexes" (1949:173). She even traced these differences to "sex differentiated reproductive strategies" (1949:160). While this is fully consistent with biological science, such state-ments tend to raise hackles among gender feminists regardless from whose pen they come.

The major proponent of gender neutrality at birth in the 1970s was psy-chologist John Money. Money's ideas about gender fluidity came under much criticism from the medical and scientific communities after informa-tion emerged about a former patient/research subject of his named David Reimer came to light (Swaab 2007). David, one of a pair of MZ twins, had his penis mutilated in a botched circumcision. David had a vagina fashioned for him, was pumped full of female hormones, and reassigned as a female with the name Brenda. Money assured Brenda's parents that since gender was primarily a matter of socialization, if they raised the child as a girl she would become a well-adjusted woman. All efforts to turn David into Brenda failed dramatically in every respect. When he learned of his medi-cal history he expressed relief and underwent further surgery to construct a penis. The whole tragic story (David committed suicide in 2004) is told in Colapinto (2006). Suffice to say that Money came to reject the notion of gender neutrality at birth: "Clearly, the brain holds the secrets of the etiology of gender identity differentiation" (Money 1986:235).

It would seem that all available credible evidence points ineluctably to the conclusion that sex/gender socialization is far from arbitrary. Parents in all cultures socialize males and females differently because they *are* different—the biological dog wags the cultural tail; the tail does not wag the dog. Socialization patterns, Sanderson (2001:198) insists, "simply represent social confirmation of a basic biological reality that is easily recognized by people in all societies." As an endless sea of parents would attest, children's sex-based behavior elicits different parental responses; this is evocative rGE in action (Leaper and Smith 2004). In accordance with the multiplier effects of evocative rGE (Dickens and Flynn 2001), gender differentiated parental responses reinforce and accentuate the behaviors that evoked them. This interplay of nature and nurture assures that gender-typical traits are not discrete categories but rather a series of continua.

Gender traits are gender-related, not gender-specific, even for the most gender-differentiated traits such as aggressiveness, sensation seeking, and risk taking—the existence of numerous Daphne Daredevils and Harvey Milquetoasts attest to this. We also saw in the last chapter that unusual cultural conditions (high rates of domestic abuse and the availability of guns) can lead females of one group (African Americans) to have consistently higher rates of homicide than males in other groups (Asians and whites), even though the homicide gap between black males and females is greater than the gap in other races. In short, robust *average* gender differences are noted in all human cultures from the earliest days of life in such things as risk taking, aggression, empathy, and self-control. These differences are dramatically underscored during the teen years, and are observed in all primate and almost all mammalian species (Archer 2006; Geary 1998) for which no one (we assume) would invoke socialization.

TYPICAL FEMINIST EXPLANATIONS FOR THE GENDER RATIO PROBLEM

The fact of huge gender differences in criminal behavior is not in dispute, although explanations of it are. Consistent with the traditional sociological view that gender differences in any behavior, aptitude, and attitude are products of differential socialization, feminist criminologists explain the gender ratio issue in like terms; i.e., males are socialized to be aggressive and dominant, and women are socialized to be nurturing and conforming. This suggests that if females were socialized in the same way as males (or vice versa) their rates of offending would be roughly the same. This idea was first seriously expressed by Freda Adler in her book *Sisters in Crime*

(1975). Adler advanced a *masculinization hypothesis* to account for the rise in female crime rates in the 1960s and 1970s marching in lockstep with rises in male rates. Adler interpreted the increase to mean that an increasing number of females were adopting male roles as the American occupational structure changed. The adoption of male roles, she asserted, leads to masculinized attitudes and eventually to greater female crime. Adler argued that "determined women are forcing their way into the world of crime," and added with apparent relish that "increasing numbers are women are using guns, knives, and wits to establish themselves as full human beings, as capable of violent aggression as any man" (1975:15).

Adler's masculinization thesis was not well received by many feminists who did not share her peculiar opinion of how women might establish their humanity, and who also perceived it as providing ammunition for those who oppose women's liberation ("We're asking for more crime by putting women in the work force"). Nor has the thesis been supported by research. Female crime in the United States has increased over the past 30 years, but as a proportion of total arrests it has not varied by more than 5 percentage points, and the male/female gap—which is the real issue—has remained essentially unchanged (Campbell 2009; Steffensmeier et al. 2006).

Darrell Steffensmeier and his colleagues (2006) compared the masculinization hypothesis with a policy change hypothesis to account for an observed diminishing ratio of female to male assault arrests documented in the Uniform Crime Reports (UCR). Using data from the UCR and the National Crime Victimization Survey (NCVS) from 1979 to 2003, they noted that the UCR data showed a 60% rise in the percentage of females arrested for simple assault over the period. At first blush this convergence of arrest rates may be considered indicative of female attitude change (masculinization), but the fly in the ointment was that there was no corresponding convergence for murder, robbery, or aggravated assault evident in the UCR data. The alternative hypothesis is that the convergence is a by-product of police policy changes (net-widening to include arrest for minor physical assaults, particularly in domestic situations).

Furthermore, the alleged increase in female violence documented in the UCR was not borne out by NCVS data collected over the same period (i.e., the ratio of female to male assaults fluctuated only randomly). NCVS data are considered more reliable to assess such matters because it is independent of criminal justice policy shifts. Steffensmeier et al. (2006) concluded that net-widening policy shifts have elevated the arrest proneness of females who commit physical attacks/threats of marginal seriousness. In other words, women are not becoming more violent; rather, official data increasingly mask differences in violent offending by men and women due to policy shifts. If women had become more masculinized, we would have observed significant increases in female arrests for other violent crimes (homicide, robbery, aggravated assault) that have not been subjected to police arrest policy changes but we have not.

264 Biology and Criminology

Efforts to account for gender differences that go beyond structural changes include that offered by Mears, Ploeger, and Warr (1998), who state the genders differ in criminal behavior because they differ in exposure to delinquent peers, and that males are more likely to be affected by delinquent peers than females. They also contend that the greater sense of morality and greater empathetic and cognitive skills among females have a strong inhibitory affect. These explanations have little substance beyond stating in a roundabout way that "boys will be boys," and "girls will be girls." What we would like to know is *why* males are more likely to be exposed (or to expose themselves) to delinquent peers and *why* they are more affected by them once exposed. We should also like to know *why* females have greater empathy, a greater sense of inhibitory morality, and greater cognitive skills. Answering these kinds of questions requires us to venture into the domains of neuroscience, endocrinology, and evolutionary biology.

Mears, Ploeger, and Warr do not venture into these realms; rather, they couch their explanations in terms of sex-role socialization. They tell us that females are not as likely to be exposed to delinquent peers because they are more strictly supervised, and they are more morally inhibited because they are socialized more strongly to conformity. These factors certainly impact decisions to misbehave for both genders, but controlling for supervision level results in the same large gender gap in offending; i.e., comparably supervised boys have higher rates of delinquency than girls (Gottfredson and Hirschi 1990). Further adding to the misery of this position is a meta-analysis of 172 studies that found a slight tendency for boys to be *more* strictly supervised than girls (Lytton and Romney 1991).

Another surprising finding from Lytton and Romney's meta-analysis was how few parent-initiated behaviors and socialization patterns were gender differentiated. One gender-differentiated behavior was that parents displayed more warmth to girls than to boys, the obvious explanation for which is that parental warmth is an evoked response to the greater warmth girls displayed toward their parents. Subsequent studies have shown that large sex differences in antisocial behavior exist regardless of level of supervision and whether or not the family is patriarchal (implying strict control of females) or egalitarian (implying more equal treatment of boys and girls) (Chesney-Lind and Shelden 1992). As Dianna Fishbein (1992:100) sums up the issue: "Cross cultural studies do not support the prominent role of structural and cultural influences of gender-specific crime rates as the type and extent of male versus female crime remains consistent across cultures."

BIOSOCIAL EXPLANATIONS

Addressing the gender-ratio issue, Steffensmeier and Allan (1996:464) write: "If the gender gap had a biological basis, it would not vary, as it does, across time and space." In other words, if biology played any role at all, the gender gap would be precisely of the same magnitude at all times and all places. This may be a logical conclusion for anyone who believes that biology is destiny. Steffensmeier and Allan do not grasp the logic that it is precisely the fact that the magnitude of the gender gap varies across time and space (indicating different cultural and other environmental influences) and yet still remains wide at all times and in all places that biological factors *must* play a very large part.

The gender gap, especially for violent crime, is so universally pervasive that the most logical explanation must lie in fundamental innate differences between the sexes. The notion that gender differences are "socially constructed" grates against so much natural science data that advocates of such a position are simply not taken seriously outside of their own circle, probably not even by the majority of nonfeminist social scientists. If gender socialization, supervision levels, and so on were important in accounting for gender differences in criminal behavior there should be a set of cultural conditions under which crime rates would be equal for both sexes, or even under which female rates would be higher, but as Fishbein pointed out earlier, no such conditions have ever been discovered. Even so, we continue to suffer the irritation of social scientists inferring the existence of gender-differentiated norms from gender-differentiated behavior that are supposedly explained by those norms. As Richard Udry (1994:563) points out: "The reason for this tautology is that we, as social scientists, can't think of any other way to explain sex differences." We never will until we break the chains of discipline parochialism and venture into the exciting worlds of neuroscience and evolutionary biology.

THE NEUROHORMONAL BASIS OF GENDER

Biosocial explanations of gender differences in many different spheres of human life rest on a foundation of differential neurological organization that reflects the influence of prenatal hormones, which in turn reflect sex-specific evolutionary pressures. There are numerous gender differences in all kinds of traits, most of which are small and inconsequential. The largest differences are those that are most salient to core gender identity, i.e., at the center of one's identity as male or female (Hines 2004). These core differences are precisely the traits that are most strongly related to criminal behavior, such as aggression, dominance, empathy, nurturance, and impulsiveness, all of which reflect evolutionary pressures that impinged on males and females differently. According to Doreen Kimura, evolutionary pressures assure that males and females come into this world with "differently wired brains," and these brain differences "make it almost impossible to evaluate the effects of experience [the socialization process] independent of physiological predisposition" (1992:119). Sarah Bennett and her colleagues (2005:273) explain the pathways from sex-differentiated brain organization to antisocial behavior:

Males and females vary on a number of perceptual and cognitive information-processing domains that are difficult to ascribe to sex-role socialization . . . the human brain is either masculinized or feminized structurally and chemically before birth. Genetics and the biological environment in utero provide the foundation of gender differences in early brain morphology, physiology, chemistry, and nervous system development. It would be surprising if these differences did not contribute to gender differences in cognitive abilities, temperament, and ultimately, normal or antisocial behavior.

The greater male susceptibility to criminogenic forces suggests that maleness per se is a potent risk for antisocial behavior, and maleness begins in the brain. All heritable traits positively associated with criminality are also related to gender, with males being much more vulnerable. Male vulnerability is largely a function of the differential exposure of male and female brains to androgens in utero (Cohen-Bendahan, van de Beek, and Berenbaum 2005). Thus if it can be said in any sense at all that there is a "crime gene," it is the SRY ("sex-determining region of the Y chromosome") gene found only in XY (male) individuals.

In all mammalian species maleness is induced from an intrinsically female form by the processes initiated by the SRY gene. All XY individuals would develop as females without the SRY gene, and XX individuals have all the material needed to make a male except this one gene. The major function of the SRY gene is to induce the development of the testes from the undifferentiated gonads rather than ovaries that would otherwise develop. The testes then produce androgens, the major one being testosterone, which, when transformed by the enzyme aromatase into estrogen, will masculinize the brain. The testes also produce Mullerian inhibiting substance (MIS), which causes the atrophy of internal female sex organs (Swaab 2004).

Hormones play tremendously important roles in sex-differentiated behaviors and traits of all kinds. The major biological factor said to underlie gender differences in aggression, violence, and general antisocial behavior is testosterone (T) (Kanazawa 2003). As noted in earlier chapters, no one claims that T is a major independent and direct cause of criminal behavior, only that it facilitates such behavior and that it is the major factor that underlies gender *differences* in criminal behavior. It is not so much the activating effect of T at puberty as it is its organizing effects in utero, because without the brain being already organized along male lines the pubescent surges of T would have nowhere to go (Sisk and Zehr 2005).

Perhaps the best way to study neurohormonal effects of gender differences in behavior is to study what happens when the process of "sexing" the brain goes awry, thus producing natural experiments that reveal the behavioral effects of prenatal hormones. On rare occasions XX (female) infants are born with ambiguous genitalia, prompting physicians to ponder what sex these infants are. Indifferent to the warring factions of social science, physicians use a classification scheme based on the appearance of the external genitalia to determine sex of rearing (Luks et al. 1988). Their only motive is a concern with the practical issue of raising children in a manner consistent with the degree of their brain virilization (how much the brain has been sculpted along male lines), and the degree of genital virilization provides an indication of the extent of this. That is, the more masculinized the genitalia, the more likely the brain has been masculinized; the more feminine the genitalia, the less brain masculinization has taken place. However, because even the most virilized of girls have internal female organs capable of reproduction, most authorities recommend female assignment via hormonal treatment and surgery despite the high risk involved of such girls later rejecting that assignment (Byne 2006).

Intersex anomalies (pseudohermaphrodites) are "deviations from the norm," and as such may provide us with important clues to the biological substrates of sex-typical behavior, just as mutants serve biologists to clarify the species norm and brain-damaged patients provide valuable information about the organization of the normal brain (Gooran 2006). Brain masculinization is not an all-or-nothing process; it is one that describes a continuum that may contain significant male/female overlap. Gender lies on a continuum from extreme femaleness (complete absence of androgens) to extreme maleness (significantly above-average levels of androgens). The female fetus is protected from the diverting effects of androgen, but not completely. However, once prenatal androgens have sensitized receptors in the male brain to their effects, they activate the brain to engage in male-typical behavior when the brain receives its second great surge of androgens at puberty.

Androgen-Insensitivity Syndrome (AIS)

AIS is a syndrome in which the receptor sites of XY males that normally bind androgens are partially (PAIS) or completely (CAIS) inoperative. If the receptors are completely inoperative, the male genotype develops a completely female phenotype. Because CAIS individuals have the SRY gene, they have androgen-producing testes (undescended), but their androgen receptors are insensitive to its effects and thus the internal male sex structures do not develop. Since the testes secrete normal amounts of MIS, CAIS individuals have neither male nor female internal sex organs, although the external genitalia are unambiguously female. Unresponsive to the masculinizing effects of androgens on the brain, CAIS individuals conform to typical behavioral patterns of normal females, often exaggeratedly so (Jurgensen et al. 2007). PAIS individuals tend to be behaviorally intermediate in terms of gendered behavior.

Congenital Adrenal Hyperplasia (CAH)

CAH is an autosomal recessive trait that prevents the synthesizing of cortisol from androgen by the adrenal cortices. In about 90% of the cases, the problem is the result is a deficient 21-hydroxlase enzyme that impairs the synthesis of cortisol, the unused precursors of which then enhance the production of androgens from the adrenal glands (Meyer-Bahlburg et al. 2006). These excess androgens result in extremely precocious sexual development in males and variable degrees of masculinization of the genitalia and brains of females. CAH females engage in more male-typical behavior and possess more male-typical traits than non-CAH females, such as a liking for rough-and-tumble play, better visual/spatial than verbal skills, lower maternal interests, less interest in marriage, a greater interest in careers, and a greater probability of bisexuality of homosexuality (about one-third of them so describe themselves [Gooren 2006]). Although there are no studies directly assessing criminal behavior among CAH women, because they are higher than hormonally normal women on traits positively associated with crime, and lower on traits negatively associated with crime (maternal interest, commitments to relationships), they are likely to be present in female criminal populations in numbers relatively greater than their number in the general population.

Klinefelter's Syndrome (KS)

KS is the most common human chromosomal anomaly, occurring in about 1 per 100 live male births (DeLisi et al. 2005). KS males have two or more X chromosomes and one or more Y chromosomes (no matter how many X chromosomes present, the SRY gene on a single Y chromosome will move the phenotype in a male direction). They tend to be taller than normal males, have smaller than normal genitalia, to develop relatively well-formed breasts at puberty, to have about half the male postpubertal amount of testosterone, a low level of sexual activity, and to have difficulty forming pair-bonds (DeLisi et al. 2005). Understandably, KS males often have sexual identity problems, and are significantly more likely than XY males to be homosexual, bisexual, and inmates in prisons and mental hospitals. KS individuals tend to be unaggressive and passive, and their crimes are almost always nonviolent or of a sexual nature (Harrison, Clayton-Smith, and Bailey 2000).

The XYY Syndrome

The XYY syndrome is the anomaly that has generated more interest than any other; it has a rate of approximately 1 per 1,000 male birth (Briken et al. 2006). The XYY male is not a "supermale" or a homicidal maniac, but he is significantly more likely than XY males to evidence exaggeration of male-typical behavioral traits. Most descriptions of the behavioral phenotype suggest aggressiveness, hyperactivity, impulsiveness, a PIQ > VIQ intellectual profile, and to show atypical brain-wave patterns relative to XY males; although these differences are not large, they are significant (Briken et al. 2006; Nicolson, Bhalerao, and Sloman 1998; O'Brien 2006). Plasma testosterone concentrations of XYY men are usually found to be significantly higher than in matched samples of normal XY men. Although most XYY males lead fairly normal lives, they are at elevated risk for a diagnosis of psychopathy and for committing sex crimes, and they are imprisoned or in psychiatric hospitals at rates greatly exceeding their incidence in the general population (Briken et al. 2006).

This brief description of some intersex anomaly syndromes show that traits associated with antisocial behavior are predictable from androgen levels across the gender continuum. CAIS individuals are the least masculine in terms of androgen levels, and are the most socially conforming of all individuals on the continuum. Normal XX females have higher androgen levels and higher levels of deviance than CAIS individuals, but as a group they are still very much at the lower end on both. CAH females have both higher androgen levels and higher deviance levels than normal females, although they have lower levels than normal males on both. KS males have significantly less testosterone than normal males but significantly more than normal females, and have a higher level of deviance than both normal females and normal males. If sexual deviance were omitted, KS males fall below the normal male average level of antisocial behavior. Normal males are next on the continuum of both androgens and antisocial behavior, and finally, XYY males have higher androgen levels and higher levels of behavioral deviance than normal males.

Although the facilitative influence of T does not appear strong enough to account for much variance in antisocial behavior when assayed solely within the normal male range, the normal male range is a restricted one in a larger distribution across the full range of human gender possibilities. Intersex anomalies reflect extreme variance in androgen levels ranging from complete insensitivity to it (CAIS) to the upper level of the normal male range (XYY males). The rarity of these anomalies renders the evidence sparse, but it is both consistent and convincing: Antisocial behavior and nonconformity increase as we move from extreme femininity to extreme masculinity as defined by androgen levels. While feminists may continue to insist that the male/female gap in antisocial behavior of all kinds is explicable in terms of sex-role socialization, it would require extraordinary sophistry to similarly explain the pattern revealed here. Social science has no theory capable of making sense of this pattern, but an understanding of the behavioral organizing effects of gonadal steroids circulating in the developing fetal brain renders it fully intelligible (Walsh 1995c).

EXPLORING MALE/FEMALE DIFFERENCES IN CORRELATES OF CRIMINAL BEHAVIOR

This section provides a brief overview of gender differences in some of the major correlates of criminal behavior. I will rely primarily of the reviews published by Bennett, Farrington, and Huesmann (2005) and Campbell (2006), adding supporting material when necessary. These reviews supply voluminous supporting references for each of their major points, but in the interests of brevity, these sources will be omitted. I will simply cite Bennett, Farrington, and Huesmann (2005) or Campbell (2006) and direct the interested reader to the original articles. I begin with general brain development and cognitive skills.

Brain Development and Cognitive Skills

It is well established that the higher brain regions (the cortex) develop sooner and faster in the female than in the male, and that males are at far greater risk for all sorts of neurological problems than females. Female brains are less lateralized than male brains, which means that female left and right hemispheres are less devoted to specialized tasks, and that both hemispheres contribute more equally to similar tasks than do the hemispheres of males (Luders et al. 2006). Active cooperation between the hemispheres leads to better social cognitive processing in females; i.e., women can more readily access and assess the rational and emotional content of social messages simultaneously.

Reflective of evolutionary pressures directed at caring for children and gathering food, females are more mindful of environmental stimuli. This is demonstrated in numerous studies of memory of spatial configurations in which females consistently outperform males. Greater female attention to environmental details may reflect greater augmentation capabilities of the reticular activating system in women. Females' greater attention to environmental stimuli may also account for their lower rates of boredom proneness relative to males, and may also constitute part of the reason why they are less prone to be sensation seekers than men.

In a book-length study of sex differences in antisocial behavior based on the Dunedin longitudinal cohort studies, it was found that family risk factors influenced both sexes roughly similarly with respect to antisocial behavior, but that neurocognitive risk factors (hyperactivity, temperamental deficits, poor impulse control, etc.) were experienced significantly more by males than females (Moffitt et al. 2001). This led Moffitt and her colleagues to conclude that "taken together, these risk factors accounted for most of the sex differences found in antisocial behavior" (2001:7). Recall from Chapter 5 that Moffitt's life-course persistent (LCP) offenders were defined by these neurological deficits. The male/female ratio of LCP offenders in the cohort was 10:1; the ratio of adolescent limited (AL) offenders was only 1.5:1 (2001:226). In a large (n = 13,000) Swedish cohort study which followed subjects to age 30, the ratios were 15.1 for LCP subjects and 4.1 for AL subjects (Kratzer and Hodkins 1999).

Psychopathy

Addressed in terms of a number of endophenotypic traits encompassing the psychopathy phenotype, Cale and Lilienfeld (2002) describe self-reports studies based on the Psychopathic Personality Inventory (PPI). The PPI contains such subscales as Machiavellian egocentricity, coldheartedness, fearlessness, impulsiveness, and stress immunity. A Cohen's *d* of .97 was found for sex differences in PPI scores averaged across all subscales, with males scoring significantly higher on all subscales. An effect size based on Cohen's *d* of .80 is considered to be "large" (Cohen 1992); thus we can say that the .97 effect size for sex differences on the PPI reported by Cale and Lilienfeld is "very large."

Empathy

Females are invariably found to more empathetic than males regardless of the tools and methods used to assess empathy (Campbell 2006), which may be traced to the effects of fetal T (Knickmeyer et al. 2006), and/or to higher oxytocin levels in females (Taylor 2006), and ultimately to sex-differentiated natural selection for nurturing behavior (Chapter 3). A study of 20 healthy females who received either a single sublingual dose of T or a placebo showed that the T administered group showed a significant reduction in empathetic responses to experimental stimuli (Hermans, Putnam, and van Honk 2006). Conversely, a study of 30 healthy males by Domes and his colleagues (2007) showed that a single intranasal dose of oxytocin significantly enhanced their ability to infer the mental states of others relative to a placebo control group. An fMRI double-bind study of 15 healthy males (Kirsch et al. 2005) showed that a single intranasal dose of oxytocin significantly reduced amygdala activity in response to angry faces and threatening scenarios relative to subjects receiving a placebo. Thus, males become more empathetic with the administration of oxytocin, and females become less so with the administration of T. All these responses took place outside conscious awareness because the target sites for both T and oxytocin are located in the limbic system.

Guilt Proneness

Guilt is a negative emotion experienced when one has committed (or contemplates committing) an act that may harm others. As one would expect, guilt is positively related to empathy (Silfver and Klaus 2007), and females are more guilt-prone than males about most, but not all, transgressions. Females are more likely to feel guilty for expressing angry aggression or for inconsiderate behavior, which is predictable both from their higher levels of empathy and from the fact that such behavior is in violation of gender roles. Sex differences in guilt are less in evidence in young children but become more pronounced with age (Bennett, Farrington, and Huesmann 2005).

Impulsiveness/Low Constraint/Low Self-Control

Although impulsiveness, low constraint, and low self-control are different constructs, they are similar enough to be treated as one construct for our purposes. All three involve disinhibited behavior in which the actor has been unable or unwilling to consider the long-term consequences of his or her behavior (Chapple and Johnson 2007). These constructs also contain elements of risk taking and sensation seeking. We have previously discussed how the various measures of behavioral inhibition (impulse control) are related to criminal and general antisocial behavior, so we will not do so again. Suffice to say that the authors of a meta-analysis that included close to 50,000 subjects concluded that it is "one of the strongest known correlates of crime" (Pratt and Cullen 2000:952). Humans learn to control their impulsive reactions to temptation and to aggressive responses to provocation via two interrelated inhibition routes. The first route is entirely visceral (sympathetic ANS arousal) in nature and is called *reactive inhibition*. Reactive inhibition is fear based and automatically impels attention to threats and initiates avoidance/withdrawal behavior in response. The second system is called *effortful control*. With the increasing ability to think before acting, effortful control develops on the superstructure of reactive inhibition and reflects the increased role of prescriptions and proscription learned in the socialization process. Effortful control enables more flexibility of inhibition and allows for reflection on the long-term negative consequences of a contemplated behavior (Rothbart 2007). This would include such things as avoiding an opportunity to steal even though there are no possible immediate negative consequences because contemplating the theft evokes thoughts of possible long-term consequences. This, of course, is what is known as conscience (see Chapter 4).

Because females are less vulnerable to a range of neurocognitive insults than males, they and are more likely to be able to exercise effortful control over their behavior (Bennett, Farrington, and Huesmann 2005). Indeed, whether it be a function of reactive or effortful control (again, these mechanisms are not mutually exclusive—the former strongly impacts the effectiveness of the later), females have, *without exception*, showed greater constraint/self-control across numerous studies regardless of differences in data, methods, culture, or ages of subjects (Chapple and Johnson 2007). A study by Helen Driscoll and her colleagues (2006:139) was designed to determine which of these two means of inhibition was stronger. They concluded that behavioral inhibition was more a function of visceral fear (harm avoidance) than other more cognitive components of inhibition such as guilt. This finding provides us with a transition to evolutionary explanations that focus on the ultimate origins of sex differences.

Evolutionary Explanations

Neurohormonal differences between the sexes provide a proximate level explanation of gender differences in the propensity to commit antisocial acts, but do not explain why these differences should exist in the first place. In contrast to the images of helpless females as the pawns of powerful males presented in many feminist writings, evolutionary theory focuses on female choice (sexual selection) as the major mechanism that drove the evolution of sex differences. The evolutionary perspective, rather than being dismissed as "sexist," may thus be viewed as offering a more affirmative view of women than the feminist perspective (Campbell 2009; Mealey 2000).

Anne Campbell (1999) surveyed a multitude of studies from anthropology, sociology, psychology, endocrinology, and primatology to provide an evolutionary explanation for sex differences in criminal behavior in her *staying alive* hypothesis. Campbell's argument has to do with evolved sex differences in basic biology relevant to parental investment and status striving. Because a female's obligatory parental investment is greater than a male's, and because of the greater dependence of the infant on the mother, a mother's presence is more critical to offspring survival (and hence to the mother's reproductive success) than is a father's. Linda Mealey (2000:341) explains that throughout most of human evolutionary history: "Desertion of one's mother means almost certain death, whereas desertion by one's father means only a reduction in resources." Surveys of the anthropological literature find that there are no human cultures in which mothers desert their children anywhere near the rate fathers do (Campbell, Muncer, and Bibel 2001). In cold, calculating terms, unlike males, females are limited in the number of children they can have, so each child represents an enormous investment in time, energy, and resources for her that is difficult to relinquish without extremely compelling reasons to do so.

Because a female's survival is more critical to her reproductive success (maximizing the probability that her offspring will survive) than is a male's survival to his, Campbell argues that females have evolved a propensity to avoid engaging in behaviors that pose survival risks. The practice of keeping nursing children in close proximity in ancestral environments posed an elevated risk to the mother and the child if the mother placed herself in risky situations. The evolved proximate mechanism to avoid doing so, Campbell proposes, is a greater propensity for females to experience more situations as fearful than men do. She surveys evidence showing that there are no sex differences in fearfulness across a number of contexts *unless* a situation contains a significant risk of physical injury. Fear of injury accounts for the greater tendency of females to avoid or remove themselves from potentially violent situations, and to employ indirect and low-risk strategies in competition and dispute resolution relative to males.

Campbell's theory provides a plausible ultimate level explanation consistent with evolutionary biological principles for why the sexes differ in fear. But what is the proximate level evidence for greater female vigilance to signals of fear?

Sex Differences in Fear

Fear is a basic primordial affective state that signals danger. It is an unpleasant state of nervous system arousal most immediately experienced as tachycardia (rapid heart rate). It is an extremely unpleasant emotion that motivates those who experience to escape the immediate threat and to avoid placing one's self in such a position in the future (Steimer 2002). Fear is thus an adaptively functional experience that facilitates the emergence of escape/avoidance behaviors that enhance an organism's chances of survival, and in the case of human females, reproductive success.

Whether assessed in early childhood (Kochanska and Knaack 2003), the middle-school years (Terranova, Morris, and Boxer 2007), or among adults across a variety of cultures (Brebner 2003), it is invariably found that females experience fear more readily and more strongly than males. A meta-analysis of 150 risk experiment studies found that when the risk involved meant actually carrying out a behavioral response, male/female differences were much greater than when subjects are asked to respond to hypothetical scenarios that merely involved cognitive appraisals of possible risk (Byrnes et al. 1999). Finally, every study of gender differences in fear of crime (as well as fear of noncrime incidents such as automobile accidents) shows that females are more fearful than males and assess their chances of victimization higher that males despite objectively being considerably less at risk for criminal victimization (reviewed in Ellis and Walsh 2000; Fetchenhauer and Buunk 2005).

As discussed in Chapter 4, the amygdala is crucially involved with processing emotional stimuli, especially fear. Numerous neuroimaging studies have shown sex-related hemispheric laterality of the amygdalae with males specializing to the right and females to the left (Cahill et al. 2004). An fMRI study by Gur and his colleagues (2002) found a highly significant greater ratio of orbital frontal cortex volume to amygdala volume in females relative to males. Because the frontal cortices play the major cognitive role in modulating behavior, this suggested to Gur and his colleagues that females would be less likely to express negative emotions in aggressive ways. This would also tend to lead more women to internalize stressful emotional experiences.

A similar study found that women activate significantly more neural systems associated with emotional experiences and with encoding experience into long-term memory than men (Canli et al. 2002). The greater tendency of females to encode emotional memories (the right amygdala specializes in detecting salient emotional stimuli and the left is involved in sustained stimulus evaluation [Williams et al. 2005]) means that past life events, both pleasant and unpleasant, are more readily available for rumination. Frequent rumination about stressful events is likely to increase their valence and lead to a number of anxiety and depressive disorders. Noting that females are about twice as likely as males to suffer from post-traumatic stress disorder (PTSD), Hamann (2005:292) states that "Because fear conditioning depends critically on the amygdala, sex differences in amygdala response may partly account for the greater prevalence of PTSD in women." Thus, while the greater tendency to experience fear may have been a female adaptation positively related to reproductive success, it comes at a mental health price.

Gender Differences in Status Seeking

The greater female concern for personal survival also has implications for sex differences in status seeking. Recall that males exhibit greater variance in reproductive success than females but less parental certainty, and thus have more to gain and less to lose than females by engaging in intrasexual competition for mating opportunities. Striving for status and dominance was a risky business in evolutionary environments, and still is in some environments today. Because dominance and status are less reproductively consequential for females than for males, there has been less evolutionary pressure for the selection of mechanisms useful in that endeavor for females than for males (Barash and Lipton 2001).

Campbell points out that although females do engage in intrasexual competition for mates, it is rarely in the form of violence and aggression in any primate species. Most of it is decidedly low key, low risk, and chronic as opposed to male competition, which is high key, high risk, and acute. A study of 20 different populations around the world found that female/female homicide constituted only 2.5% of all homicides (Daly and Wilson 2001). When females kill they typically kill males (spouse, boyfriend, ex-spouse) in self-defense situations (Daly and Wilson 2001). The female assets most pertinent to reproductive success are youth and beauty, which cannot be won in intrasexual competition; one either has them or one does not. Male assets are the resources that females desire for their reproductive success, and unlike youth and beauty, can be achieved in competition with other males. Males are willing to incur high risks to achieve the status and dominance that bring them resources and thus access to more females and have evolved neurohormonal mechanisms to enable them to do so.

Campbell shows that when females engage in crime they almost always do so for instrumental reasons, and the crimes themselves rarely involve risk of physical injury. Both robbery and larceny theft involve expropriating resources from others, but as seen in Figure 10.1 females constitute about 38% of arrests for larceny/theft and only about 11% of arrests for robbery (almost always with and at the instigation of male partners [Miller 1998]), a crime carrying a relatively high risk for personal injury. There is no mention in the literature that female robbers crave the additional payoffs of dominance that male robbers do, or seek reputations as "badasses." Dominant and aggressive females are not particularly desirable as mates, and certainly a woman with a reputation as a "badass" would be most unattractive. Campbell (1999:210) notes that while women do aggress and do steal, "they rarely do both at the same time because the equation of resources and status reflects a particularly masculine logic."

It is important to realize that sex differences in aggression, dominance seeking, and sexual promiscuity across species are related to parental investment, rather than biological sex per se. It is the level of parental investment that exerts pressure for the selection of the neurohormonal mechanisms that underlie these behaviors. There are species in which females do not carry the primary burden of parental investment. In a number of bird and fish species, males contribute greater parental investment (e.g., incubating the eggs and feeding the young), and in these species it is the female who takes the risks, who is promiscuous and the aggressor in courtship, and who engages in intrasexual competition for mates (Barash and Lipton 2001; Betzig 1999). Males and females in these species thus assume traits that are opposite those of males and females in species in which the females assume all or most of the burden of parenting. This sex-role reversal provides support for Campbell's thesis and underlines the usefulness of cross-species comparisons.

Tending and Befriending

While taking a somewhat different approach, Shelly Taylor and her colleagues (Taylor 2006; Taylor et al. 2000) augment Campbell's (1999) position in their *tending-and-befriending* hypothesis. Taylor and her colleagues surveyed the voluminous rodent, primate, and human literature on coping strategies in response to stress and found robust and consistent sex differences; females typically "tend and befriend" as opposed to the more male-typical "fight or flight." This is not to say that males do not tend and befriend in response to threat, but rather that they do not do so in the same ways, to the same degree, or in response to the same stresses and threats. Likewise, women do fight or flee when conditions demand it, but they do so in different ways to different degrees.

Tending and befriending essentially refers to the tendency to respond to environmental stressors by drawing closer to offspring and intensifying their care and by drawing closer to social support networks of other females. The tending-and-befriending model is a biobehavioral model of stress response, not a model of parenting, but the evolutionary pressures underlying it are clearly the result of the female parenting role. The female inclination to tend and befriend is assumed to have coevolved from the more primitive attachment and nurturing systems which are chemically organized and sustained by oxytocin, which is modulated by estrogens. The tendency among females apparently arose to protect themselves and their children in situations in which the male responses of fighting or fleeing were not viable options. When there are no other options, then of course females will flee with their young and fight fiercely to protect them.

Most male threat stress encountered in evolutionary environments came from other males in the competition for status and access to females or from predators in the hunting/hunted nature of primitive food acquisition.

We may characterize this as "fight or forfeit mating" in the first case, and "take risks or starve" in the second. Whether an organism fights or flees depends on if it perceives the threat as surmountable. These male typical responses are driven by T, which, as we have seen, works antagonistically to oxytocin. When men befriend in the face of threat, T tends to rise in anticipation of aggressive responses (Geary and Flynn 2002).

In most situations of threat females faced in evolutionary environments, the fight-or-flight response would have made very little sense. If threat came from an aggressive male, a woman was highly unlikely to either out-fight or outrun him. If the threat came from a nonhuman predator, fleeing would mean abandoning her offspring to their fate. A much more adaptive response in both cases was to hunker down and tend to her offspring and to turn to her social network of other females to help with her defense. The initial physiological stress response is identical in males and females, i.e., arousal of the sympathetic ANS and the HPA axis, but is apparently down-regulated in females by oxytocin (Geary and Flynn 2000; Taylor 2006).

The tend-and-befriend hypothesis is useful to us in understanding the ultimate evolutionary and proximate hormonal mechanisms that underlie the tendency for women to be less physically aggressive when competitively threatened. Numerous studies indicate either no sex differences in the fre-quency or intensity of anger or a small insignificant tendency for women to experience anger more frequently than men. However, men are more likely to express it via confrontation and verbal and/or physical aggression against those who have threatened and angered them. This is part of the fight/flight response. The female tendency is to discuss the situation (vent-ing) with a friend or other uninvolved individual or to cry and seek solace (Campbell 2006). This is part of the tend/befriend response.

In sum, when anger exceeds inhibition, overt behavioral aggression ensues. From a meta-analysis of differential gender-regulation abilities, Knight et al. (2002) concluded that the greater inhibitor ability of females underlies gender differences in aggressive behavior. Subsequent studies con-tinue to find large gender differences in the behavioral inhibition system (BIS) (Mardaga and Hansenne 2007). Parsey and his colleagues' (2002) PET scan studies suggest that the greater female ability to inhibit reactive aggression in response to provocation lies in the greater binding potential of serotonin in areas of the PFC and amygdala in female brains (for a review of gender, serotonin, and inhibition studies, see Driscoll et al. 2006).

Sexual Selection and the Sexually Dimorphic Brain

Charles Darwin proposed the theory of sexual selection to complete his theory of evolution, noting that while natural selection accounted for dif-ferences *between* species, it did not account for the often profound male/female differences *within* species. Sexual selection involves competition for reproductive partners and favors traits that lead to reproductive success, even though those characteristics may not be favored overall by natural selection; i.e., while traits such as bright plumage may attract females by indicating "good genes," they also attract predators by indicating a good meal. Sexual selection, like natural selection, causes changes in the relative frequency of alleles in populations in response to environmental challenges, but in response to sex-specific mating challenges rather than general sex-neutral challenges (Qvarnstrom, Brommer, and Gustafsson 2006).

There are two primary paths by which sexual selection proceeds: Intrasexual selection and epigamic or mate choice selection. Both sexes engage in both processes, but intrasexual selection is primarily male competition for access to females, and epigamic selection is more a process of females choosing with whom they will mate. The more sexual selection operates on a species (the more competition there is for mates) the more sexual dimorphism will be selected for. In species where intrasexual selection is paramount, there are large differences between males and females in size, strength, and aggression; in species where epigamic selection is stressed, males are more striking in their appearance than females (Andersson and Simmons 2006). I bring up sexual selection to show that while males and females inhabit the same physical environments in which natural selection operates in sex-neutral ways, they inhabit quite different mating environments that lead to sexual selection operating differently on them.

In Chapter 4 I addressed the possibility that selection for intelligence was driven by the cognitive demands of group life. Recent studies of the primate brain have shown that group life has produced striking differences between the sexes in brain mechanisms related to carrying out the different demands placed on males and females in evolutionary environments (Dunbar 2007; Lindenfors 2005; Lindenfors, Nunn, and Barton 2007). Because of the competitive demands of sexual selection for males, we should observe greater development of subcortical (limbic) brain structures involved in sensory-motor skills and aggression. Conversely, female fitness depends more on acquiring male resources and navigating social networks ("staying alive"/"tending and befriending"), and thus we should expect greater development of neocortical areas, particularly of the frontal lobe structures.

The studies by Patrik Lindenfors and his colleagues (2005, 2007) have shown this to be the case among 21 nonhuman primate species ranging from chimpanzees to rhesus monkeys. Specifically, they found that the more affiliative sociality of females is related to greater neocortex volume, and that the more competitive male sociality is more closely related to subcortical volume. The authors suggest that this neural sexual dimorphism should extend beyond the primate species that have been studies so far. Indeed, the fMRI study by Gur et al. (2002) discussed in the context of gender differences in fear showing the greater ratio of orbital frontal cortex volume to amygdala volume in human females relative to males fits in nicely with these findings. They also slot in nicely with evidence from genomic imprinting data. Genomic imprinting is an epigenetic phenomenon that takes place in a small number of genes whereby alleles are switched off (methylated) according to their parental origin. These data show that neocortical (the "social brain") volume is inherited maternally and that limbic system volume is inherited paternally (Dunbar 2007; Goos and Silverman 2001).

Evolutionary, hormonal, genetic, and neurological data all converge on the conclusion that we should strongly *expect* to see the low rate of criminal offending among females relative to males. There is certainly no mystery about it; indeed, a profound mystery would exist if we actually found a culture in which we *did not* find large sex differences in antisocial behavior. Criminology can ill afford to remain oblivious to the robust and consistent evidence coming from mature sciences if it is ever to reach maturity itself. If we continue to call upon differential gender socialization as our only explanation for gender differences in antisocial behavior we will go nowhere except into scientific oblivion. We must realize that gender socialization rests on the solid bedrock of sex-differentiated biology forged by countless thousands of years of contrasting selection pressures.

EVOLUTIONARY CONTRIBUTIONS TO UNDERSTANDING RAPE

This section addresses an issue at the forefront of feminist criminology: rape. Women are overwhelmingly the victims of this horrible crime, and thus would benefit most from a better understanding of its origins. Naturalistic explanations for crimes such as rape tend to evoke strong emotions in those who believe that such explanations seek to excuse them (the naturalistic fallacy) when they do no such thing. Although evolutionary theory differs from feminist theory in terms of explaining rape, in many respects they are surprisingly consistent.

Feminist Theories of Rape

The fundamental assumption of most feminist theories of rape is that it is motivated by power, not sexual desire. It is viewed as a crime of violence and degradation designed to intimidate and keep women "in their place" (Gilmartin 1994). Most feminist theorists feel that to understand rape we have to understand that there are large social, legal, and economic power differentials between men and women, that these differentials affect all social interactions between men and women. Because men enjoy the power advantage, they are able to use any and all means to control women, which includes rape.

A central assertion of feminist theory is that males are socialized to rape, not directly, of course, but via the many gender-role messages society sends them asserting their authority and dominance over women (Gilmartin 1994). Rape is the major weapon males have used to establish and maintain culture-wide patriarchy and the dominance of individual men over individual women because the threat of rape forces a woman to seek the protection of a man from the predations of other men, thus forcing her into permanent subjugation. In this politicized view feminists tend not to see rape as an act committed by a few psychologically unhealthy men. As the master symbol of women's oppression, it is considered an act that all men may commit and one that is indicative of the general hatred of women that characterizes the behavior of "normal" adult men (Herman 1991:178). The most extreme statement of this position was made by Susan Brown-miller, who asserted that rape "is nothing more or less than a conscious process of intimidation by which *all* men keep *all* women in a state of fear" (1975:5, emphasis added). Feminists provide evidence for this by pointing to "rape proclivity" studies indicating that between 35 and 69% of surveyed males admit that they would commit rape given assurances of never being exposed and punished (Skinner et al. 1995), and the widespread use of rape in war where potential victims are plentiful and the likelihood of punishment is near zero.

The major problem with feminist theories of rape is the insistence that it is a nonsexual act. While it is obvious that rape is a violent act, it is just as obvious that it is a sexual act. Most clinicians engaged in the treatment of sex offenders insist that rape is primarily sexually motivated (Barbaree and Marshall 1991), and very few men (about 6%) or women (about 18%) believe that it is motivated by anything other than sex (Hall 1987). After all, sex is *the* primordial motivator of behavior, and the treatment literature is replete with studies that explicitly or implicitly view rape as sexually motivated, as evidenced by treatment modalities that emphasize cognitive restructuring for deviant *sexual* fantasies and/or medications designed to reduce *sexual* arousal (Dreznick 2003; Giotakos et al. 2003; Grubin 2007; Harvard Mental Health Letter 2004; Howard 2002; Maletzky and Field 2003). Science should be the guide to unraveling the nature of this horrible crime, not mantras that thwart this goal.

In a plea to depoliticize rape, Craig Palmer asserts that the "not sex" explanation prevents researchers from learning more about the phenomenon of rape, and this occurs "at the expense of an increased number of rape victims" (1994:59). Palmer's point is that if we misidentify the causes of rape we compromise treatment plans for rapists as well as efforts at rape prevention. This is an important point, because perpetuating the notion that rape is about power and control and not about sex misrepresents reality for young women who are bombarded with such messages, and it confuses their rape-avoidance tactics. Pinker (2002:370) points out additional dangers of this position: "The flight from reality of the rape-is-not-sex doctrine warps not just advice to women but policies for deterring rapists . . . Savvy offenders who learn to mouth the right psychobabble of feminist slogans can be seen as successfully treated, which can win them early release and the opportunity to prey on women anew."

An increasing number of feminist writers now disavow the idea that rape is not about sex, stating that the "not sex" claim was necessary earlier to combat the idea that women saw rape as an act of sex and that they enjoy it. Brownmiller herself has said that while women did not see rape as a sex act and were humiliated by it: "But obviously we didn't think it had nothing to do with the sex act—of course it is, sexual organs are used" (in Palmer and Thornhill 2003:253). Despite countless examples in the literature to the contrary, some feminists even claim that it was never the feminist position that sex was not a major motivator for rape. For example, Natalie Angier has stated that "neither Brownmiller nor any other thoughtful person ever claimed that rape was only about power and aggression, not about sex" (in Palmer and Thornhill 2003:254).

The Evolutionary Perspective

Evolutionists argue among themselves as to whether the general male disposition to rape is an adaptation or a by-product or other adaptations (e.g., aggression, dominance striving) that may promote sexual assault in the absence of internal or external constraints. In fact, the coauthors of a book on rape from the evolutionary perspective admit that they differ on this point (Thornhill and Palmer 2000). The adaptationist hypothesis avers that ancestral males who were most inclined to pursue multiple sexual opportunities, forcefully or otherwise, would have enjoyed greater reproductive success than males who did not, thus leaving modern males with a genetic legacy inclining them to do the same. In this view men are monomorphic for genes inclining them to rape, which is exactly what many feminists have long argued (Brownmiller 1975).

The by-product hypothesis contends that rape would not have occurred frequently enough or successfully enough (in terms of resulting in pregnancy) to induce an evolutionary trajectory that resulted in a mental mechanism dedicated to rape. However, even minuscule fitness advantages can eventually lead to an adaptation over the vast stretches of evolutionary time, and it seems to be the case that pregnancy rates are twice as common for rape (about 6%) than they are for consensual sex (about 3%) (Gottschall and Gottschall 2003). Nevertheless, it is difficult to view rape as an adaptation since the vast majority of copulations in any mammalian species are voluntary, and when force is used to secure compliance, it tends to be a tactic of last resort pursued only after other tactics have failed (Ellis 1991; Mealey 2003). Rape is common enough to be considered an unfortunate part of our evolutionary baggage (a by-product), however. Whatever the case may be, both positions view rape as a high-risk behavior most likely to be committed by males lacking the status to acquire consenting partners, that is, by low-status males living in environments in which high rates of other forms of violence are common (Ellis and Walsh 1997; Mckibben et al. 2008; Mills, Anderson, and Kroner 2004; Ward and Siegert 2002).

The likely key to understanding the evolutionary origins of rape is the huge disparity in parental investment between males and females. The energy expended in copulation is the only necessary male investment in reproduction; thus male reproductive success in ancestral environment rested on the ability to obtain access to as many partners as possible. Males who were most inclined to pursue multiple sex partners, which occasionally may have included forced copulation, probably enjoyed greater fitness than other males, thus passing on these inclinations to their male offspring (Thornhill and Palmer 2000; White et al. 2008).

Because of greater obligatory investment, female fitness rested on the ability to secure paternal investment in exchange for exclusive sexual access. Female promiscuity is only evolutionarily viable in species requiring no paternal investment and very little maternal investment after weaning, or in species in which the parental burden rests almost entirely on males. Promiscuous mating leads to paternal uncertainty and thus would have been maladaptive for human females because paternal uncertainty is not likely to result in male investment (Buss 2005; Wright 1994). Females have thus evolved a tendency to resist casual copulation with multiple partners. The conflict between the reckless and indiscriminate male strategy and the careful and discriminating female strategy, according to this view, sometimes results in rape.

Most of us find this line of reasoning distasteful because of our propensity to subscribe to the naturalistic fallacy, but I am not aware of any evolutionary theorist, male or female, who excuses, condones, justifies, or considers rape any less onerous on the basis of its hypothesized natural origins. Rape is a maladaptive consequence of a mating strategy that *may* have been adaptive in ancestral environments (Alcock 2005). It is a morally reprehensible crime that requires strong preventative legal sanctions. Calling something "natural" does not dignify it or place it beyond the power of culture to modify, as manifestly it is not (Walsh 2006). Behavior must be judged by its consequences, not by its origins.

The evolutionary theory of rape has also been criticized on the grounds that it cannot explain the rape of males, children, or postmenopausal women, or sexual attacks not including vaginal intercourse because such acts do not enhance reproductive success. This criticism also reveals a misunderstanding of evolutionary logic. As repeatedly stressed, organisms are not adapted to seek ultimate goals; they are adapted to directly seek proximate goals that generally blindly serve ultimate goals. More often than not we consciously attempt to thwart our fitness via the use of contraception even as we continue to enjoy the mechanism that promotes it. Just as a hammer can be used for purposes other than those for which it was designed, we can use and misuse our adaptations in various ways. Nonreproductive sex is not adaptive, and neither is the nurturing of pets, but both are examples of the nonadaptive diffusion of mechanisms that *are* adaptive (Lykken 1995). Nurturing pets or nongenetically related children is nonadaptive, but it calls on the same mechanisms that give us pleasure in nurturing our own offspring, which is adaptive, just as nonreproductive sexual acts (nonadaptive) call on the same mechanisms that lead males to want to copulate with fertile females (adaptive). Evolution is a mindless algorithmic process; it does not instruct us to use our nurturing and sexuality for one thing but not another; human cultures have constructed moral norms to instruct us in these matters.

An adequate theory of rape must go beyond claims that all men are monomorphic for rape-inclining genes, or that all males learn cultural messages that lead them to disvalue women. The evolutionary claim that rape behavior is expressed facultatively is true, but we should also like to know why some men experience a situation as conducive to rape and others do not. We also want to go beyond the obvious social demographics that tell us that men most likely to rape are found in the same places as men who will take anything they want, be it your car, your wallet, or your sex by any means necessary (White et al. 2008). In short, we need a theory that integrates biological, psychological, and cultural factors (Ward and Siegert 2002).

Lee Ellis (1989, 1991) provides such a theory in his *synthesized biosocial theory*. Ellis pulls together many lines of evidence to integrate the most empirically supported concepts from evolutionary, neurohormonal, psychological, feminist, and cultural theories and finds a great deal of similarity, as well as many differences, among them. It is the most comprehensive and most value-neutral theory of rape to be found in the literature. The theory is briefly presented following in the form of four propositions.

1. The sex drive and the drive to possess and control motivates rape.

All sexually producing organisms possess an unlearned sex drive, although the manner in which it is expressed is open to learning. The contention that rape is sexually motivated rests on many kinds of evidence, the most compelling being that nonstranger rapists are likely to use force only when other tactics have failed. The rapist's preliminary use of nonforceful tactics to gain sexual access makes it difficult to claim that rape is "nonsexual." Additionally, the existence of forced copulation in other animal species makes it difficult to claim that similar behavior among humans is motivated by hatred of females, that rape occurs because males and females are differentially socialized, or that rapists are subconsciously protecting the privileged political and economic positions of males.

Animals also possess a strong drive to possess and control, as evidenced by the catching, hoarding, burying, and protection of resources needed to survive and attract mates. The drive to possess and control is especially strong where sex partners are concerned. Among humans there is plentiful evidence that men and women are extremely possessive of one another. We can be sure that ancient humans who were inclined to be nonchalant about who had access to their resources and mates are not among our ancestors.

Jealousy and male sexual proprietariness is responsible for a huge percentage of domestic violence and spousal and lover homicides around the world (Daly and Wilson 1988a; Lepowsky 1994). It has also been found that the rape of one's spouse or lover is positively related to her infidelity (or suspected infidelity), and is a tactic apparently used to reassert "ownership" (Mckibben et al. 2008).

2. The average sex drive of men is stronger than the average sex drive of women.

Among the many facts called upon to support this proposition are that males commit the vast majority of sexual crimes, consume the vast majority of pornography, constitute practically all the customers of both male and female prostitutes, masturbate more frequently, and have a much greater interest in casual sex with multiple partners (Geary 2000; Oliver and Hyde 1993). Not only are these and many other indices of gender differences in strength of sex drive found across cultures but also among other species. Opposing the male tendency to readily learn forceful tactics to obtain sex is the evolved female tendency to resist them.

3. Although the motivation for rape is unlearned, the specific behavior surrounding it is learned.

Rape is learned via operant conditioning, and individuals with the strongest sex drives will learn rape behavior more readily. Men who have successfully used "pushy" (but not necessarily forceful) tactics to gain sexual favors in the past have learned that those tactics may pay off. A male's initial payoff may be little more than a necking or petting session, but if he finds that each time he escalates his pushiness he succeeds in gaining greater sexual access, his behavior may be gradually shaped by reinforcement in ways that could lead to rape.

4. Because of neurohormonal factors, people differ in the strength of their sex drives and in their sensitivity to threats of punishment.

The sex drive is primarily a function of T, and different levels of T lead to varying intensities of the sexual drive. Exposure to fetal androgens results in lessened male sensitivity to environmental stimuli. Males in many species are less sensitive on average to noxious stimuli than females, leading them to be more likely to discount the consequences of antisocial behavior for themselves and for their victims. Individuals with the strongest sex drives are usually the same individuals who are the most insensitive to environmental stimuli because the same neurohormonal factors are responsible for both. These are the individuals who are most likely to rape and to be engaged in criminal activity in general.

Ellis's theory illustrates for us once again the utility of theories that are informed by the subject matter of all disciplines that study human behavior. His theory draws on insights from evolutionary biology, endocrinology, genetics, neuroscience, psychology, and sociology, and is thus greatly to be preferred over any theory that limits itself to a single discipline for its insights. Indeed, as I have opined time and again throughout this work, to refuse the intellectual nourishment of more advanced, sophisticated, and robust sciences when such nourishment is there for the taking is not only scientific malfeasance; it is scientific suicide.

11 Retrospect and Prospect

RETROSPECT

This brief chapter looks back on the variety of evidence supporting the view that biosocial criminology is the storm that will stir the stagnant waters of criminology and forward to its prospects of becoming the paradigm for the 21st century when the storm abates. I began by attempting to convince sociologically trained criminologists to at least examine the possibility that the biosocial sciences can provide valuable insight and promote further understanding of the phenomena of criminality. I hope that I have demonstrated that the key to progress in other sciences has been to maintain consistency with the methods, theories, and concepts of the more fundamental sciences. History has shown that initial opposition to such integration based on supposed threats to the younger disciplines' autonomy soon dissipated when the benefits the older disciplines offered became apparent.

Another point I hope that I have highlighted is that I have made clear that there is nothing antihumanistic about injecting biology into the study of human nature. Recognizing this, we can stop the endless and senseless ideological debates that often as not end up attacking individual researchers and throwing at his or her work epithets with hissing suffixes such as "racist," "sexist," "fascist" that inflame one side and intimidate the other. Only when such immature, shameful, and counterproductive tactics are no longer tolerated can we begin a cooperative endeavor that will eventually lead to a truly scientific criminology. There will still be arguments enough, for the data rarely speak unequivocally for themselves, but hopefully the arguments will be less personal and more useful.

Genetics

The primary lesson from our discussion of genetics is that the accusation of genetic determinism is a red herring of the ripest kind. I have yet to hear any biosocial scientist make the kind of statements they are accused of making. Neither, I'll wager, have any of their critics. There are no genes "for" any kind of nontrivial behavior. Geneticists are acutely aware that genes simply make proteins that interact with other proteins and with the physical and social environments in which their carriers find themselves in complicated ways. Genes do not cause behavior; rather, they simply bias the organism's trait values in one direction or another. It cannot be emphasized enough that the genes of most interest to criminologists are not self-activated; rather, they are activated by environmental information, including information that originates in our own thoughts and feelings. There is a determinism in all this (what would *not* being a determinist imply for a scientist?), but biosocial determinism is a determinism more respectful of human dignity than cultural determinism because it implies self-determinism. After all, our genes are *ours*; they are at the beck and call of our purposes whether they are morally good, bad, or neutral. They manufacture the proteins that facilitate the traits that modulate our personal responses to the variety of situations we encounter or fashion for ourselves.

Evolutionary Psychology

Evolutionary psychology focuses on why we have the traits and characteristics we do, and is more interested in their universality than in their variability. The major lesson from this chapter is the notion that crime and criminality, while immoral, are normal responses to environmental contingencies. Evolutionary psychology avers that many human adaptations forged by natural selection in response to survival and reproductive pressures in ancestral environments are easily co-opted to serve morally obnoxious purposes. In common with all sexually producing species, humans are designed to be selfish—to be preeminently concerned with our own survival and reproductive success. There are bright and dark sides to selfishness: one side leads to mutual affection and support (reciprocal altruism) and the other to a crabbed selfishness, an egoism shorn of concern for others (a cheater strategy). Certainly, evolutionary psychology reminds us of humanity's dark side, but it also reminds us of our nobler side, and how both sides aided in our ancestors' reproductive success. We are alive and kicking today because of aspects of our species characteristics we would like to disown, as well as those of which we are justly proud.

The Neurosciences

The neurosciences best capture how wrong critics are to accuse "biological" perspectives of hardwired determinism. The takeaway lesson from the neurosciences is that the genes have surrendered control of the human organism to that wondrous organ of adaptation we call the brain. It was noted that following some elementary genetic wiring to jump-start the process, the brain literally wires itself in response to environmental input. Neurons eventually make trillions of connections with each another, with each connection reflecting the organism's experiences (experience-dependent brain development). But this does not mean that the human mind is a blank slate; there are certain built-in algorithms guiding responses to the environment that are too important to be left to the vagaries of learning (experience-expected brain development). Neural Darwinists have long claimed that the brain's self-programming function can be viewed as the basis of human freedom, and they may well be right.

Anomie and Status Striving

Anomie/strain theory was examined with emphasis placed on the central concept of this tradition—SES and status striving. Status striving is central to both evolutionary psychology and anomie/strain theory. Perhaps no concept in this book generates more opposition among sociological criminologists than the idea that in an open society a person's SES is caused by individual traits and characteristics. The correlations between IQ and income, occupation, and education are consistent and robust, much more so than between parental SES and those variables. The other important individual trait linked to occupational success is conscientiousness, which like intelligence is highly heritable. Without conscientiousness, intelligence won't get you very far, but high conscientiousness, even if paired with somewhat less than average IQ, will lift you up the status hierarchy. An intelligent and conscientious person is able to embark on long-term strategies to obtain status and resources legitimately. Those less well endowed need not be abandoned to "innovation" and "retreatism" if programs recognizing and attacking these deficiencies can be implemented.

Social Learning and Adolescence

The main theme in this chapter is adolescence and the antisocial behavior it generates. Trying to understand the age-crime curve—why many adolescents suddenly become antisocial at puberty why they overwhelmingly desist in the early 20s—without understanding the physiological changes that occur during puberty is like trying to understand automotive locomotion without understanding combustion. At puberty teens are subjected to a huge surge of hormones, which in turn signal the brain to begin sculpting itself into its eventual adult form. During this period there is also a shift in the ratio of excitatory to inhibitory neurotransmitter, all of which favor the kinds of immature and often antisocial behavior we observe among adolescents.

Control Theories and the Family

Both social- and self-control theory emphasize the importance of the family and that antisocial behavior is the default option that occurs in the absence of adequate socialization. Neither version of control theory has much to say about *why* the family is so important to human development in any ultimate sense. Our discussion of the family (the reproductive team of male, female, and offspring) focused on its evolutionary origins and claimed that the family is an adaptation in the fullest sense of the word. If this is so, then we can expect problems when the family is disrupted. It was acknowledged that family problems could have preceded family disruption, and that the nuclear family of today's industrialized societies places tremendous burdens on parents. Nevertheless, we must be aware of the many disadvantages that accrue when children are reared in evolutionarily unexpected family arrangements.

Human Ecology/Social Disorganization and Race

The major concern of the human ecology/social disorganization tradition is the influence of the physical and cultural environment on antisocial behavior. Social disorganization is greatest in America's inner cities inhabited overwhelmingly by African Americans, and the proportion of blacks in a city has long been the best predictor of a city's crime rate. The inner cities suffer from multiple disabilities ranging from deindustrialization to the presence of crack, and to the spawning of an explicitly hostile oppositional culture. These areas are overflowing with unsupervised youths born to single mothers who have formed themselves into gangs and who wreak havoc in their neighborhoods. It was also shown that the cultural practices in inner-city environments may alter the biology of a person in ways conducive to criminal behavior. These practices include high levels of substance abuse among pregnant females that can lead to conditions such as fetal alcohol syndrome in their offspring, low levels of experience-expected breastfeeding, and high levels of environmental lead.

Critical Theories and Conflict

Critical theorists tend to be radical leftists and vehemently opposed to applying biological thinking to criminal behavior. It was emphasized that Marxism is not inherently antibiological since Marx often spoke of human nature, although what was meant by it is hotly contested. Conflict is certainly inherent in social life, and is necessary to resolve competing interests. The Marxist vision that conflict will cease with the coming of the communist state in which all people will be equal is a biological fiction. Max Weber, who had a more nuanced understanding of class and conflict, was well aware of this. The opinion was expressed that a Darwinian view of human nature has more to offer the left than it does the right. The message in Peter Singer's (2000) plea for a "Darwinian left" is that leftists must make their knowledge commensurate with their compassion if they are to avoid the horrendous mistakes of the past.

Feminist Theory and Gender

The major concern of feminist criminology is to understand why everywhere and always males commit far more antisocial acts than females. Explanations in terms of environmental variables just won't wash. The proximate answer to this central concern appears to be neurohormonal. Male and female brains are different because male brains are diverted from their default female form by androgen surges in utero. The male brain is further sensitized to the affects of androgens at puberty. Male-typical behaviors (status striving, dominance, aggressiveness) are facilitated by androgen-activated brain areas that subserve reproductive efforts. These same behaviors also facilitate criminal behavior in certain situations.

In terms of the ultimate origins of male/female differences in the propensity to commit crime, we explored Anne Campbell's (1999) staying-alive hypothesis and Shelly Taylor's tending-and-befriending hypothesis. Both theories posit that the mothering role is so important to a female's reproductive success that natural selection favored women who had a fear of placing themselves in physical danger, and when they are in danger they resort to different strategies than males do. It was noted that as the potential for physical harm becomes greater and greater in the commission of a crime, female participation becomes less and less frequent. Greater female parental investment also accounts for the lower level of status striving, an activity that can lead to criminal behavior in some instances.

BIOSOCIAL CRIMINOLOGY AND ETHICS

Whenever someone writes about biology and human behavior, he or she is inevitably expected to defend it on ethical grounds. Biological explanations have long provided ammunition for those with an axe to grind, probably because they are seen as more mysterious and powerful, and thus more threatening, than environmental explanations. Theorists writing from a purely environmental perspective are never burdened with the task of defending their theories as being open to abuse, despite history's sad catalogue of inquisitions, gulags, pogroms, genocides, and wars fought in the name of religious and secular ideologies far removed from any whiff of the demon biology.

Poor biology: little more than a century old as a respectable science, it has become the whipping boy of the heartfelt protectors of the disadvantaged. I appreciate their concern, and agree that findings in biology relevant to human nature can be used against disadvantaged and disvalued people.

But bigots and hatemongers will ride any vehicle that will give their preju-
dices a free ride, and have done so for centuries before anyone ever heard
of genes.

Nazi Germany is frequently evoked as an example of the dangers of biology, although I have yet to understand why. Surely we do not believe that
if Hitler and his cronies had never heard of eugenics the holocaust would
not have happened. The nightmares of racial purity and ethnic cleansing
that have bedeviled us throughout human history did not wait for Gregor
Mendel or Charles Darwin to sanctify them. Nationalism, not "biology,"
was used to hypnotize the German people. The Nazis created a myth—
Aryan superiority—and a monster—Jews, Slavs, anyone not belonging to
the "master race"—and set the myth to devour the monster. While the
Nazis tapped into ancient biological underpinnings of tribalism and xeno-
phobia, the mechanisms allowing them to access these ancient traits were
social and psychological. Control of the media, the destruction of institutions mediating the relationship between the individual and the state, the
frighteningly magnificent rallies, and the cult of the Führer were among the
many mechanisms employed.

The communist terror, unlike the Nazi terror, was based squarely on a
well-articulated theory of causation and was both longer-lived and quan-
titatively more heinous than the Nazi terror. The theory on which it was
based was purely environmental. Marxist ideals sound admirable in theory,
but in practice they reduced almost everyone to a numbing equality of prop-
ertylessness and powerlessness. An egalitarian ideology that goes beyond
the ideals of equality of justice and opportunity is biologically doomed. It is
doomed because it is founded on the assumption that all people are essentially equal in their capacities, and if they are not, by God we will make
them so. Only force and repression can mold human being in ways alien to
their nature. Repression and terror are inherent in *any* ideology that brooks
no dissent.

Does the potential for misuse exist in biosocial theories? Of course it
does; but it exists in any kind of systematic account of what humans are
and can become. Misuse of knowledge should be opposed when and where
is it misused, but knowledge must never be swept under the rug because
some feel that it could be misused. As Bryan Vila (1994:329) has remarked:
"Findings can only be used for racist or eugenic ends only if we allow perpetuation of the ignorance that underlies these arguments." For instance,
social Darwinism is stone dead, wounded by science when it was realized
that it was a perverse misinterpretation of Darwinian concepts, and killed
off by an evolving sense that in some ways we are our brother's keeper.
Could such a philosophy return? Yes, but it will receive no sustenance from
a modern understanding of Darwinism.

We hear less of social Darwinism these days, but we continue to hear of
eugenics. Perhaps eugenics is feared more because it is activist and inter-
ventionist whereas social Darwinism promoted only a laissez-faire attitude

toward the unfortunate. As we have seen, early eugenic programs in the United States and Britain were endorsed as sensible and humane by many ethical and progressive individuals. The basic principle of eugenics was that it is best to prevent suffering by sterilizing individuals who were likely to give birth to "defective" children.

A form of "back door" eugenics is practiced today by parents who elect to abort the fetus after being informed that it has some form of defect. Most of us probably see no ethical issue here since the decision was made by individuals in consultation with their physicians and not by some oppressive government. Thus it is perhaps government control that most bothers critics of "biological" approaches to crime. These critics tend to view such approaches as ultimately leading to a kind of therapeutic tyranny, with parolees running around with wires sticking out of their heads, sex offenders with their testes in a jar, and former gangbangers reduced to pliant zombies cleaning off the tables at Burger King. That these hypothetical measures might reduce crime is not the issue. The issue is that treating individuals diverts attention away from the social factors that critics "know" are responsible for criminal behavior.

At one level I agree with this assessment in that I acknowledge that environmental interventions are the best crime control methods since such interventions reduce the prevalence of crime. Unfortunately, such environmental interventions are hard to find, and those that have been implemented (for example, the Chicago Area Project) have produced little in the way of results, or even negative results (Rosenbaum, Lurigio, and Davis 1998). On another level I disagree. Flesh-and-blood people commit crimes, not disembodied "social factors," the influence of which depends on individual vulnerability to them. Proponents of the treatment approach view treatment as humane and superior to years of soul-destroying imprisonment. Advocating a therapeutic approach for individuals with a demonstrated need for it is a far cry from advocating it as a general method of social control. They would also point out the success of their methods with syndromes that were also once "known" to be caused exclusively by social factors, such as schizophrenia and depression. I for one would love to see a form of long-acting implant that calms the violent, shames the psychopath, brings tears to the eyes of the heartless, makes the impulsive contemplate, and causes a man contemplating rape to have intensely graphic visions of pit bulls devouring his nether parts. Yes, this is "treating symptoms not causes," but medicine treats symptoms rather than elusive causes all the time. It is the symptoms that cause the patient's pain, and it is the criminal's manifest symptoms that cause his or her victim's suffering, not the supposed "root causes" of those symptoms, whatever they may be.

We may have the technology to implement such interventions one day, but if we do it will be a political decision, not a scientific one. Biology can no more be blamed for these uses we may decide to put it to than can electricity be blamed for electrocutions. Whatever biological therapies we eventually come up with, they probably will not be genetic. As has been constantly emphasized, there are genes that bias traits in certain directions, but there are no genes that lead directly to any kind of nontrivial behavior. Genes act in concert with other genes, and collectively they act in concert with the environment.

Suppose we did find a polymorphism that coded "for" sensation seeking/risk taking. Because we know that this trait is significantly related to antisocial behavior, should we counsel aborting fetuses carrying such an allele? If we did we would be depriving society of millions of vibrant people, for the same sensation-seeking/risk-taking trait that gets some people into trouble with the law is also a characteristic of explorers, police officers, firefighter, soldiers, inventors, and entrepreneurs.

What if the hypothetical polymorphism was "for" something with no apparent social usefulness, such as extreme impulsiveness, and what if we could exorcise it through genetic engineering? Our extremely impulsive people might just become prudent and restrained, but what else might they become? Because genes interact with each other and with the environment in ways far from perfectly predictable, the elimination of the "impulsive gene" might upset a whole fleet of genetic apple carts to the benefit of neither the individual nor society. All genes have potential pleiotropic effects, meaning that they can have multiple effects on the organism via the proteins they produce. Thus we would potentially eliminate positive as well as negative affects if we were able to engineer the allele out of existence. We do nothing less than play with fire when we image we can eliminate a particular allele and thereby change behavior in the directions we desire.

The real ethical dilemma accompanying our increasing genetic knowledge is the use of genetic information to fire employees or refuse people medical insurance on the basis of genetic screening that reveals some genetic anomaly. Thousands of such cases occur every year, a fact that often prevents individuals from undergoing potentially lifesaving genetic testing (Martindale 2001). Is this a genetic or a social problem? It is, of course, the latter, and so must be the solution. We can benefit from the positive fruits of genetic research without fear of negatives consequences simply by passing legislation forbidding access to personal genetic information to anyone but the individual and his or her physician. Neither genetics, nor any other branch of biology for that matter, is going to go away because there is potential for misuse.

I am not so naive as to think that anything I have written will change the mind of the true believer. Such people are so ideologically driven that they cannot imagine that others could conduct their research not similarly driven. For those with open minds, I suggest that any remaining fear of "biology" as it pertains to explanations of criminal behavior will be assuaged by learning some.

PROSPECT

In the first paragraph of the first chapter I quoted eminent criminologist Francis Cullen's prediction that the biosocial approach will be criminology's paradigm in the 21st century. With Cullen, I am cautiously optimistic that biosocial criminology will be "normal science" within one or two more decades. Criminological journals are increasingly featuring biosocial articles, and journals outside of criminology such as *Social Biology, Behavior and Brain Function, Human Nature,* and *Neuroscience* all frequently contain articles pertaining to criminology. Numerous books on biosocial criminology are arriving on the market every year, and most criminology textbooks now contain chapters at least making reference to genetics and neuroscience without the sneering that used to be prevalent (although many such references are still woefully off target).

It was noted in Chapter 1 that Walsh and Ellis's (2004) 1997 study of the state of criminology found that the theories criminologists were attracted to were highly influenced by their ideological leanings. This study was repeated 10 years later by Cooper, Walsh, and Ellis (submitted) who found that ideology, while still maintaining its grip on the discipline, has weakened considerably. Favored theory broken down by ideology in the 1997 sample yielded a chi-square of 177.23 and a strong Cramer's V of 0.65; in the 2007 sample the chi-square was 134.6 and Cramer's V was a weak to moderate 0.34. A number of other correlations in which ideology was the independent variable pointed to the decreasing role of ideology (or at least an increasing realization that biology is not the ogre many criminologists once thought it to be).

An issue not addressed in the 1997 sample but addressed in the 2007 sample was the opinions of criminologists regarding the usefulness of psychology and biology to the discipline and whether those disciplines should be integrated with criminology. There was no significant disagreement among the ideological groups regarding the statement "Psychology has a lot to offer criminology," with 81% either strongly agreeing or agreeing. However, there was significant disagreement regarding integrating psychology with criminology ($\chi^2 = 28.26$, df = 12, p < .01), with conservatives (66.6%) most enthusiastically supporting integration and radicals (56.6%) either disagreeing or strongly disagreeing. One wonders about the logic involved where a slight majority of the overall sample (51.3%) rejected integration with a discipline which 81% of the same sample thought had a lot to offer!

There was more disagreement along ideological lines with the statement "Biology has a lot to offer criminology" ($\chi^2 = 40.06$, df = 12, p < .001). Most conservatives and moderates agreed or strongly agreed, and most radicals disagreed or strongly disagreed. While more than one-fourth of the liberals sat on the neutral "don't know" position, there were more who agreed or strongly agreed (42.9%) than disagreed or strongly disagreed (31.9%).

However, contemporary criminologists of all ideological persuasions are less than enthusiastic about integrating biology with criminology regardless of how useful they think it may be to do so. Only 27.9% of the overall sample either agreed or strongly agreed that the disciplines should be integrated and 52.5% either disagreed or strongly disagreed. The extent of the agreement or disagreement was again significantly related to ideology with agreement falling linearly from conservative to radical ($\chi^2 = 27.5$, df = 12, p < .01). Nevertheless, overall the results of this study give me reason for optimism. While I cannot document a trend regarding the acceptance of biology in criminology because trend data do not exist, anyone familiar with the discipline in the 1960s through 1990s would bet the rent if not the mortgage that attitudes about discipline integration have changed considerably. The mere mention of biology in my graduate student days was to invite hostility and derision from professors.

Discipline integration is the answer to progress in criminological theory and research. Biosocial research, dealing as it does with the quicksilver of human behavior, has certainly been bedeviled by poor research. But all the faults of this research have also bedeviled other approaches to the study of human behavior, and more seriously so. A truly multidisciplinary criminology would serve a "check and balance" function to minimize research flaws. We have seen throughout this book that when biosocial theory, methods, technology, and data have been brought to bear on important criminological issues, criminology has benefited.

For individual criminologists to benefit, however, they must leave the comfort of their own stockade to explore the wonders awaiting them in the new world. Apparently intellectual adventurism is not all that popular among the current crop of criminologists, however. The modal number of biology classes taken by Cooper, Walsh, and Ellis's (submitted) 770 respondents was a fat zero, and the modal number of sociology classes was 10. Future criminologists cannot afford to be so hermetically sealed in their tight sociological boxes.

Leaving behind the familiar is exciting for some and disquieting for others, but necessary for all. Kurt Gödel's famous *incompleteness theorem*, although aimed at the inherent incompleteness of mathematics, has been interpreted as demonstrating that no science can be complete in itself and must venture beyond itself (Fitting 2000). All social sciences eventually reduce to psychology, psychology to biology, biology to chemistry, chemistry to physics, and physics to mathematics. What then of mathematics; to what does it reduce? Perhaps it was the inherent incompleteness of mathematics that moved Gödel (who has been called the greatest logician since Aristotle [Fitting 2000]) to formulate his mathematical proof of God. Now that is really "venturing beyond"! I bring this up not to introduce a metaphysical argument at this late juncture, but rather to make the point once again that it is perverse to believe that we can understand criminology's subject matter within the highly circumscribed world in which it presently operates. To believe that we can is to contradict the most brilliant minds that the natural and physical sciences have produced in the past two centuries.

Bibliography

Abu El-Haj, M. (2007). The genetic reinscription of race. *Annual Review of Anthropology*, 36:283–300.

Adler, F. (1975). *Sisters in crime: The rise of the new female criminal*. New York: McGraw-Hill.

Adler, F., G. Mueller, & W. Laufer (2001). *Criminology and the criminal justice system*. Boston: McGraw-Hill.

Agnew, R. (1992). Foundations for a general strain theory of crime and delinquency. *Criminology*, 30, 47–87.

Agnew, R.(1995). The contribution of social-psychological strain theory to the explanation of crime and delinquency. In F. Adler & W. Laufer (Eds.), *The legacy of anomie theory*, pp. 113–137. New Brunswick, NJ: Transaction.

Agnew, R. (1997). Stability and change in crime over the lifecourse: A strain theory explanation. In T. Thornberry (Ed.), *Developmental theories of crime and delinquency*, pp. 101–132. New Brunswick, NJ: Transaction.

Agnew, R. (2005). *Why do criminals offend? A general theory of crime and delinquency*. Los Angeles: Roxbury.

Agnew, Z., K. Bhakoo, & B. Puri (2007). The human mirror system: A motor resonance theory of mind reading. *Brain Research Reviews*, 54:286–293.

Akers, R. (1985). *Deviant behavior: A social learning approach*. Belmont, CA: Wadsworth.

Akers, R. (1994). *Criminological theories: Introduction and evaluation*. Los Angeles: Roxbury.

Akers, R. (1998). *Social learning and social structure: A general theory of crime and deviance*. Boston: Northeastern University Press.

Akers, R. (1999). Social learning and social structure: Reply to Sampson, Morash, and Krohn. *Theoretical Criminology*, 3:477–493.

Akers, R., & G. Jensen (2006). The empirical status of social learning theory of crime and deviance: The past, present, and future. In F. Cullen, J. Wright, & K. Blevins (Eds.), *Taking stock: The status of criminological theory*, pp. 37–76. New Brunswick, NJ: Transaction.

Albanese, J. & R. Pursley (1993). *Crime in America: Some existing and emerging issues*. Englewood Cliffs, NJ: Prentice-Hall.

Alcock, J. (2001). *The triumph of sociobiology*. New York: Oxford University Press.

Alcock, J. (2005). *Animal behavior: An evolutionary approach*. Sunderland, MA: Sinauer Associates.

Alexander, R., & K. Noonan (1979). Concealment of ovulation, parental care, and human social evolution. In N. Chagnon & W. Irons (Eds.), *Evolutionary biology and human social behavior*, pp. 436–453. North Scituate, MA: Duxbury.

Allman, William (1994). *The stone age present*. New York: Simon & Schuster.

Almgren, G. (2005). The ecological context of interpersonal violence. *Journal of Interpersonal Violence*, 20:218–224.

Al-Samarrai, S. (2006). Achieving education for all: How much does money matter? *Journal of International Development*, 18:179–206.

Altukhov, Y., & E. Salmenkova (2002). DNA polymorphisms in population genetics. *Russian Journal of Genetics*, 38:1173–1195.

Amato, P., & B. Keith (1991a). Parental divorce and well-being of children: A meta-analysis. *Psychological Bulletin*, 110:26–46.

Amato, P., & B. Keith (1991b). Parental divorce and adult well-being: A meta-analysis. *Journal of Marriage and the Family*, 53:43–58.

Anderson, A. (2002). Individual and contextual influences on delinquency: The role of the single-parent family. *Journal of Criminal Justice*, 30:575–587.

Anderson, E. (1994). The code of the streets. *The Atlantic Monthly*, 5:81–94.

Anderson, E. (1999). *Code of the street: Decency, violence, and the moral life of the inner city*. New York: W.W. Norton.

Anderson, W., & C. Summers (2007). Neuroendocrine mechanisms, stress coping strategies, and social dominance: Comparative lessons about leadership potential. *Annals of the American Academy of political and Social Science*, 614:102–130.

Andersson, M., & L. Simmons (2006). Sexual selection and mate choice. *Trends in Ecology and Evolution*, 21:296–302.

Anestis, S. (2006). Testosterone in juvenile and adolescent chimpanzees (Pan troglodytes): Effects of dominance, rank, aggression, and behavioral style. *American Journal of Physical Anthropology*, 130:536–545.

Archer, J. (2006). Testosterone and human aggression: An analysis of the challenge hypothesis. *Neuroscience and Biobehavioral Reviews*, 30:319–345.

Aron, A., H. Fisher, D. Mashek, G. Strong, H. Li, & L. Brown (2005). Reward, motivation, and emotion systems associated with early-stage intense romantic love. *Journal of Neurophysiology*, 94:327–337.

Axelrod, R. (1984). *The evolution of cooperation*. New York: Basic Books.

Badcock, C. (2000). *Evolutionary psychology: A critical introduction*. Cambridge, England: Polity Press.

Bailey, M. (1997). Are genetically based individual differences compatible with species-wide adaptations? In N. Segal, G. Weisfeld, & C. Weisfeld (Eds.), *Uniting psychology and biology* (pp. 81–100). Washington, DC: American Psychological Association.

Baker, L., S. Bezdjian, & A. Raine (2006). Behavior genetics: The science of antisocial behavior. *Law and Contemporary Problems*, 69:7–46.

Bamshad, M., S. Wooding, W. Watkins, C. Ostler, M. Batzer, & L. Jorde (2003). Human population genetic structure and inference of group membership. *American Journal of Human Genetics*, 72:578–589.

Barak, G. (1998). *Integrating criminologies*. Boston, Allyn & Bacon.

Barash, D., & J. Lipton (2001). Making sense of sex. In D. Barash (Ed.), *Understanding violence*, pp. 20–30. Boston: Allyn & Bacon.

Barbaree, H., & W. Marshall (1991). The role of male sexual arousal in rape: Six models. *Journal of Consulting and Clinical Psychology*, 59:621–630.

Barber, N. (2000a). On the relationship between country sex ratios and teen pregnancy: A replication. *Cross-Cultural Research*, 34:26–37.

Barber, N. (2000b). The sex ratio as a predictor of cross-national variation in violent crime. *Cross-Cultural Research*, 34:264–282.

Barber, N. (2003). Paternal investment prospects and cross-national differences in single parenthood. *Cross-Cultural Research*, 37:163–177.

Barkow, J. (1989). *Darwin, sex and status: Biological approaches to mind and culture*. Toronto: University of Toronto Press.

Barkow, J. (1992). Beneath new culture is an old psychology: Gossip and social stratification. In J. Barkow, L. Cosmides, & J. Tooby (Eds.), *The adapted mind:*

Evolutionary psychology and the generation of culture, pp.627–637. New York: Oxford University Press.

Barkow, J. (1997). Happiness in evolutionary perspective. In N. Segal, G. Weisfeld, & C. Weisfeld (Eds.), *Uniting Psychology and Biology*, pp. 397–418. Washington, DC: American Psychological Association.

Barkow, J. (2006). Introduction: Sometimes the bus does wait. In J. Barkow (Ed.), *Missing the revolution: Darwinism for social scientists*, pp. 1–59. Oxford: Oxford University Press.

Barnett, R., L. Zimmer, & J. McCormack (1989). P>V sign and personality profiles. *Journal of Correctional and Social Psychiatry*, 35:18–20.

Bartels, A., & S. Zeki (2004). The neural correlates of maternal and romantic love. *NeuroImage*, 21: 1155–1166.

Bartels, M., & J. Hudziak (2007). Genetically informative designs in the study of resilience in Developmental psychopathology. *Child and Adolescent Psychiatric Clinics of North America*, 16:323–339.

Bartollas, C. (2005). *Juvenile delinquency* (7th ed.). Boston: Allyn & Bacon.

Bateson, C. (1997). Self-other merging and the empathy-altruism hypothesis. *Journal of Personality and Social Psychology*, 73:517–612.

Baumeister, R., L. Smart, & J. Boden (1996). Relation of threatened egoism to violence and aggression: The dark side of self-esteem. *Psychological Review*, 103:5–33.

Baumrind, D. (1993). The average expected environment is not good enough: A response to Scarr. *Child Development*, 64:1299–1317.

Beaver, K. (2009). Molecular genetics and crime. In A. Walsh & K. Beaver, *Biosocial criminology: New directions in theory and research*, pp. 50–72. New York: Routledge.

Beaver, K., J. Wright, & M. DeLisi (2008). Delinquent peer group formation: Evidence of a gene x environment correlation. *The Journal of Genetic Psychology*, 169:227–244.

Beaver, K., J. Wright, & A. Walsh (2008). A gene-based evolutionary explanation for the association between criminal involvement and number of sex partners. *Social Biology*, 54:47–55.

Beaver, K., J. Wright, M. DeLisi, A. Walsh, M. Vaughn, D. Boisvert, & J. Vaske (2007). A gene X gene interaction between DRD2 and DRD4 in the etiology of conduct disorder and antisocial behavior. *Behavioral and Brain Functions*. 30:1–8.

Beatty, M. (2002). Do we know a vector from a scalar? Why measures of association (not their squares) are appropriate indices of effect. *Human Communication Research*, 25:605–611.

Beckman, M. (2004). Crime, culpability, and the adolescent brain. *Science*, 305:596–599.

Beirne, P., & J. Messerschmidt (2000). *Criminology*. Boulder, CO: Westview.

Bell, M., & K. Deater-Deckard (2007). Biological systems and the development of self-regulation: Integrating behavior, genetics, and psychophysiology. *Journal of Developmental and Behavioral Pediatrics*, 28:409–420.

Bellinger, D. (2008). Neurological and behavioral consequences of childhood lead exposure. *PLoS Medicine*, 5:690–692.

Belsky, J. (1999). Conditional and alternative reproductive strategies: Individual differences in susceptibility to rearing experiences. In J. Rodgers & D. Rowe (Eds.), *Genetic influences on fertility and sexuality*. Boston: Klumer.

Belsky, J., & P. Draper (1991). Childhood experience, interpersonal development, and reproductive strategy: An evolutionary theory of socialization. *Child Development*, 62:647–670.

Bem, S. (1987). Androgyny and gender schema theory: A conceptual and empirical integration. In R. Neinstbeir & T. Donderroger (Eds.), *Nebraska symposium on motivation: Psychology and gender*, pp. 179–226. Lincoln: University of Nebraska Press.

Bennett, A., K. Lesch, A. Heills, J. Long, J. Lorenz, S. Shoaf, M. Champoux, S. Suomi, M. Linnoila, & J. Higley (2002). Early experience and serotonin transporter gene variation interact to influence primate CNS functioning. *Molecular Psychiatry*, 7:118–122.

Bennett, S., D. Farrington, & L. Huesman (2005). Explaining gender differences in crime and violence: The importance of social cognitive skills. *Aggression and Violent Behavior*, 10:263–288.

Berg, S., & K. Wynne-Edwards (2001). Changes in testosterone, cortisol, and estradiol levels in men becoming fathers. *Mayo Clinic Proceedings*, 76:582–592.

Berger, B., & P. Berger (1984). *The war on the family: Capturing the middle ground*. Garden City, NY: Anchor.

Bernard, T. (1981). The distinction between conflict and radical criminology. *Journal of Criminal Law and Criminology*, 72:362–379.

Bernard, T. (1987). Testing structural strain theories. *Journal of Research in Crime and Delinquency*, 24, 264–270.

Bernard, T. (1990). Angry aggression among the "truly disadvantaged." *Criminology*, 28: 73–96.

Bernhardt, P. (1997). Influences of serotonin and testosterone in aggression and dominance: Convergence with social psychology. *Current Directions in Psychological Science*, 6:44–48

Betzig, L. (1999). When women win. *Behavioral and Brain Sciences*, 22:217.

Birger, M., M. Swartz, D. Cohen, Y. Alesh, C. Grishpan, & M. Kotelr (2003). Aggression: The testosterone–serotonin link. *Israel Medical Association Journal*, 5:653–658.

Bjorklund, D. (1997). The role of immaturity in human development. *Psychological Bulletin*, 122:153–169.

Bjorklund, D., & A. Pellegrini (2000). Child development and evolutionary psychology. *Child Development*, 71:1687–1708.

Black, J., & W. Greenough (1997). How to build a brain: Multiple memory systems have evolved and only some of them are constructivist. *Behavioral and Brain Sciences*, 20:558–559.

Blair, R. (2005a). Responding to the emotions of others: Dissociating forms of empathy through the study of typical and psychiatric populations. *Consciousness and Cognition*, 14:698–718.

Blair, R. (2005b). Applying a cognitive neuroscience perspective to the disorder of psychopathy. *Development and Psychopathology*, 17:865–891.

Blankenship, M., & S. Brown (1993). Paradigm or perspective? A note to the discourse community. *Journal of Crime and Justice*, 16:167–175.

Blumstein, A. (1995). A LEN interview with Professor Alfred Blumstein of Carnegie Mellon University. *Law Enforcement News*, 21:10–13.

Bohm, R. (2001). *A primer on crime and delinquency* (2nd ed.). Belmont, CA: Wadsworth.

Bond, A. (2005). Antidepressant treatments and human aggression. *European Journal of Pharmacology*, 526:218–225.

Bond, R., & P. Saunders (1999). Routes of success: Influences on the occupational attainment of young British males. *British Journal of Sociology*, 50:217–240.

Bonger, Willem (1969). *Criminality and Economic Conditions*. Bloomington: Indiana University Press.

Booth, A., K. Carver, & D. Granger (2000). A biosocial perspective on the family. *Journal of Marriage and the Family*, 62:1018–1034.

Booth, A., D. Granger, A. Mazur, & Katie Kivligan (2006). Testosterone and social behavior. *Social Forces*, 85:167–191.

Bottomore, T. (1956). *Karl Marx: Selected writings in sociology and social philosophy.* London: Watts & Company.

Bouchard, T., & McGue, M. (1981). Familial studies of intelligence: A review. *Science,* 250:1055–1059.

Bouchard, T., & Segal, N. (1985). Environment and IQ. In B. Wolman (Ed.), *Handbook of intelligence: Theories, measurements, and applications,* pp. 391–464. New York: John Wiley.

Bouchard, T., Segal, N., Tellegen, A., McGue, M., Keyes, M., & Krueger, R. (2003). Evidence for the construct validity and heritability of the Wilson-Patterson conservatism scale: A reared-apart twins study of social attitudes. *Personality and Individual Differences,* 34:959–969.

Boyce, W., J. Quas, A. Alkon, N. Smider, M. Essex, & D. Kupfer (2001). Autonomic reactivity and psychopathology in middle childhood. British Journal of Psychiatry. 179:144–150.

Boyd, T. (1996). A small introduction to the "G" funk era: Gangsta rap and black masculinity in contemporary Los Angeles. In M. Dear, E. Schockman, & G. Wise (Eds.), *Rethinking Los Angeles,* pp. 127–146. Thousand Oaks, CA: Sage.

Box, S. (1987). *Recession, crime and punishment.* Totowa, NJ: Barnes & Noble.

Brammer, G., M. Raleigh, & M. McGuire (1994). Neurotransmitters and social status. In L. Ellis (Ed.), Social stratification and socioeconomic inequality. Vol. 2: Reproductive and interpersonal aspects of dominance and status, pp. 75–91. Westport, CT: Praeger.

Brannigan, A. (1997). Self-control, social control and evolutionary psychology: Towards an integrated perspective on crime. *Canadian Journal of Criminology,* Oct.:403–431.

Brebner, J. (2003). Gender and emotions. *Personality and Individual Differences,* 34:387–394.

Bremner, D. (2002). *Does stress damage the brain? Understanding trauma-related disorders from a neurological perspective.* New York: Norton.

Brennan, P., A. Raine, F. Schulsinger, L. Kirkegaard-Sorenen, J. Knop, B. Hutchings, R. Rosenberg, & S. Mednick (1997). Psychophysiological protective factors for male subjects at high risk for criminal behavior. *American Journal of Psychiatry,* 154:853–855.

Briken, P., N. Habermann, W. Berner, & A. Hill (2006). XYY chromosome abnormality in sexual homicide perpetrators. *American Journal of Medical Genetics,* 141b:198–200.

Brodie, B. (1880). *Ideal chemistry: A lecture.* London: Macmillan and Co.

Bromage, T. (1987). The biological and chronological maturation of early hominids. *Journal of Human Evolution,* 16:257–272.

Bromhall, C. (2003). *The eternal child: An explosive new theory of human origins and behavior.* London: Ebury Press.

Bronfenbrenner, U., & S. Ceci (1994). Heredity, environment, and the question "how"—A first approximation. In R. Plomin & G. McClearn (Eds.), *Nature, nurture, and psychology,* pp.313–324. Washington, DC: American Psychological Association.

Brown, D. (1991). *Human universals.* New York: McGraw-Hill.

Brown, R., & J. Frank (2006). Race and officer decision making: Examining differences in arrest outcomes between black and white officers. *Justice Quarterly,* 23:96–126.

Brownmiller, S. (1975). *Against our will: Men, women, and rape.* New York: Simon & Schuster.

Brunero, J. (2002). Evolution, altruism and internal reward explanations. *Philosophical Forum,* 33:413–424.

Buck, R. (1999). The biological affects: A typology. *Psychological Review*, 106:301–336.

Buck, K., & D. Finn (2000). Genetic factors in addiction: QTL mapping and candidate gene studies implicate GABAergic genes in alcohol and barbiturate withdrawal in mice. *Addiction*, 96:139–149.

Buckley, P. 2004. Pharmacological options for treating schizophrenia with violent behavior. *Psychiatric Times* (supplement), October: 1–8.

Bukowski, W., L. Sippola, & A. Newcomb (2000). Variations in patterns of attraction to same-and other-sex peers during early adolescence. *Developmental Psychology*, 36: 147–154.

Burgess, R., & R. Akers (1966). A differential association-reinforcement theory of criminal behavior. *Social Problems*, 14:128–147.

Buss, D. (1994). *The evolution of desire*. New York: Basic Books.

Buss, D. (2005). *The murderer next door: Why the mind is designed to kill*. New York: Penguin.

Buss, D., & J. Duntley (2006). The evolution of aggression. In M. Schaller, J. Simpson, & D. Kenrick, *Evolution and social psychology*, pp. 263–285. New York: Psychology Press.

Butcher, L., E. Meaburn, L. Liu, C. Fernandes, L. Hill, A. Al-Chalabi, R. Plomin, L. Schalkwyk, & I. Craig (2004). Genotyping pooled DNA on microarrays: A systematic genome screen of thousands of SNPs in large sample to detect QTLs for complex traits. *Behavior Genetics*, 34:549–555.

Button, T., R. Corley, S. Rhee, J. Hewitt, S. Young, & M. Stallings (2007). Delinquent peer affiliation and conduct problems: A twin study. *Journal of Abnormal Psychology*, 116:554–564.

Byne, W. (2006). Developmental endocrine influences on gender identity. *The Mount Sinai Journal of Medicine*, 73:950–959.

Byrne, J. (1986). Cities, citizens, and crime: The ecological/nonecological debate revisited. In J. Byrne & R. Sampson (Eds.), *The Social Ecology of Crime*, pp. 116–130. London: Springer-Verlag.

Byrnes, J., D. Miller, & W. Schafer (1999). Gender differences in risk taking: A meta-analysis. *Psychological Bulletin*, 125:367–383.

Cadoret, R., W. Yates, E. Troughton, G. Woodworth, & M. Stewart (1995). Genetic-environmental interaction in the genesis of aggressivity and conduct disorders. *Archives of General Psychiatry*, 52:916–924.

Cahill, L., N. Uncapher, L. Kilpatrick, M. Alkire, & J. Turner (2004). Sex-related hemispheric lateralization of amygdala function in emotionally-influenced memory: An fMRI investigation. *Learning and Memory*, 11:261–266.

Cale, E., & S. Lilienfeld (2002). Sex differences in psychopathy and antisocial personality disorder: A review and integration. *Clinical Psychology Review*, 22:1179–1207.

Campbell, A. (1999). Staying alive: Evolution, culture, and women's intrasexual aggression. *Behavioral and Brian Sciences*, 22:203–214.

Campbell, A. (2006). Sex differences in direct aggression: What are the psychological mediators? *Aggression and Violent Behavior*, 6:481–497.

Campbell, A. (2009). Gender and crime: An evolutionary perspective. In A. Walsh & K. Beaver (Eds.), *Criminology and biology: New directions in theory and research*, pp. 117–136. New York: Routledge.

Campbell, A., S. Muncer, & D. Bibel (2001). Women and crime: An evolutionary approach. *Aggression and Violent Behavior*, 6:481–497.

Canli, T. (2006). *Biology of personality and individual differences*. New York: Guilford.

Canli, T., J. Desmond, Z. Zhao, & J. Gabrieli (2002). Sex differences in the neural basis of emotional memories. *Proceedings of the National Academy of Sciences*, 99:10789–10794.

Cao, L. (2004). Major criminological theories: Concepts and measurement. Belmont, CA: Wadsworth.

Carey, G. (2003). *Human genetics for the social sciences.* Thousand Oaks, CA: Sage

Carmen, I. (2007). Genetic configurations of political phenomena: New theories, new methods. *Annals of the American Academy of Political and Social Sciences,* 614:34–54.

Carnagey, N., C. Anderson, & B. Bushman (2007). The effect of video game violence on physiological desensitization to real violence. *Journal of Experimental Social Psychology,* 43:489–496.

Carrasco, X., P. Rothhammer, M. Moraga , H. Henríquez, R. Chakraborty, F. Aboitiz, & F. Rothhammer (2006). Genotypic interaction between DRD4 and DAT1 loci is a high risk factor for attention-deficit/hyperactivity disorder in Chilean families. *American Journal of Medical genetics (Neuropsychiatric Genetics),* 141B:51–54.

Casear, P. (1993). Old and new facts about perinatal brain development. *Journal of Child Psychology and Psychiatry,* 34:101–109.

Caspi, A. (2000). The child is the father of the man: Personality continuities from childhood to adulthood. *Journal of Personality and Social Psychology,* 78:158–172.

Caspi, A., D. Bem, & G. Elder (1989). Continuities and consequences of interaction styles across the lifecourse. *Journal of Personality,* 57:375–406.

Caspi, A., B. Henry, R. McGee, T. Moffitt, & P. Silva (1995). Temperamental origins of child and adolescent behavior problems: From age three to age fifteen. *Child Development,* 66:55–68.

Caspi, A., McClay, J., Moffitt, T. E., Mill, J., Martin, J., Craig., I. W., Taylor, A., & Poulton, R. (2002). Role of genotype in the cycle of violence in maltreated children. *Science,* 297, 851–854.

Caspi, A., & T. Moffitt. (1995). The continuity of maladaptive behavior: From description to explanation in the study of antisocial behavior. In D. Cicchetti & D. J. Cohen (Eds.), *Developmental Psychopathology.* Vol. 2: pp. 472-511. *Risk, disorder, and adaptation.* New York: Wiley.

Caspi, A., T. Moffitt, P. Silva, M. Stouthamer-Loeber, R. Krueger, & P. Schmutte. (1994). Are some people crime-prone? Replications of the personality-crime relationship across countries, genders, races, and methods. *Criminology,* 32:163–194.

Catalano, S. (2006). Criminal victimization, 2005. Washington, DC: Bureau of Justice Statistics.

Cauffman, E., L. Steinberg, & A. Piquero, A. (2005). Psychological, neuropsychological, and psychophysiological correlates of serious antisocial behavior in adolescence. *Criminology,* 43, 133–176.

Cavalli-Sforza, L. (2000). *Genes, peoples, and languages.* New York: North Point Press.

Cecil K., C. Brubaker, C. Adler, K. Dietrich, M. Altaye, J. Egelhoff, S. Wessel, I. Elangovan R. Hornung, K. Jarvis, & B. Lanphear (2008). Decreased brain volume in adults with childhood lead exposure. *PLoS Medicine,* 5:742–750.

Chagnon, N. (1988). Life histories, blood revenge, and warfare in a tribal population. *Science,* 239: 985–992.

Chagnon, N. (1996). Chronic problems in understanding tribal violence and warfare. In G. Bock & J. Goode (Eds.), *Genetics of criminal and antisocial behavior,* pp. 202–236. Chichester, England: Wiley.

Chakraborty, B., H. Lee, M. Wolujewicz, J. Mallik, G. Sun, K. Dietrich, A. Bhattacharya, R. Deka, & R. Chakraborty (2008). Low dose effect of chronic lead exposure on neuromotor response impairment in children is moderated by genetic polymorphisms. *Journal of Human Ecology,* 23:183–194.

Chambliss, W. (1976). *Criminal law in action.* Santa Barbara, CA: Hamilton.

Chamorro-Premuzic, T. & A. Furnham (2005). Intellectual competence. *The Psychologist*, 18:352-354.

Chapple, C., & K. Johnson (2007). Gender differences in impulsivity. *Youth Violence and Juvenile Justice*, 5:221-234.

Chesney-Lind, M., & R. Sheldon (1992). *Girls' delinquency and juvenile justice.* Pacific Grove, CA: Brooks/Cole.

Child Trends Data Bank (2000a). Infant homicide. http://www.childtrendsdatabank.org.

Child Trends Data Bank (2000b). Child maltreatment. http://www. childtrendsdatabank.org.

Chilton, R. (1986). Urban crime rates: Effects of inequality, welfare dependency, region, and race. In J. Byrne & R. Sampson (Eds.), *The social ecology of crime*, pp. 116-130. New York: Springer-Verlag.

Ciotti, P. (1998). Money and school performance: Lessons from the Kansas City desegregation experiment. *Policy Analysis*, 298:1-25.

Clark, C., & N. Gist (1938). Intelligence is a factor in occupational choice. *American Sociological Review*, 3:683-694.

Clark, K. (1965). *Dark Ghetto: Dilemmas of social power.* New York: Harper & Row.

Clarke, J. (1998). *The lineaments of wrath: Race, violent crime, and American culture.* New Brunswick, NJ: Transaction Publishers.

Cleveland, H., R. Wiebe, & D. Rowe (2005). Sources of exposure to smoking and drinking friends among adolescents. *The Journal of Genetic Psychology*, 166:153-169.

Cleveland, H., R. Wiebe, E. van den Oord, & D. Rowe (2000). Behavior problems among children from different family structures: The influence of genetic self-selection. *Child Development*, 71:733-751.

Cloward, R., & L. Ohlin (1960). *Delinquency and opportunity.* New York: Free Press.

Clutton-Brock, T. (1991). *The Evolution of Parental Care.* Princeton, NJ: Princeton University Press

Clutton-Brock, T., & Parker, G. (1995). Punishment in animal societies. *Nature*, 373:209-216.

Cohen, A. (1955) *Delinquent boys.* New York: Free Press.

Cohen, J. (1992). A power primer. *Psychological Bulletin*, 112:155-159.

Cohen, L. (1987). Throwing down the gauntlet: A challenge to the relevance of sociology for the etiology of criminal behavior. *Contemporary Sociology*, 16:202-205.

Cohen, L., & R. Machalek (1994). The normalcy of crime: From Durkheim to evolutionary ethology. *Rationality and Society*, 6: 286-308.

Cohen-Bendahan, C., C. van de Beek, & S. Berenbaum (2005). Prenatal sex hormone effects on child and adult sex-typed behavior: Methods and findings. *Neuroscience and Biobehavioral Reviews*, 29:353-384.

Colapinto, J. (2006). *As nature made him: The boy who was raised as a girl.* New York: Harper Perennial.

Cole, S. (Ed.). (2001). *What's wrong with sociology?* New Brunswick, NJ: Transaction.

Collier, S. (2005). Evolutionary perspectives on human development. [Review of the book by the same name]. *Human Ethology Bulletin*, 20:9-12.

Collins, R. (2004). Onset and desistence in criminal careers: Neurobiology and the age-crime relationship. *Journal of Offender Rehabilitation*, 39:1-19.

Collins, W., E, Maccoby, L. Steinberg, M. Heatherington, & M. Bornstein (2000). Contemporary research on parenting: The case for nature *and* nurture. *American Psychologist*, 55:218-232.

Comings, D., T. Chen, K. Blum, J. Mengucci, S. Blum, & B. Meshkin (2005). Neurogenetic interactions and aberrant behavioral co-morbidity of attention deficit hyperactivity disorder (ADHD): Dispelling myths. *Theoretical Biology and Medical Modelling*, 2:1–15.

Congdon, E., & T. Canli (2005). The endophenotype of impulsivity: Reaching consilience through behavioral genetic and neuroimaging approaches. *Behavioral and Cognitive Neuroscience Reviews*, 4:262–281.

Cooley-Quille, M., R. Boyd, E. Frantz, & J. Walsh (2001). Emotional and behavioral impact of exposure to community violence in inner-city adolescents. *Journal of Clinical Child Psychology*, 30:199–206.

Coolidge, F., L. Thede, & S. Young (2000). Heritability and the comorbidity of attention deficit hyperactivity disorder with behavioral disorders and executive function deficits: A preliminary investigation. *Developmental Neuropsychology*, 17:273–287.

Cooper, J., A. Walsh, & L. Ellis (submitted). *Is criminology ripe for a paradigm shift? Evidence from a survey of American criminologists.*

Corwin, E. (2004). The concept of epigenetics and its role in the development of cardiovascular disease. *Biological Research for Nurses*, 6:11–16.

Coser, L. (1971). *Masters of sociological thought.* New York: Harcourt Brace Jovanovich.

Cota-Robles, S., M. Neiss, & D. Rowe (2002). The role of puberty in violent and nonviolent delinquency among Anglo American, Mexican American, and African American boys. *Journal of Adolescent Research*, 17:364–376.

Cote, S. (2002). *Criminological theories*, pp. 5–13. Thousand Oaks, CA: Sage.

Crabbe, J. (2002). Genetic contributions to addiction. *Annual Review of Psychology*, 53:435–462.

Crano, W., & R. Prislin (2006). Attitudes and persuasion. *Annual Review of Psychology*, 57:345–374.

Crawford, C. (1998). Environments and adaptations: Then and now. In C. Crawford & D. Krebs (Eds.), *Handbook of evolutionary psychology: Ideas, issues, and applications* pp. 275–302. Mahwah, NJ: Lawrence Erlbaum.

Crippen, T. (1994). Neo-Darwinian approaches in the social sciences: Unwarranted concerns and misconceptions. *Sociological Perspectives*, 37:391–401.

Crow, T. (2007). How and why genetic linkage has not solved the problem of psychosis: Review and hypothesis. *American Journal of Psychiatry*, 164:13–21.

Cullen, F. (2003). Foreword to J. Crank, *Imagining justice.* Cincinnati, OH: Anderson.

Cullen, F. (2005). Challenging individualistic theories of crime. In S. Guarino-Ghezzi & J. Trevino (Eds.), *Understanding crime: A multidisciplinary approach*, pp. 55–60. Cincinnati, OH: Anderson.

Cullen, F. (2009). Foreword to A. Walsh & K. Beaver, *Biosocial Criminology: New directions in theory and research.* New York: Routledge.

Cummins, D., R. Cummins, & P. Poirier (2003). Cognitive evolutionary psychology without representational nativism. *Journal of Experimental and Theoretical Artificial Intelligence*, 15:143–159.

Curtis, J., & Z. Wang (2003). The neurochemistry of pair bonding. *Current Directions in Psychological Sciences*, 12: 49–53.

D'Alessio, S., & L. Stolzenberg (2003). Race and the probability of arrest. *Social Forces*, 81:1381–1397.

Daly, M. (1996). Evolutionary adaptationism: Another biological approach to criminal and antisocial behavior. In G. Bock & J. Goode (Eds.), *Genetics of criminal and antisocial behaviour*, pp. 183–195. Chichester, England. Wiley.

Daly, M., & M. Wilson (1985). Child abuse and other risks of not living with both parents. *Ethology and Sociobiology*, 6:197–210.

Daly, M., & M. Wilson (1988a). *Homicide*. New York: Aldine De Gruyter.

Daly, M., & M. Wilson (1988b). Evolutionary social psychology and family homicide. *Science*, 242:519–524.

Daly, M., & M. Wilson (1994). Some differential attributes of lethal assaults on small children by stepfathers versus genetic fathers. *Ethology and Sociobiology*, 15:207–217.

Daly, M., & M. Wilson (1996). Violence against stepchildren. *Current Directions in Psychological Science*, 5:77–81.

Daly, M., & M. Wilson (2001). Risk-taking, intersexual competition, and homicide. *Nebraska Symposium on Motivation*, 47:1–36.

Daly M., & M. Wilson (2005). Human behavior as animal behavior. In J. J. Bolhuis & L. A. Giraldeau (Eds.), *Behavior of animals: Mechanisms, function, and evolution*, pp. 393–408. Oxford: Blackwell Publishing.

Davidson, R., K. Putman, & C. Larson (2000). Dysfunction in the neural circuitry of emotion regulation—a possible prelude to violence. *Science*, 289:591–594.

Davis, D., P. Webster, H. Stainthorpe, J. Chilton, L. Jones, & R. Dio (2007). Declines in sex ratios at birth and fetal deaths in Japan, and in U.S. whites but not African Americans. *Environmental Health Perspectives*, 115:941–946.

Dawkins, R. (1982). *The extended phenotype*. Oxford: Oxford University Press.

Dawson, D. (2002). The marriage of Marx and Darwin. *History and Theory*, 41:43–59.

Day J., & R. Carelli (2007). The nucleus accumbens and Pavlovian reward learning. *The Neuroscientist*, 13:148–159.

De Quervain, D., U. Fischbacher, T. Valerie, M. Schellhammer, U. Schnyder, A. Buch, & E. Fehr (2004). The neural basis of altruistic punishment. *Science*, 305:1254–1259.

de Rougemont (1973). Love. In P. Weiner (Ed), *Dictionary of ideas*, pp. 94–108. New York: Charles Scribner.

de Waal (1996). *Good natured: The origins of right and wrong in humans and other animals*. Cambridge, MA: Harvard University Press.

de Waal, F. (2002). Evolutionary psychology: The wheat and the chaff. *Current Directions in Psychological Science*, 11:187–191. New York: Charles Scribner.

de Waal, F. (2008). Putting the altruism back into altruism: The evolution of empathy. *Annual Review of Psychology*, 59:279–300.

Deary, I., F. Spinath, & T. Bates (2006). Genetics of intelligence. *European Journal of Human Genetics*, 14:690–700.

Decety, J., & P. Jackson (2006). A social-neuroscience perspective on empathy. *Current Directions in Psychological Science*, 15:54–58.

DeFina, R., & T. Arvanites (2002). The weak effect of imprisonment on crime: 1971–1998. *Social Science Quarterly*, 83:635–653.

Degler, C. (1991). *In search of human nature: The decline and revival of Darwinism in American social thought*. New York: Oxford University Press.

DeLisi, L., A. Maurizo, C. Svetina, B. Ardekani, K. Szulc, J. Nierenburg, J. Leonard, & P. Harvey (2005). Klinefelter's syndrome (XXY) as genetic model for psychotic disorders. *American Journal of Medical Genetics*, 135:15–23.

DeLisi, M. (2003). Conservatism and common sense: The criminological career of James Q. Wilson. *Justice Quarterly*, 20:661–674.

Dellarosa-Cummins, D., & R Cummins (1999). Biological preparedness and evolutionary explanations. *Cognition*, 73:37–53.

DeLozier, P. (1982). Attachment theory and child abuse. In C. Parks & J. Stevenson-Hinde (Eds.), *The place of attachment in human behavior*, pp. 95–117. New York: Basic.

Demir, B.,G. Ucar, B. Ulug, S. Ulusoy, I. Sevinc & S. Batur (2002). Platelet monoamine oxidase activity in alcoholism subtypes: relationship to personality traits and executive functions. *Alcohol and Alcoholism*, 37:597–602

Demuth, S., & S. Brown (2004). Family structure, family processes, and adolescent delinquency: The significance of parental absence versus parental gender. *Journal of Research in Crime and Delinquency*, 41:56–81.

Dennett, D. (1995). *Darwin's dangerous idea: Evolution and the meanings of life.* New York: Simon & Schuster.

Department of Health and Human Services (2004). Breastfeeding practices—results from the National Immunization Survey. http://www.cdc.gov/breastfeeding/data/NIS_2004.htm.

Depue, R., & P. Collins (1999). Neurobiology of the structure of personality: Dopamine, facilitation of incentive motivation, and extraversion. *Behavioral and Brain Sciences*, 22:491–569.

Dickens, W., & J. Flynn (2001). Heritability estimates versus large environmental effects: The IQ Paradox resolved. *Psychological Review*, 108: 346–349.

Dilalla, L., & I. Gottesman (1989). Heterogeneity of causes for delinquency and criminality: Lifespan perspectives. *Development and psychopathology*, 1:339–349.

Ding, Y., H. Chi, D. Grady, A. Morishima, J. Kidd, K. Kidd, P. Flodman, M. Spence, S. Schuck, J. Swanson, Y. Zhang, & R. Moyziz (2002). Evidence of positive selection acting at the Human dopamine receptor D4 gene locus. *Proceedings of the National Academy of Science*, 99:309–314.

DiRago, A., & G. Viallant (2007). Resilience in inner city youth: Childhood predictors of occupational status across the lifespan. *Journal of Youth Adolescence*, 36:61–70.

Dobzhansky, T. (1973). Nothing in biology makes sense except in light of evolution. *The American Biology Teacher*, 35:125–129.

Dobzhansky, T., F. Ayala, G. Stebbins, & J. Valentine (1977). *Evolution.* San Francisco: W. H. Freeman.

Dolan, M. (2004). Psychopathic personality in young people. *Advances in Psychiatric Treatment*, 10:466–473.

Dollard, J. (1988). *Caste and class in a southern town.* Madison: University of Wisconsin Press.

Domes, G., M. Heinrichs, A. Michel, C. Berger, & S. Herpertz (2007). Ocytocin improves "mind-reading" in humans. *Biological Psychiatry*, 61:731–733.

Drake, R. (1995). A neuropsychology of deception and self-deception. *Behavioral and Brain Sciences*, 18:552–553.

Dreznick, M. (2003). Heterosexual competence of rapists and child molesters: A meta-analysis. *The Journal of Sex Research*, 40:170–178.

Drigotas, S., & Udry, J. (1993). Biosocial models of adolescent problem behavior: Extensions to panel design. *Social Biology*, 40:1–7.

Driscoll, H., A. Zinkivskay, K. Evans, & A. Campbell (2006). Gender differences in social representations of aggression: The phenomenological experience of differences in inhibitory control. *British Journal of Psychology*, 97:139–153.

D'Sousa, D. (1995a). *The end of racism: Principles for a multiracial society.* New York: Free Press.

Du Boise, W. (1899/1967). *The Philadelphia Negro: A social study.* Millwood, NY: Kraus-Thompson.

Du Bois, W. (1903/1969). *The souls of black folk.* New York: New American Library.

Dugatkin, L.(1992). The evolution of the con artist. *Ethology and Sociobiology,* 13:3-18

Dunbar, R. (2007). Male and female brain evolution is subject to contrasting selection pressures in primates. *BioMedCentral Biology,* 5:1–3.

Dunbar, R., & S. Shultz (2007). Evolution of the social brain. *Science,* 317:1344–1347.

DuPont, R. (1997). *The selfish brain: Learning from addiction.* Washington, DC: American Psychiatric Press

Dupre, J. (1992). Blinded by "science": How not to think about social problems. *Behavioral and Brain Sciences,* 15:382–383.

Durant, W. (1952). *The story of philosophy.* New York: Simon & Schuster.

Durkheim, E. (1951a). *The division of labor in society.* Glencoe, IL: Free Press.

Durkheim, E. (1982). *Rules of sociological method.* New York: Free Press.

Durston, S. (2003). A review of the biological bases of ADHD: What have we learned from imaging studies? *Mental Retardation and Developmental Disabilities,* 9:184–195.

Edelman, G. (1987). *Neural Darwinism. The theory of neuronal group selection.* New York: Basic Books.

Edelman, G. (1992). *Bright air, brilliant fire.* New York: Basic Books.

Egan M., M. Kojima, J. Callicott, T. Goldberg, B. Kolachana, A. Bertolino, E. Zaitsev, B. Gold, D. Goldman, M. Dean, B. Lu, & D. Weinberger (2003). The BDNF val66met polymorphism affects activity-dependent secretion of BDNF and human memory and hippocampal function. *Cell,* 112:257–269

Elder, L. (2005). *Children having children.* http://www.townhall.com/columnists/larryelder/printle20041223.shtml.

Elbert, T., & B. Rockstroh (2004). Reorganization of human cerebral cortex: The range of changes follow use and injury. *The Neuroscientist,* 10:129–141.

Elliot, D., D. Huizinga, & S. Menard (1989). *Multiple problem youth: Delinquency, substance abuse, and mental health problems.* New York: Springer-Verlag.

Ellis, L. (1977). The decline and fall of sociology: 1975–2000. *American Sociologist,* 12, 56–66.

Ellis, L. (1987). Criminal behavior and r/K selection: An extension of gene-based\ evolutionary theory. *Deviant Behavior,* 8:149-176

Ellis, L. (1989). *Theories of rape: Inquiries into the causes of sexual aggression.* New York: Hemisphere.

Ellis, L. (1991). A synthesized (biosocial) theory of rape. *Journal of Consulting and Clinical Psychology,* 59:631–642.

Ellis, L. (1996). A discipline in peril: Sociology's future hinges on curing its biophobia. *American Sociologist,* 27:21–41.

Ellis, L. (1998). Neo-Darwinian theories of violent criminality and antisocial behavior: Photographic evidence from nonhuman animals and a review of the literature. *Aggression and Violent Behavior,* 3:61–110.

Ellis, L. (2005). A theory explaining biological correlates of criminality. *European Journal of Criminology,* 2:287–315.

Ellis, L., Beaver, K., & Wright, J. P. (2009). *Handbook of Crime Correlates.* San Diego, CA: Elsevier.

Ellis, L., & H. Hoffman (1990). Views of contemporary criminologists on causes and theories of crime. In L. Ellis & H. Hoffman (Eds.), *Crime in biological, social, and moral contexts,* pp. 50–58. New York: Praeger.

Ellis, L., & H. Nyborg (1992). Racial/ethnic variations in male testosterone levels: A probable contributor to group differences in health. *Steroids,* 57:72–75.

Ellis, L., & A. Walsh (1997). Gene-based evolutionary theories in criminology. *Criminology,* 35:229–276.

Ellis, L., & A. Walsh (2000). *Criminology: A global perspective.* Boston: Allyn & Bacon.

Ellwood, D., & J. Crane (1990). Family changes among black Americans: What do we know? *Journal of Economic Perspectives.* 4:65–84.

Elman, J., E. Bates, M. Johnson, A. Karmiloff-Smith, D. Parisi, & K. Plunkett (1996). *Rethinking innateness: A connectionist perspective on development.* Cambridge, MA: MIT Press.

Ember, M., & C. Ember (1998). Facts of violence. *Anthropology Newsletter,* October:14–15.

Emlen, S. (1995). An evolutionary theory of the family. *Proceedings of the National Academy of Sciences,* 92:8092–8099.

Endler, J. (1986). *Natural selection in the wild.* Princeton, NJ: Princeton University Press.

Engels, F. (1884/1988). Engels on the origin and evolution of the family. *Population and Development Review,* 14:705–729. Originally published as *The origin of the family, private property, and the state.*

Ernst, M., D. Pine, & M. Hardin (2006). Triadic model of the neurobiology of motivated behavior in adolescence. *Psychiatric Medicine,* 36:299–312.

Esch, T., & G. Stefano (2005). Love promotes health. *Neuroendocrinology Letters,* 3:264–267.

Eshel, N., E. Nelson, R. Blair, D. Pine, & M. Ernst (2007). Neural substrates of choice selection in adults and adolescents: Development of the ventrolateral prefrontal and anterior cingulated cortices. *Neuropsychologia,* 45:1270–1279.

Evans, P., S. Gilbert, N. Mekel-Bobrov, E. Vallender, J. Anderson, L. Vaez-Azizi, S. Tishkoff, R. Hudson, & B. Lahn (2005). *Microcephalin,* a gene regulating brain size, continues to evolve adaptively in humans. *Science,* 309:1717–1720.

Evans, G., P. Kim, A. Ting, H. Tesher, & D. Shannis (2007). Cumulative risk, maternal responsiveness, and allostatic load among young adolescents. *Developmental Psychology,* 43:341–351

Eysenk, H. (1982). The sociology of psychological knowledge, the genetic interpretation of the IQ and Marxist-Leninist ideology. *Bulletin of the British Psychological Society,* 35:449–451.

Eysenck, H., & G. Gudjonsson (1989). *The causes and cures of criminality.* New York: Plenum.

Farley, J. (1990). *Sociology.* Englewood Cliffs, NJ: Prentice-Hall.

Farley, R. (1996). *The new American reality.* New York: Russell Sage Foundation.

Farrington, D. (1996). The explanation and prevention of youthful offending. In J. Hawkins (Ed.), *Delinquency and crime: Current theories,* pp 68–148. Cambridge: Cambridge University Press.

Feder, K., & M. Park (1989). *Human antiquity.* Mountain View, CA: Mayfield.

Federal Bureau of Investigation (2007). Crime in the United State: 2006. Washington, DC: U.S. Government Printing Office.

Fehr, E., & S. Gachter (2002). Altruistic punishment in humans. *Nature,* 415:137–140.

Felson, R. (2001). Blame analysis: Accounting for the behavior of protected groups. In S. Cole (Ed.), *What's wrong with sociology?* pp. 223–245. New Brunswick, NJ: Transaction.

Felson, R., & D. Haynie (2002). Pubertal development, social factors, and delinquency among adolescent boys. *Criminology,* 40:967–988.

Ferber, S. (2004). The nature of touch in mothers experiencing maternity blues: The contribution of parity. *Early Human Development,* 79:65–75.

Fergusson, D., & L. Horwood (1995). Early disruptive behavior, IQ, and later school achievement and delinquent behavior. *Journal of Abnormal Child Psychology,* 23:183–199.

Fessler, D. (2006). The male flash of anger: Violent responses to transgression as an example of the intersection of evolved psychology and culture. In J. Barkow

(Ed.), *Missing the revolution: Darwinism for social scientists*, pp. 101–117. Oxford: Oxford University Press.

Fetchenhauer, D., & B. Buunk (2005). How to explain gender differences in fear of crime: Towards an evolutionary approach. *Sexualities, Evolution and Gender,* 7:95–113. Fields, R. (2005). Myelination: An overlooked mechanism of synaptic plasticity. *The Neuroscientist,* 11:528–531.

Fields, S. (2002). White guilt and affirmative action. http//www.townhall.com/columnists/suzannefields/printsf20021205.shtml

Figueredo, A., B. Sales, K. Russel, J. Becker, & M. Kaplan (2000). A Brunswickian evolutionary-developmental theory of adolescent sex offending. *Behavioral Sciences and the Law,* 18:309–329.

Figueredo, A., G. Vasquez, B. Broumbach, S. Schneider, J. Sefcek, I. Tal, D. Hill, C. Wenner, & W. Jacobs (2006). Consilience and life history theory: From genes to brain to reproductive strategy. *Developmental Review,* 26:243–275.

Fincher, J. (1982). *The human brain: Mystery of matter and mind.* Washington, DC: U.S. News Books.

Fishbein, D. (1992). The psychobiology of female aggression. *Criminal Justice and Behavior,* 19:99–126.

Fishbein, D. (1998a). Building bridges. *Academy of Criminal Justice Sciences ACJS Today,* 17:1–5.

Fishbein, D. (1998b). Differential susceptibility to comorbid drug abuse and violence. *Journal of Drug Issues,* 28:859–891.

Fishbein, D. (2001). *Biobehavioral perspectives in criminology.* Belmont, CA: Wadsworth.

Fisher, H., A. Aron, & L. Brown (2005). Romantic love: An fMRI study of a neural mechanism for mate choice. *The Journal of Comparative Neurology,* 493: 58–62.

Fisher, H., A. Aron, & L. Brown (2006). Romantic love: A mammalian brain system for mate choice. *Philosophical Transactions of the Royal Society,* 361: 2173–2186.

Fitting, M. (2002). *Types, tableaus, and Godel's god.* Norwell, MA: Kluwer.

Fleming, A., C. Corter, J. Stallings, & M. Steiner (2002). Testosterone and prolactin are associated with emotional responses to infant cries in new fathers. *Hormones and Behavior,* 42:399–413.

Fletcher, (1991). Mating, the family, and marriage: A sociological view. In Reynolds, V. & Kellett, J. (Eds.), *Mating and Marriage,* pp. 111-162. Oxford. Oxford University Press.

Foley, D., L. Eaves, B. Wormley, J. Silberg, H. Maes, J. Kuhn & B. Riley (2004). Childhood adversity, monoamine oxidase a genotype, and risk for conduct disorder, *Archives of General Psychiatry,* 61: 738–744.

Fortune, W. (1939). Apapesh warfare. *American Anthropologist,* 41:22–41.

Fox, R. (1991) *Encounter with anthropology.* New Brunswick, NJ: Transaction.

Fraga, M., et al. (2005). Epigenetic differences arise during the lifetime of monozygotic twins. *Proceedings of the National Academy of Sciences,* 102:10604–10609.

Frank, D., P. Klass, F. Earls, & L. Eisenberg (1996). Infants and young children in orphanages: One view from pediatrics and child psychiatry. *Pediatrics,* 97:569–578.

Franken, I., P. Muris, & E. Rassin (2005). Psychometric properties of the Dutch BIS/BAS scales. *Journal of Psychopathology and Behavioral Assessment,* 27, 25–30.

Freedman, D. (1997). Is nonduality possible in the social and behavioral sciences? Small essay on holism and related issues. In N. Segal, G. Weisfeld, & C. Weisfeld (Eds.), *Uniting psychology and biology,* pp. 47–80. Washington, DC: American Psychological Association.

Freud, S. (1961). *Civilization and its discontents.* New York: Norton.

Friedman, M., N. Chhabildas, N., Budhiraja, E. Willcutt, & B. Pennington (2003). Etiology of the comorbidity between reading disability and ADHD: Exploration of the non-random mating hypothesis. *American Journal of Medical Genetics Part B (Neuropsychiatric genetics)*, 120b:109–115.

Friedman, M., & R. Freidman (1980). *Free to choose: A personal statement.* New York: Harcourt Brace Jovanovich.

Friedman, N., A. Miyake, S. Young, J. DeFries, R. Corely, & J. Hewitt (2008). Individual differences in executive functions are almost entirely genetic in origin. *Journal of Experimental Psychology*, 137:201–225.

Galernter, J., H. Kranzler, & J. Cubells (1997). Serotonin transporter protein (SLC6A4) allele and haplotype frequencies and linkage disequilibria in African- and European-American and Japanese populations in alcohol-dependent subjects. *Human Genetics*, 101:243–246.

Galvan, A., T. Hare, C. Parra, J. Penn, H. Voss, G. Glover, & B. Casey (2006). Earlier development of the accumbens relative to orbitofrontal cortex might underlie risk-taking behavior in adolescents. *The Journal of Neuroscience*, 26:6885–6892.

Gane, N. (2005). Max Weber as social theorist: 'Class, status, party.' *European Journal of Social Theory*, 8: 211–226.

Gatzke-Kopp, L., A. Raine, R. Loeber, M. Stouthamer-Loeber, & S. Steinhauer (2002). Serious delinquent behavior, sensation seeking, and electrodermal arousal. *Journal of Abnormal Child Psychology*, 30:477–486.

Gaulin, S., & D. McBurney (2001). *Psychology: An evolutionary approach.* Upper Saddle River, NJ: Prentice-Hall.

Gazzaniga, M. (1998). The split brain revisited. *Scientific American*, July: 50–55.

Ge, X, R. Conger, R. Cadoret, J. Neiderhiser, W. Yates, E. Troughton, & M. Stewart (1996). The developmental interface between nature and nurture: A mutual influence model of child antisocial behavior and parental behavior. *Developmental Psychology*, 32:574–589.

Geary, D. (1998). Functional organization of the human mind: Implications for behavioral genetic research. *Human Biology*, 70:185-198.

Geary, D. (2000). Evolution and proximate expression of human paternal investment. *Psychological Bulletin*, 126:55–77.

Geary, D. (2005). *The origin of mind: Evolution of brain, cognition, and general intelligence.* Washington, DC: American Psychological Association.

Geary, D., & M. Flinn (2002). Sex differences in behavioral and hormonal response to social threat: Commentary on Taylor et al. (2000). *Psychological Review*, 109:745–750.

Gelles, R. (1991). Physical violence, child abuse, and child homicide: A continuum of violence or distinct behaviors? *Human Nature*, 2:59–72.

Gewertz, C. (2000). A hard lesson for Kansas City's troubled schools. *Education Week*, April 22, 1–5.

Gewertz, D. (1981).A historical reconsideration of female dominance among the Chambri of Papua New Guinea. *American Ethnologist*, 8:94–106.

Gewertz, D., & Errington, F. (1991). *Twisted histories, altered contexts: Representing the Chambuli in a world system.* Cambridge: Cambridge University Press.

Giddens, A. (1977). Capitalism and modern social theory: An analysis of the writings of Marx, Durkheim, and Weber. Cambridge: Cambridge University Press.

Giedd, J. (2004). Structural magnetic resonance imaging of the adolescent brain. *Annals of the New York Academy of Science*, 1021:77–85.

Gilmartin, P. (1994). *Rape, incest, and child sexual abuse: Consequences and recovery.* New York: Garland.

Giotakos, O., M. Markianos, N. Vaidakis, & G. Christodoulou (2003). Aggression, impulsivity, plasma sex hormones, and biogenic amine turnover in a forensic population of rapists. *Journal of Sex and Marital Therapy*, 29:215–225.

Glahn, D., P. Thompson, & J. Blangero (2002). Neuroimaging endophenotypes: Strategies for finding genes influencing brain structure and function. *Human Brain Mapping*, 28:488–501.

Glaser, D. (2000). Child abuse and neglect and the brain—a review. *Journal of Child Psychology and Psychiatry*, 41:97–116.

Glueck, S. (1956). Theory and fact in criminology: A criticism of differential association theory. *British Journal of Criminology*, 7:92–109.

Glueck, S., & E. Glueck (1950). *Unraveling juvenile delinquency.* New York: Commonwealth Fund.

Goldberg, E. (2001). *The executive brain: Frontal lobes and the civilized mind.* New York: Oxford University Press.

Goldsmith, H., & R. Davidson (2004). Disambiguating the components of emotion regulation. *Child Development*, 75: 361–365.

Goldstein, D., & I. Kopin (2007). Evolution of the concept of stress. *Stress*, 10:109–120.

Goldstein, H. (1990). *Problem-oriented policing.* New York: McGraw-Hill.

Gomez-Smith, Z., & A. Piquero (2005). An examination of adult-onset offending. *Journal of Criminal Justice*, 33:515–525.

Goodlett, C., K. Horn, & F. Zhou (2005). Alcohol teratogenesis: Mechanisms of damage and strategies for intervention. *Developmental Biology and Medicine*, 230:394–406.

Gooren, L. (2006). The biology of human psychosexual development. *Hormones and Behavior*, 50:589–601.

Goos, L., & I. Silverman (2001). The influence of genomic imprinting on brain development and behavior. *Evolution and Human Behavior*, 22:385–407.

Gottesman, I., & D. Hanson. (2005). Human development: biological and genetic processes. *Annual Review of Psychology*, 56, 263–286.

Gottfredson, L. (1986). Social consequences of the g factor in employment. *Journal of Vocational Behavior* 29:379–410.

Gottfredson, L. (1997). Why g matters: The complexity of everyday life. *Intelligence*, 24:79–132.

Gottfredson, M. (2006). The empirical status of control theory in criminology. In F. Cullen, J. Wright, & K. Blevins (Eds.), *Taking stock: The status of criminological theory*, pp. 77–100. New Brunswick, NJ: Transaction.

Gottfredson, M,. & T. Hirschi (1990). A general theory of crime. Stanford, CA: Stanford University Press.

Gottfredson, M., & Hirschi, T. (1997). National crime control policies. In M. Fisch (Ed.), *Criminology, 97/98*, pp. 27–33. Guilford, CT: Dushkin Publishing.

Gottlieb, G. (2007). Probabilistic epigenesis. *Developmental Science,* 10 (1)1–11.

Gottschall, J., & T. Gottschall (2003). Are per-incident rape-pregnancy rates higher than per-incident consensual pregnancy rate? *Human Nature*, 14:1–20.

Gould, S. (1976). Darwin's untimely death. *Natural History*, 85:24–30.

Gould, S. (1991). Exaptation: A crucial tool for an evolutionary psychology. *Journal of Social Issues*, 47:43–65.

Gouldner, A. (1973). Foreword to *The new criminology: For a social theory of deviance.* I. Taylor, P. Walton, & J. Young. London: Routledge & Kegan Paul.

Gove, W. (1985). The effect of age and gender on deviant behavior: A biopsychosocial perspective. In A. Rossi (Ed.), *Gender and the life course*, pp. 115–144. Chicago: Aldine.

Gove, W. & C. Wilmoth (2003). The neurophysiology of motivation and habitual criminal behavior. In A. Walsh & L. Ellis (Eds), Biosocial criminology: Challenging environmentalism's supremacy, pp. 227-245. Hauppauge, NY: Nova Science.

Gray, J. (1994). Three fundamental emotional systems. In P. Ekman & R. Davidson (Eds.), *The nature of emotion: Fundamental questions*, pp. 243–247. New York: Oxford University Press.

Gray, P., J. Chapman, T. Burnham, M. McIntyre, S. Lipson, & P. Ellison (2004). Human male pair bonding and testosterone. *Human Nature*, 15: 119–131.

Green, M. (1983). Marx, utility, and right. *Political Theory*, 11:433–446.

Greenberg, D. (1980). *Crime and capitalism*. Palo Alto, CA: Mayfield.

Greenberg, D. (1981). *Crime and capitalism: Readings in Marxist criminology*. Palo Alto, CA: Mayfield.

Greenberg, D. (1985). Age, crime, and social explanation. *American Journal of Sociology*, 91:1–21.

Griffiths, P. (1990). Modularity and the psychoevolutionary theory of emotion. *Biology and Philosophy*, 5:175-196.

Gross, M. (2005). The evolution of parental care. *The Quarterly Review of Biology*, 80:37–46.

Grossman, W., H. Hylton, & R. McCulley (2006). What happened to the gangs of New Orleans? *Time*, May 22:54–61.

Grosvenor, P. (2002). Evolutionary psychology and the intellectual left. *Perspectives in Biology and Medicine*, 45:433–448.

Groves, C. (1989). *A theory of human and primate evolution*. Oxford, Claredon Press.

Grubin, D. (2007). Sexual offending: The treatment of sex offenders. *Psychiatry*, 6:439–443.

Gubernick, D., D. Sengelaub, & E. Kurtz (1993). A neuroanatomical correlate: Paternal and maternal behavior in the biparental California mouse (*Peromyscus californicus*). *Behavioral Neuroscience*, 107:194–201.

Gunnar, M., & K. Quevedo (2007). The neurobiology of stress and development. *Annual Review of Psychology*, 58:145–173.

Guo, G. (2006). The linking of sociology and biology. *Social Forces*, 85:145–149.

Guo, G., M. Roettger, & J. Shih (2007). Contributions of the DAT1 and DRD2 genes to serious and violent delinquency among adolescents and young adults. *Human Genetics*, 121:125–136.

Guo, G., Y. Tong, C.-W. Xie, & L. Lange (2007). Dopamine transporter, gender, and number of sexual partners among young adults. *European Journal of Human Genetics*, 15:279–287

Gur, R., C. F. Gunning-Dixon, W. Bilker, & R. E. Gur (2002). Sex differences in temporo-limbic and frontal brain volumes of healthy adults. *Cerebral Cortex*, 12:998–1003.

Guttentag, M., & P. Secord (1983). *Too many women: The sex ratio question*. Beverly Hills, CA: Sage.

Haberstick, B. C., J. M. Lessem, C. J. Hopfer, A, Smolen, M. A., Ehringer, D. Timberlake, D., & J. K. Hewitt. (2005). Monoamine oxidase A (MAOA) and antisocial behaviors in the presence of childhood and adolescent maltreatment. *American Journal of Medical Genetics Part B (Neuropsychiatric Genetics)*, 135B, 59–64.

Hall, E. (1987). Adolescents' perceptions of sexual assault. *Journal of Sex Education and Therapy*, 13:37–42.

Hall, S. (2002). Daubing the drudges of fury: Men, violence and the piety of the 'hegemonic masculinity' thesis. *Theoretical Criminology*, 6:35–61.

Hamann, S. (2005). Sex differences in the responses of the human amygdala. *The Neuroscientist*, 11:288–293.

Hamburg, D. (1993). The American family transformed. *Society*, 30:60–69.

Hamilton, W. (1964). The evolution of social behavior. *Journal of Theoretical Biology*, 7:1–52.

Hampton, S. (2004). Domain mismatches, scruffy engineering, exaptations and spandrels. *Theory and Psychology*, 14:147–166.

Hare, R. (1993). *Without conscience: The disturbing world of the psychopaths among us.* New York: Pocket Books.

Harlow, H., & M. Harlow (1962). Social Deprivation in monkeys. *Scientific American*, 206:137–144.

Harpending, H., & G. Cochran (2002). In our genes. *Proceedings of the National Academy of Science*, 99:10–12.

Harpending, H., & P. Draper (1988). Antisocial behavior and the other side of cultural evolution. In T. Moffitt, & S. Mednick (Eds.), *Biological contributions to crime causation*, pp. 293–307. Dordrecht, Netherlands: Martinus Nyhoff.

Harris, J. (1998). *The nurture assumption: Why children turn out the way they do.* New York: Free Press.

Harris, A., & J. Shaw (2001). Looking for patterns: Race, class, and crime. In J. Sheley (Ed.), *Criminology: A contemporary handbook*, pp. 129–163. Belmont, CA: Wadsworth.

Harris, G., M. Rice, & V. Quinsey (1994). Psychopathy as a taxon: Evidence that psychopaths are a discrete class. *Journal of Consulting and Clinical Psychology*, 62:387–397.

Harrison, L., J. Clayton-Smith, & S. Bailey (2001). Exploring the complex relationship between adolescent sexual offending and sex chromosome abnormality *Psychiatric Genetics*, 11:5–10.

Harvard Mental Health Letter (2004). *Pedophilia*, 20:1–4.

Hasin, D., M. Hatzenbuehler & R. Waxman (2006). Genetics of substance use disorders. In William R. Miller & Kathleen Carroll (Eds.), *Rethinking substance abuse: What the science shows, and what we should do about it*, pp. 61–77. New York: Guilford Press.

Hawley, A. (1944). Ecology and human ecology. *Social Forces*, 22:398–405.

Hawley, A. (1950). *Human ecology: A theory of community structure.* New York: Ronald.

Hayner, N. (1933). Delinquency areas in the Puget Sound region. *American Journal of Sociology*, 39:314–28.

Haynes, E., H. Kalkwarf, R. Hornung, R. Wenstrup, K. Dietrich, & P. Lanphear (2003). Vitamin receptor Fok1 polymorphism and blood lead concentration in children. *Environmental Health Perspectives*, 111:1665–1669.

Haynie, D. (2003). Contexts of risk? Explaining the link between girls' pubertal development and their delinquent involvement. *Social Forces*, 82:355–397.

Hayslett-McCall, K., & T. Bernard (2002). Attachment, masculinity, and self-control: A theory of male crime rates. *Theoretical Criminology*, 6:5–33.

Hazard, J., W. Butler, & P. Maggs (1977). *The Soviet legal system.* Dobbs Ferry, NY: Oceana.

Heaney, R. (2006). Low calcium intake among African Americans: Effects on bones and body weight. *Journal of Nutrition*, 136:1095–1098.

Heck, G., & A. Walsh (2000). The effects of maltreatment and family structure on minor and serious delinquency. *International Journal of Offender Therapy and Comparative Criminology*, 44:178–193.

Hendricks, S. (2006). The frame problem and theories of belief. *Philosophical Studies*, 129:317–333.

Hensch, T. (2004). Critical period regulation. *Annual Review of Neuroscience*, 27:549–579.

Herbert, W. (1997). The politics of biology. *U.S. News & World Report*, April 21, 72–80.

Herman, J. (1990). Sex offenders: A feminist perspective. In W. Marshall, D. Laws, & H. Barbaree (Eds.), *Handbook of sexual assault: Issues, theories, and treatment of the offender*, pp. 177–193. New York: Plenum.

Hermans, E., P. Putnam, & J. van Honk (2006). Testosterone administration reduces empathetic behavior: A facial mimicry study. *Psychoneuroendocrinology*, 31:859–866.

Hickey, E. (2006). *Serial murderers and their victims* (4th ed.). Belmont, CA: Wadsworth.

Hiller, J. (2004). Speculations on the links between feelings, emotions and sexual behaviour: Are vasopressin and oxytocin involved? *Sexual and Relationship Therapy*, 19:1468–1479.

Himmelfarb, G. (1994). A de-moralized society: The British/American experience. *The Public Interest*, Fall:57-80.

Hines, M. (2004). *Brain gender*. Oxford: Oxford University Press.

Hirschi, T. (1969). *The causes of delinquency*. Berkeley: University of California Press.

Hirschi, T. (1977). Causes and prevention of juvenile delinquency. *Sociological Inquiry*, 47:322–341.

Hirschi, T. (1995). The family. In J. Wilson& J. Petersilia (Eds.), *Crime*, pp.121–140. San Francisco, ICS Press.

Hirschi, T. (2004). Self-control and crime. In R. Baumeister & K. Vohs (Eds.), *Handbook of self-regulation research, theory, and applications*, pp. 537–552, New York: Guilford.

Hirschi, T., & M. Gottfredson (1983). Age and the explanation of crime. *American Journal of Sociology*, 89:552–584.

Hirschi, T., & M. Hindelang (1977). Intelligence and delinquency: A revisionist review. *American Sociology Review*, 42:571–587.

Holzer, H., & P. Offner (2004). The puzzle of black male unemployment. *The Public Interest*, 154:74-84.

Horgan, J. (1995). The new social Darwinists. *Scientific American*, October:174-182.

Horowitz, I. (1993). *The decomposition of sociology*. New York: Oxford University Press.

Hosking, G. (1985). *The first socialist society: A history of the Soviet Union from within*. Cambridge, MA: Harvard University Press.

Howard, R. (2002). Brain waves, dangerousness and deviant desires . *The Journal of Forensic Psychiatry*, 13:367–384.

Hrdy, S. (1999). *Mother Nature: A history of mothers, infants, and natural selection*. New York: Pantheon.

Hublin, J., & H. Coqueugniot (2006). Absolute or proportional brain size: That is the question. *Journal of Human Evolution*, 50:109–113.

Humes, K., & J. McKinnon (2000). The Asian and Pacific Islander in the United States. U.S. Census Bureau. Washington, DC: U.S. Government Printing Office.

Hur, Y., & T. Bouchard (1997). The genetic correlation between impulsivity and sensation-seeking traits. *Behavior Genetics*, 27:455–463.

Hurford, J., S. Joseph, S. Kirby, & A Reid (1997). Evolution might select constructivism. *Behavioral and Brain Sciences*, 20:567–568.

Hurst, C. (1995). *Social inequality: Forms, causes, and consequences*. Boston: Allyn & Bacon.

Hutchinson, E. (2001). Behind, beside, in front of him? Black women talk about their men. In N. Benokraitis (Ed.), *Contemporary ethnic families in the United States*, pp. 58–70. Upper Saddle River, NJ: Prentice-Hall.

Ingham, J., & D. Spain (2005). Sensual attachment and incest avoidance in human evolution and child development. *Journal of the Royal Anthropological Institute*, 11:677–701.

Jacobson, J., & S. Jacobson (2002). Effects of prenatal alcohol exposure on child development. *Alcohol Research and Health*, 26:282–286.

Jafee, S., T. Moffitt, A. Caspi, & A. Taylor (2003). Life with (or without) father: The benefits of living with two biological parents depend on the father's antisocial behavior. *Child Development*, 74:109-126.

James, W. (1986). Hormonal control of the sex ratio. *Journal of Theoretical Biology*, 118:427–441.

James, W. (1987). The human sex ratio. Part 1: A review of the literature. *Human Biology*, 59:721–752.

Jeffery, C. (1977). Criminology—Whither or wither? *Criminology*, 15:283-286.

Jeffery, C. R. (1993). Obstacles to the development of research in crime and delinquency. *Journal of Research in Crime and Delinquency*, 30:491–497.

Jensen, A. (1998). *The g factor*. Westport, CT: Praeger.

Jensen, A. (1998b). Adoption data and two g-related hypotheses. *Intelligence*, 25:1–6.

Jianhui, L., & H. Fan (2003). Science as ideology: The rejection and reception of sociobiology in China. *Journal of the History of Biology*, 36:567–578.

Jockin, V., V. McGue & D. Lykken (1996). Personality and divorce: A genetic analysis. *Journal of Personality and Social Psychology*, 71: 288–299.

Johnson, L. (1996). Rap, misogyny, and racism. *Radical American*, 26:7–19.

Jorde, L., W. Watkins, M. Barnshad, M. Dixon, C. Ricker, M. Seielstad, & M. Batzer (2000). The distribution of human genetic diversity: A comparison of mitochondrial, autosomal, and Y-chromosome data. *American Journal of Human Genetics*, 66:979–988.

Jost, L., & J. Jost (2007). Why Marx left philosophy for social science. *Theory and Psychology*, 17:297–322.

Judge, T., C. Higgins, C. Thoresen, & M. Barrick (1999). The big five personality traits, general mental ability, and career success across the lifespan. *Personnel Psychology*, 52:621–652.

Jurgensen, M., O. Hiort, P. Holterhus, & U. Thyen (2007). Gender role behavior in children with XY karyotype and disorders of sex development. *Hormones and Behavior*, 51:443–453.

Kagan, J., & N. Snidman (2007). Temperament and biology. In D. Coch, K. Fischer, & G. Dawson (Eds.), *Human behavior, learning, and the developing brain: Typical development*, pp. 219–246. New York: Guilford.

Kalekin-Fishman, D. (2006). Studying alienation: Toward a better society? *Kybernetes*, 35:522–530.

Kanazawa, S. (2003). A general evolutionary psychological theory of criminality and related male-typical behavior. In A. Walsh & L. Ellis (Eds.), *Biosocial criminology: Challenging environmentalism's supremacy*, pp. 37–60. Hauppauge, NY: Nova Science.

Katz, J. (1988). *Seductions of crime: Moral and sensual attractions in doing evil*. New York: Basic Books.

Kelly, R. (2005). The evolution of lethal intergroup violence. *Proceedings of the National Academy of Sciences*, 102:15294–15298.

Kelly, S., N. Day & A. Streissguth (2000). Effects of prenatal alcohol exposure on social behavior in humans and animals, *Neurotoxicology and Teratology*, 22: 43–149.

Kendler, K. (1983). Overview: A current perspective on twin studies of schizophrenia. *American Journal of Psychiatry*, 140:1413–1425.

Kendler, K., K. Jacobson, C. Gardner, N. Gillespie, S. Aggen, & C. Prescott (2007). Creating a social world: A developmental twin study of peer group deviance. *Archives of General Psychiatry.*, 64:958–965.

Kennelly, I., S. Mertz, & J. Lorber (2001). What is gender? *American Sociological Review*, 66:598–605.

Kenrick, D., & J. Simpson (1997). Why social psychology and evolutionary psychology need one another. In J. Simpson & D. Kenrick (Eds.), *Evolutionary social psychology*, pp. 1–20. Mahwah, NJ: Lawrence Erlbaum Associates.

Ketelar, T. & B. Ellis (2000). Are evolutionary explanations unfalsifiable? Evolutionary psychology and the Lakatosian philosophy of science. *Psychological Inquiry*, 11:1-21.

Kim-Cohen, J., A. Caspi, A. Taylor, B. Williams, R. Newcombe, I. Craig, & T. Moffitt (2006). MAOA, maltreatment, and gene-environment interaction predicting children's mental health: New evidence and a meta-analysis. *Molecular Psychiatry*, 11, 903–913.

Kimura, D. (1992). Sex differences in the brain. *Scientific American*, 267:119–125.

King, A. (1999). African American females' attitudes toward marriage: An exploratory study. *Journal of Black Studies*. 29:416–437.

Kinner, S. (2003). Psychopathy as an adaptation: Implications for society and social Policy. In R. Bloom & N. Dass (Eds.), *Evolutionary psychology and violence*, pp. 57–81. Westport, CT: Praeger.

Kirsch P., C. Esslinger, Q. Chen, D. Mier, S. Lis, S. Siddhanti, H. Gruppe, V. Mattay, B. Gallhofer, & A. Meyer-Lindenberg (2005). Oxytocin modulates neural circuitry for social cognition and fear in humans. *Journal of Neuroscience*, 25:11489–11493.

Kleber, H. (2003) Pharmacological treatments for heroin and cocaine dependence. *The American Journal on Addictions*, 12:S5–S18.

Klein, J., & N. Takahata (2002). *Where do we come from? The molecular evidence of human descent*. Berlin: Springer-Verlag.

Kling, J., J. Liebman, & L. Katz (2005). Neighborhood effects on crime for female and male youth: Evidence from a randomized housing voucher experiment. *The Quarterly Journal of Economics*, 120:87–130.

Kling, J., J. Liebman, & L. Katz (2007). Experimental analysis of neighborhood effects. *Econometrica*, 75:83–119.

Knickmeyer, R., S. Baron-Cohen, P. Raggatt, K. Taylor, & G. Hackett (2006). Fetal testosterone and empathy. *Hormones and Behavior*, 49:282–292.

Knight, D. (1992). Ideas in chemistry: A history of the science. New Brunswick, NJ: Rutgers University Press.

Knight, G., I. Guthrie, M. Page, & R. Fabes (2003). Emotional arousal and gender differences in aggression: A meta-analysis, *Aggressive Behavior*, 28:366–393.

Kochanska, G., & N. Aksan (2004). Conscience in childhood: Past, present, and future. *Merrill-Palmer Quarterly*, 50:299–310.

Kochanska, G., & A. Knaack (2003). Effortful control as a personality characteristic of young children: Antecedents, correlates, and consequences. *Journal of Personality*, 71:1087–1112.

Kochanska, M. (1991). Socialization and temperament in the development of guilt and conscience. *Child Development*, 62:1379-1392.

Koller, K., T. Brown, A. Spurfeon, & L. Levy (2004). Recent developments in low-level lead exposure and intellectual impairment in children. *Environmental Health Perspectives*, 112:987–994.

Kornhauser, R. (1978). *Social sources of delinquency: An appraisal of analytical methods*. Chicago: University of Chicago Press.

Kraemer, G. (1992). A psychobiological theory of attachment. *Behavioral and Brain Sciences*, 15:493–541.

Kraemer, G., M. Ebert, D. Schmidt, & W. McKinney (1998). A longitudinal study of the effect of different social rearing conditions on cerebrospinal fluid norepinephrine and biogenic amine metabolites in rhesus monkeys. *Neuopsychopharmacology*, 2:175–189.

Kramer, D. (2005). Commentary: Gene-environment interplay in the context of genetics, epigenetics, and gene expression. *Journal of the American Academy of Child and Adolescent Psychiatry*, 44:19–27.

Kramer, M., F. Aboud, E. Mironova, I. Vanilovich, R. Platt, L. Matush, S. Igumnov, E. Fombonne, N. Bogdanovich, T. Ducruet, J. P. Collet, B. Chalmers, E. Hodnett, S. Davidovsky, O. Skugarevsky, O. Trofimovich, L. Kozlova, S. Shapiro, for the Promotion of Breastfeeding Intervention Trial (PROBIT) Study Group (2008). Breastfeeding and child cognitive development: New evidence from a large randomized trial. *Archives of General Psychiatry*, 65:578–584.

Kratzer, L., & S. Hodgins (1999). A typology of offenders: A test of Moffitt's theory among males and females from childhood to age 30. *Criminal Behavior and Mental Health*, 9:57–73.

Krebs, D. (1998). The evolution of moral behavior. In C. Crawford & D. Krebs (Eds.), *Handbook of evolutionary psychology: Ideas, issues, and applications*, pp. 337–368, Hillsdale, NJ: Lawrence Erlbaum Associates.

Krebs, J., & N. Davies (1993). *An introduction to behavioral ecology*. London: Blackwell.

Kreek, M. J., D. A. Nielsen, E. R. Butelman, & S. K. LaForge (2005). Genetic influences on impulsivity, risk taking, stress responsivity and vulnerability to drug abuse and addiction. *Nature: Neuroscience*, 8, 1450–1457.

Krueger, R., K. Markon, & T. Bouchard, Jr. (2003). The extended genotype: The heritability of personality accounts for the heritability of recalled family environments in twins reared apart. *Journal of Personality*, 71: 809–833.

Krueger, R., T. Moffitt, A. Caspi, A. Bleske, & P. Silva (1998). Assortative mating for antisocial behavior: Developmental and methodological implications. *Behavior Genetics*, 28:173–185.

Krueisi, M., H. Leonard, S. Swedo, S. Nadi, S. Hamburger, J. Lui, & J. Rapoport (1994). Endogenous opioids, childhood psychopathology, and Quay's interpretation of Jeffrey Gray. In D. Routh (Ed.), *Disruptive behavior disorders in childhood*, pp. 207–219. New York: Plenum.

Kruger, D. (2003). Evolution and altruism: Combining psychological mediators with naturally selected tendencies. *Evolution and Human Behavior*, 24:118–125.

Kruk, M., J. Halasz, W. Meelis, & J. Haller (2004). Fast positive feedback between the adrenocortical stress response and a brain mechanism involved in aggressive behavior. *Behavioral Neuroscience*, 118:1062–1070.

Kuhn, T. (1970). *The structure of scientific revolutions*. Chicago: University of Chicago Press.

Kumar, A., K. Choi, W. Renthal, N. Tsankova, D. Theobald, H. Truong, S. Russo, Q. Laplant, T. Sasaki, K. Whistler, R. Neve, D. Self, & E. Nestler (2005). Chromatin remodeling is a key mechanism underlying cocaine-induced plasticity in striatum. *Neuron*, 20:303–314.

Kubrin, C., & R. Weitzer (2003). New directions in social disorganization theory. *Journal of Research in Crime and Delinquency*, 40:374–402.

Kyl-Heku, L., & D. Buss (1996). Tactics as units of analysis in personality psychology: An illustration using tactics of hierarchy negotiation. *Personality and Individual Differences*, 21: 497–517.

Lacourse, E., D. Nagin, F. Vitaro, S. Côté, L. Arseneault, & R. E. Tremblay (2006). Prediction of early-onset deviant peer group affiliation: A 12-year longitudinal study. *Archives of General Psychiatry*, 63:562–568.

LaFree, G. (1996). Race and crime trends in the United States, 1946–1990. In D. Hawkins (Ed.), *Ethnicity, race, and crime: Perspectives across time and space*, pp. 169–193. Albany, NY: State University of New York Press.

LaFree, G., & K. Russell (1993). The argument for studying race and crime. *Journal of Criminal Justice Education*, 4:273–289.

Lahey, B., & R. Loeber (1994). Framework for a developmental model of oppositional defiant disorder and conduct disorder. In D. Routh (Ed.), *Disruptive behavior disorders in childhood*, pp.139–180. New York: Plenum.

Lamanna, M. (1985). *Abortion: Understanding differences*. New York: Plenum.

Lancaster, J., & Lancaster, C. (1987). The watershed: Changes in parental investment and family formation strategies in the course of human evolution. In J. Lancaster, J. Altman, A. Rossi, & L. Sherrod (Eds.), *Parenting across the lifespan: Biosocial perspectives*, pp. 187–205. New York: Aldine De Gruyter.

Lanier, M., & S. Henry (1998). *Essential criminology*. Boulder, CO: Westview.

Laub, J. (1983), Urbanism, race, and crime. *Journal of research in crime and delinquency*, 20:183–198.

Leaper, C., & T. Smith (2004). A meta-analytic review of gender variations in children's language use: talkativeness, affiliative speech, and assertive speech. *Developmental Psychology*, 40:993–1027.

Lee, V., & P. Hoaken (2007). Cognition, emotion, and neurobiological development: Mediating the relation between maltreatment and aggression. *Child Maltreatment*, 12: 281–298.

Lehmann, L., & L. Keller (2006). The evolution of cooperation and altruism—a general framework and classification of models. *European Society for Evolutionary Biology*, 19:1365–1376.

Lemery, K., & H. Goldsmith (2001). Genetic and environmental influences on preschool sibling cooperation and conflict: Associations with difficult temperament and parenting style. Marriage and Family Review, 33:77–99.

Lenski, G. (1978). Marxist experiments in destratification: An appraisal. *Social Forces*, 57:364–383.

Lepowsky, M. (1994). Women, men, and aggression in egalitarian societies. *Sex Roles*, 30:199–211.

Leslie, C. (1990). Scientific racism: Reflections on peer review, science and ideology. *Social Science and Medicine*, 31:891–912.

Levine, D. (1993). Survival of the synapses. *The Sciences*, 33:46–52.

Levine, D. (2006). Neural modeling of the dual motive theory of economics. *The Journal of Socio-Economics*, 35:613–625.

Levy, F. (2004). Synaptic gating and ADHD: A biological theory of comorbidity of ADHD and anxiety. *Neuropsychopharmacology*, 29:1589–1596.

Levy, F., D. Hay, M. McStephen, C. Wood, & I. Waldman (1997). Attention-deficit hyperactivity disorder: A category or a continuum? Genetic analysis of a large-scale twin study. *Journal of the American Academy of Child and Adolescent Psychiatry*, 36:737–744.

Lewis, D. (1991). Conduct disorder. In M. Lewis (Ed.), *Child and adolescent psychiatry: A comprehensive textbook*, pp. 561–583. Baltimore: Williams & Wilkins.

Lewontin, R. (1982). *Human diversity*. New York: Scientific American.

Li, C., X. Mao, & L. Wei (2008). Genes and (common) pathways underlying drug addiction. *PLoS Computational Biology*, 4:28–34.

Li, R., Z, Zhao A. Mokdad, L. Barker, & L. Grummer-Strawn (2003). Prevalence of breastfeeding in the United States: The 2001 National Immunization Survey. *Pediatrics*, 111:1198–1201.

Lilly, J., F. Cullen, & R. Ball (2007). *Criminological theory: Context and consequences*. Thousand Oaks, CA: Sage.

Lin, K. (2001). Biological differences in depression and anxiety across races and ethnic groups. *Journal of Clinical Psychiatry*, 62:13–19.

Lindenfors, P. (2005). Neocortex evolution in primates: The 'social brain' is for females. *Biology Letters*, 1:407–410.

Lindenfors, P., C. Nunn, & R. Barton (2006). Primate brain architecture and selection in relation to sex. *BioMedCentral Biology*, 5:1–9.

Lipsitt, D., S. Buka, & L. Lipsett (1990). Early intelligence scores and subsequent behavior. *American Journal of Family Therapy*, 18:197–208.

Loeber, J. (1994). *The paradoxes of gender.* New Haven, CT: Yale University Press.

Lodi-Smith, J., & B. Roberts (2007). Social investment and personality: A meta-analytic analysis of the relationship of personality traits to investment in work, family, religion, and volunteerism. *Personality and Social Psychology Review*, 11:68–86.

Lopez-Rangel, E., & M. Lewis (2006). Loud and clear evidence for gene silencing by epigenetic mechanisms in autism spectrum and related neurodevelopmental disorders. *Clinical Genetics*, 69:21–25.

Lopreato, J., & T. Crippen (1999). *Crisis in sociology: The need for Darwin.* New Brunswick, NJ: Transaction.

Lovell, D. (2004). Marx's utopian legacy. *The European Legacy*, 9:629–640.

Low, B. (1998). The evolution of life histories. In C. Crawford & D. Krebs (Eds.), Handbook of evolutionary psychology: Ideas, Issues, and applications. pp. 131–161. Mahwah, NJ: Lawrence Erlbaum

Lu, B. (2003). BDNF and activity-dependent synaptic modulation. *Learning and Memory*, 10: 86–98.

Lubinski, D., & Humphreys, L. (1997). Incorporating intelligence into epidemiology and the social sciences. *Intelligence*, 24:159–201.

Lubman, D., M. Yucel, & W. Hall (2007). Substance abuse and the adolescent brain: A toxic combination? *Journal of Psychopharmacology*, 21:792–794.

Luders, E., K. Narr, E. Zaidel, P. Thompson, L. Jancke, & A. Toga (2006). Parasagittal asymmetries of the corpus callosum. *Cerebral Cortex*, 16:346–354.

Ludwig, J., G. Duncan, & P. Hirschfield (2001). Urban poverty and juvenile crime: Evidence from a randomized housing-mobility. *The Quarterly Journal of Economics*, 116:655–679.

Luks, F., F. Hansbrough, D. Klotz, P. Kottmeier, & F. Tolete-Valcek (1988). Early gender assignment in true hermaphrodism. *Journal of Pediatric Surgery*, 23:1122–1126.

Lykken, D. (1995). *The antisocial personalities.* Hillsdale, NJ: Lawrence Erlbaum Associates.

Lynam, D. (1996). Early identification of chronic offenders: Who is the fledgling psychopath? *Psychological Bulletin*, 120:209–234.

Lynn, R. (1990). Testosterone and gonadotropin levels and r/K reproductive strategies. *Psychological Reports*, 67:1203–1206.

Lynn, R. (1996). *Dysgenics: Genetic deterioration in modern populations.* Westport, CT: Greenwood Press.

Lyons, M., W. True, S. Eusen, J. Goldberg, J. Meyer, S. Faraone, L. Eaves, & M. Tsuang (1995). Differential heritability of adult and juvenile antisocial traits. *Archives of General Psychiatry*, 53:906–915.

Lytton, H., & D. Romney (1991). Parents' differential socialization of boys and girls: A meta-analysis. *Psychological Bulletin*, 109:267–296.

Maccoby, E. (2000). Parenting and its effects on children: On reading and misreading behavior genetics. *Annual Review of Psychology*, 51:1–27.

MacDonald, K. (1992). Warmth as a developmental construct: An evolutionary analysis. *Child Development*, 63:753–773.

MacDonald, K. (1997). Life history theory and human reproductive behavior: Environmental/contextual influences and heritable variation. *Human Nature*, 8:327–359.

Machalek, R. (1996). The evolution of social exploitation. *Advances in Human Ecology*, 5:1–32.

Machalek, R. & L. Cohen (1991). The nature of crime: Is cheating necessary for cooperation? *Human Nature*, 2:215-233.

Macionis, J. (1989). Sociology. Englewood Cliffs, NJ: Prentice-Hall.

Mackey, W. (1997). Single-parent families contribute to violent crime. In K. Swisher, (Ed.), *Single-parent families*, pp. 49–52. San Diego, CA: Greenhaven Press.

MacLean, P. (1990). *The triune brain in evolution: Role in paleocerebral functions.* New York: Plenum.

Maletzky, B., & G. Field (2003). The biological treatment of dangerous sexual offenders, a review and preliminary report of the Oregon Pilot Depo-Provera Program. *Aggression and Violent Behavior*, 8:391–412.

Manica, A., & R. Johnstone (2004). The evolution of paternal care with overlapping broods. *The American Naturalist*, 164:517–530.

Mann, C. (1990). Black female homicides in the United States. *Journal of Interpersonal Violence*, 5:176–201.

Mann, C. (1995). Women of color and the criminal justice system. In Price, B. & N. Sokoloff (Eds.). *The Criminal Justice System and Women: Offenders,Victims, and Workers*, pp. 118-119. New York: McGraw-Hill.

Manning, M., & H. Hoyme (2007). Fetal alcohol syndrome disorders: A practical clinical approach to diagnosis. *Neuroscience and Biobehavioral Review*, 31:230–238.

Mardaga, S., & M. Hansenne (2007). The relationship between Cloniger's biosocial model of personality and the behavioral inhibition/approach systems (BIS/BAS). *Personality and Individual Differences*, 42:715–722.

Marks, J. (1996). Science and race. *American Behavioral Scientist*. 40:123-133.

Martindale, D. (2001). Pink slips in your genes. *Scientific American*, 284:19–20.

Martin, S. (2001). The links between alcohol, crime and the criminal justice system: Explanations, evidence and interventions. *The American Journal on Addictions*, 10:136–158.

Marx, K. (1967). *Capital* (volume 1). New York: International Publishers.

Marx, K. (1977). *Karl Marx: Selected writings.* (D. McLellan, Ed.). Oxford: Oxford University Press.

Marx, K. (1978a). Theses on Feuerbach. R. Tucker (Ed.), *The Marx-Engels reader*, pp. 143–145. New York: W.W. Norton.

Marx, K. (1978b). Economic and philosophical manuscripts of 1844. R. Tucker (Ed.), *The Marx-Engels reader*, pp. 66–123. New York: W.W. Norton.

Marx, K., & F. Engels (1948). *The communist manifesto.* New York: International.

Marx, K., & F. Engels (1956). *The holy family.* Moscow: Foreign Languages Publishing House.

Marx, K., & F. Engels (1965). The *German ideology.* London: Lawrence & Wishart.

Marx, K., & F. Engels (1978c). The German ideology: Part 1. R. Tucker (Ed.), *The Marx-Engels reader*, pp. 146–200. New York: W.W. Norton.

Marx, K., & F. Engels (1988). *Economic and philosophical manuscripts of 1844.* Amherst, NY: Prometheus.

Massey, D. (2002). A brief history of human society: The origin and role of emotions in social life. *American Sociological Review*, 67, 1–29.

Massey, D. (2004). Segregation and stratification: A biosocial perspective. *Du Bois Review*, 1:7–25.

Masters, R. (1991). Naturalistic approaches to the concept of justice. *American Behavioral Scientist*, 34:289-313.

Matsueda, R., & K. Anderson (1996). The dynamics of delinquent peers and delinquent behavior. *Criminology*, 36:269–308.

Matsueda, R., & K. Anderson (1998). The Dynamics of delinquent peers and delinquent behavior. *Criminology* 36:269-308.

Matsueda, R., K. Drakulich, & C. Kubrin (2006). Race and neighborhood codes of violence. In R. Peterson, L. Krivo, & J. Hagan (Eds.), *The many colors of crime:*

Inequalities of race, ethnicity, and crime in America, pp. 334–356. New York: New York University Press.

Matthen, M., & A, Ariew (2005). How to understand causal relations in natural selection: Reply to Rosenberg and Bouchard. *Biology and Philosophy*, 20:355–364.

Mauer, M., & R. King (2007). *Uneven justice: State rates of incarceration by race and ethnicity.* Washington, DC: The Sentencing Project.

Maughan, B. (2005). Developmental trajectory modeling: A view from developmental psychopathology. *The Annals of the American Academy of Political and Social Science*, 602:118–130.

Maughan, B., & A. Pickles (1990). Adopted and illegitimate children growing up. In L. Robins & M. Rutter (Eds.), *Straight and devious pathways from childhood to adulthood*, pp. 36–61. Cambridge: Cambridge University Press.

Maughan, B., R. Rowe, J. Messer, R. Goodman, & H. Meltzer (2004). Conduct disorder and oppositional defiant disorder in a national sample: Developmental epidemiology. *Journal of Child Psychology and Psychiatry*, 43:609–621.

Mayes, L., M. Bornstein, K. Chawarska, O. Haynes, & R. Granger (1995). Informational processing and developmental assessment in 3-month-old infants exposed prenatally to cocaine. *Pediatrics*, 95:539–545.

Mayr, E. (2002). The biology of race and the concept of equality. *Daedalus*, Winter:89–94.

Mazur, A. (2005). *Biosociology of dominance and deference.* Lanham, MD: Rowman & Littlefield.

Mazur, A., & A. Booth (1998). Testosterone and dominance in men. *Behavioral and Brain Sciences*, 21:353–397.

Mazur, A., & J. Michalek (1998). Marriage, divorce, male testosterone. *Social Forces*, 77:315–330.

McBurnett, K., B. Lahey, P. Rathouz, & R. Loeber (2000). Low salivary cortisol and persistent aggression in boys referred for disruptive behavior. *Archives of General Psychiatry*, 57:38–43.

McDermott, R. (2004). The feeling of rationality: The meaning of neuroscientific advances for political science. *Perspectives on Politics*, 2:691–706.

McGue, M. (1999). The behavioral genetics of alcoholism. *Current Directions in Psychological Science*, 8:109–115.

McGue, M., S. Bacon, & D.Lykken (1993). Personality stability and change in early adulthood: A behavioral genetic analysis. *Developmental Psychology* 29:96-109.

Mckibbin, W., T. Shackelford, A. Goetz, & V. Starratt (2008). Evolutionary psychological perspectives on rape. In J. Duntley & T. Shackelford (Eds.), *Evolutionary forensic psychology: Darwinian foundations of crime and law*, pp. 101–120. Oxford: Oxford University Press.

McKinnon, J., & K. Humes (2000). *The black population in the United States.* U.S. Census Bureau: Washington, DC.

McWhorter, J. (2000). *Losing the race: Self-sabotage in black America.* New York: Free Press.

Mead, M. (1935). *Sex and temperament in three primitive societies.* New York: Morrow.

Mead, M. (1949). *Male and female: A study of the sexes in a changing world.* New York: Morrow.

Mead, V. (2004). New model for understanding the role of environmental factors in the origins of chronic illness: A case study of type 1 diabetes mellitus. *Medical Hypotheses*, 63:1035–1046.

Mealey, L. (1995). The sociobiology of sociopathy: An integrated evolutionary model. *Behavioral and Brain Sciences*, 18:523–559.

Mealey, L. (2000) *Sex differences: Developmental and evolutionary strategies*. London: Academic Press.

Mealey, L. (2003). Combating Rape: Views of an Evolutionary Psychologist. In R. Bloom and N. Dess, eds., *Evolutionary Psychology and Violence*, pp. 83-113. Westport, CT: Praeger.

Meaney, M. (2001). Maternal care, gene expression, and the transmission of individual differences in stress reactivity across generations. *Annual Review of Neuroscience*, 24:1161–1192.

Mears, D., M. Ploeger, & M. Warr (1998). Explaining the gender gap in delinquency: Peer influence and moral evaluations of behavior. *Journal of Research in Crime and Delinquency*, 35:251–266.

Mednick, S., W. Gabrielli, & B. Hutchings (1984). Genetic influences in criminal convictions: Evidence from an adoption cohort. *Science*, 224: 891–894.

Mehta, P., & R. Josephs (2006). Testosterone change after losing predicts the decision to compete again. *Hormones and Behavior*, 50:684–692.

Mekel-Bobrov, N., S. Gilbert, P. Evans, E. Vallender, J. Anderson, R. Hudson, S. Tishkoff, & B. Lahn (2005). Ongoing adaptive evolution of *ASPM*, a brain size determinant in *Homo sapiens*. *Science*, 309:1720–1722

Menard, S., & S. Mihalic (2001). The tripartite conceptual framework in adolescence and adulthood: Evidence from a national sample. *Journal of Drug Issues*, 31:905–940.

Messner, S., & R. Rosenfeld (2001). *Crime and the American dream*. (3rd ed.). Belmont, CA: Wadsworth.

Messner, S., & R. Sampson (1991). The sex ratio, family disruption, and rates of violent crime: The paradox of demographic structure. *Social Forces*, 69:693–723.

Merton, R. (1938). Social structure and anomie. *American Sociological Review*, 3:672–682.

Merton, R. (1968). *Social theory and social structure*. Glencoe, IL: Free Press.

Messerschmidt, J. W. (1993). *Masculinities and crime*. Lanham, MD: Rowman & Littlefield.

Methwin, E. (1997). Mugged by reality. *Policy Review*, July/August:32–39.

Meyer-Bahlburg, H., C. Dolezal, S. Baker, A. Ehrhardt, & M. New (2006). Gender development in women with congenital adrenal hyperplasia as a function of disorder severity. *Archives of Sexual Behavior*, 35:667–684.

Meyer-Lindenberg, A., J., Buckholtz, B. Kolachana, A. Hariri, L. Pezawas, G. Blasi, A. Wabnitz, R. Honea, B. Verchinski, J. Gallicott, M. Egan, V. Mattay, & D. Weinberger (2006). Neural mechanisms of genetic risk for impulsivity in violence in humans. *Proceedings of the National Academy of Sciences*, 103:6269–6274.

Miles, D., & G. Carey (1997). Genetic and environmental architecture of human aggression. *Journal of Personality and Social Psychology*, 72:207–217.

Miller, J. (1998). Up it up: Gender and the accomplishment of street robbery. *Criminology*, 36:37–65.

Miller, L. (1987). Neuropsychology of the aggressive psychopath: An integrative review. *Aggressive Behavior*, 13:119–140.

Miller-Butterworth, C., J. Kaplan, J. Shaffer, B. Devlin, S. Manuck, & R. Ferrell (2008). Sequence variation in the primate dopamine transporter gene and its relationship to social dominance. *Molecular Biology and Evolution*, 25:18–28

Millner, D., & N. Chiles (1999). *What brothers think, what sisters know: The real deal on love and relationships*. New York: Morrow.

Mills, J., D. Anderson, & D. Kroner (2004). The antisocial attitudes of sex offenders. *Criminal Behavior and Mental Health*, 14:134–145.

Millstein S. G., & Halpern-Felsher B. L. (2002). Judgments about risk and perceived invulnerability in adolescents and young adults. *Journal of Research on Adolescence*, 12:399–422.

Milner, C., & R. Milner (1972). *Black players*. Boston: Little, Brown.

Mitchell, K. (2007). The genetics of brain wiring: From molecule to mind. *PLoS Biology*, 4:690–692.

Moffitt, T. (1993). Adolescent-limited and life-course-persistent antisocial behavior: A developmental taxonomy. *Psychological Review*, 100:674–701.

Moffitt, T. (1996). Childhood-onset versus adolescent-onset antisocial conduct problems in males: Natural history from ages 3 to 18. *Development and Psychopathology*, 8:399–424.

Moffitt, T. (1996). The neuropsychology of conduct disorder. In P. Cordella & L. Siegel (Eds.), *Readings in contemporary criminology*, pp. 85–106. Boston: Northeastern University Press.

Moffitt, T. (2005). The new look of behavioral genetics in developmental psychopathology: Gene-environment interplay in antisocial behavior. *Psychological Bulletin*, 131:533–554.

Moffitt, T. E., A. Caspi, N. Dickson, P. A. Silva, & W. Stanton (1996). Childhood-onset versus adolescent-onset antisocial conduct in males: Natural History from age 3 to 18. *Development and Psychopathology*, 8, 399–424.

Moffitt, T., A. Caspi, H. Harrington, & B. Milne (2002). Males on the life-course persistent and adolescence-limited antisocial pathways: Follow-up at age 26. *Development and Psychopathology*, 14, 179–206.

Moffitt, T., A. Caspi, M. Rutter, & P. Silva (2001). *Sex differences in antisocial behaviour: Conduct disorder, delinquency and violence in the Dunedin longitudinal study*. Cambridge: Cambridge University Press.

Moffit, T., & the E-Risk Study Team (2002). Teen-aged mothers in contemporary Britain. *Journal of Child Psychology and Psychiatry*, 43:1–16.

Moffitt, T., D. Lynam, & P. Silva (1994). Neuropsychological tests predicting persistent male delinquency. *Criminology*, 32:277–300.

Moffitt, T., & P. Silva (1988). IQ and delinquency: A test of the differential detection hypothesis. *Journal of Abnormal Psychology*, 97:330–333.

Moffitt, T., & A. Walsh (2003). The adolescence-limited/life-course persistent theory of antisocial behavior: What have we learned? In A. Walsh & L. Ellis (Eds.), *Biosocial criminology: Challenging environmentalism's supremacy*, pp. 125–144. Hauppauge, NY: Nova Science.

Moir, A., & D. Jessel (1995). *A mind to crime*. London: Michael Joseph.

Molenaar, P., D. Boomsma, & C. Dolan (1993). A third source of developmental differences. *Behavior Genetics*, 23:519–524.

Molles, M. (2008). *Ecology: Concepts and applications* (4th ed.). New York: McGraw-Hill.

Money, J. (1986). *Venuses Penuses: Sexology, Sexosophy, and Exigency Theory*. Buffalo, NY: Prometheus.

Monk, C., R. Klein, E. Telzer, E. Schroth, S. Mannuzza, J. Moulton, M. Guardino, C. Masten, E. McClure-Tone, S. Fromm, R. Blair, D. Pine, & M. Ernst (2008). Amygdala and nucleus accumbens activation to emotional facial expressions in children and adolescents at risk for major depression *American Journal of Psychiatry*, 165:90–98

Montagu, A. (1978). Touching: The human significance of the skin. New York: Harper & Row.

Montagu, A. (1981). *Growing young*. New York: McGraw-Hill.

Moore, C., & M. Rose (1995). Adaptive and nonadaptive explanations of sociopathy. *Behavior and Brain Sciences*, 18:566–567.

Mouse Genome Sequencing Consortium (2002). Initial sequencing and comparative analysis of the mouse genome. *Nature*, 420:520–562.

Moynihan, D. (1965). *The Negro family: The case for national action*. Washington, DC: U.S. Department of Labor.

Mumola, C. (2000). *Incarcerated parents and their children.* Washington, DC: Bureau of Justice Statistics.

Murray, C. (1997). IQ and economic success. *Public Interest,* 128:21–35.

Murray, J., M. Liotti, H. Mayberg, Y. Pu, F. Zamarripa, & Y. Liu (2006). Children's brain activations while viewing televised violence revealed by fMRI. *Media Psychology,* 8:25–37.

Nair, H., & L. Young (2006). Vasopressin and pair-bond formation: Genes to brain to behavior. *Physiology,* 21:146–152.

Nasser, A. (1975). Marx's ethical anthropology. *Philosophy and Phenomenological Research,* 35:484–500.

Needleman, H. (2004). Lead poisoning. *Annual Review of Medicine,* 55:209–222.

Neisser, U., G. Boodoo, T. Bouchard, A. Boykin, N. Brody, S. Ceci, D. Halpern, J. Loehlin, R. Perloff, R. Sternberg, & S. Urbina (1995). *Intelligence: Knowns and unknowns. Report of a task force established by the board of scientific affairs of the American Psychological Association.* Washington, DC: American Psychological Association.

Nelson, E., E. McClure, C. Monk, E. Zahran, E. Leibenluft, D. Pine, & M. Ernst (2003). Developmental differences in neuronal engagement during implicit encoding of emotional faces: An event-related fMRI study. *Journal of Child Psychology and Psychiatry and Allied Disciplines,* 44:1015–1024.

Nelson, K., & J. White (2002). Androgen receptor CAG repeats and prostate cancer. *American Journal of Epidemiology,* 155:883–890.

Nesse, R., & A. Lloyd (1992). The evolution of psychodynamic mechanisms. In J. Barkow, L. Cosmides, & J. Tooby (Eds.), *The adapted mind: Evolutionary psychology and the generation of culture,* pp. 601–620. New York: Oxford University Press.

Nettle, D. (2003). Intelligence and class mobility in the British population. *British Journal of Psychology,* 94:551–561.

Nettler, G. (1978). *Explaining crime.* New York: McGraw-Hill.

Nettler, G. (1984). *Explaining crime* (3rd ed.). New York: McGraw-Hill.

Newton, S., & R. Duman (2006). Chromatin remodeling: A novel mechanism of psychotropic drug action. *Molecular Pharmacology,* 70:440–443.

Nicholson, S. (2002). On the genealogy of norms: A case for the role of emotion in cultural evolution. *Philosophy of Science,* 69: 234–255.

Nicolson, R., S. Bhalerao, & L. Sloman (1998). Karyotype and pervasive developmental disorders. *Canadian Journal of Psychiatry,* 43:619–622.

Niehoff, D. (2003). A vicious circle: The neurobiological foundations of violent behavior. *Modern Psychoanalysis,* 28:235–245.

Nielsen, F. (1994). Sociobiology and sociology. *Annual Review of Sociology,* 20:267–303.

Nielsen, F. (2006). Achievement and ascription in educational attainment: Genetic and environmental influences on adolescent schooling. *Social Forces,* 85:193–216.

Noble, M., M. Mayer-Proschel, & R. Miller (2005). The oligodendrocyte. In M. Rao & M. Jacobson (Eds.), *Developmental neurobiology,* pp. 151–196. New York: Kluwer/Plenum.

Norton, E. (1987). Restoring the traditional black family. In L. Barnes (Ed.), *Social problems.* Guilford, CT: Dushkin.

Nowak, M., & K. Sigmund (2005). Evolution of indirect reciprocity. *Nature,* 437:1291–1298

Oberwittler, D. (2006). A multilevel analysis of neighborhood contextual factors on serious juvenile offending. *European Journal of Criminology,* 1:201–235.

O'Brien, G. (2006). Behavioural phenotypes: Causes and clinical implications. *Advances in Psychiatric Treatment,* 12:338–348.

O'Brien, R. (2001). Crime facts: Victim and offender data. In Sheley, J. (Ed.). *Criminology: A contemporary handbook* (pp. 59-83). Belmont: CA, Wadsworth.

O'Connor, T., K. Deater-Deckard, D. Fulker, M. Rutter, & R. Plomin (1998). Genotype-environment correlations in late childhood and early adolescence: Antisocial behavioral problems and coercive parenting. *Developmental Psychology*, 34:970–981.

O'Leary, M., B. Loney, & L. Eckel (2007). Gender differences in the association between psychopathic personality traits and cortisol response to induced stress. *Psychoneuroendocrinology*, 32:183–191.

Oliver, M., & Hyde, J. (1993). Gender differences in sexuality: A meta-analysis. *Psychological Bulletin*, 14:29–51.

Olson, J., P. Vernon, & J. Harris (2001). The heritability of attitudes: A study of twins. *Journal of Personality and Social Psychology*, 80:845–860.

O'Manique, J. (2003). *The origins of justice: The evolution of morality, human rights, and law*. Philadelphia: University of Philadelphia Press.

Oniszczenko, W., & U. Jakubowski (2005). Genetic determinants and personality correlates of sociopolitical attitudes in a Polish sample. *Twin Research and Human Genetics*, 8:47–52.

Osgood, D., & J. Chamber (2003). Community correlates of rural youth violence. *Juvenile Justice Bulletin*, May. U.S. Department of Justice.

Osofsky, J. (1995). The effects of exposure to violence on young children. *American Psychologist*, 50:782–788.

Palmer, C. (1994). Twelve reasons why sex is not sexually motivated: A skeptical explanation. In R. Francoeur (Ed.), *Taking sides: Clashing views on controversial issues in human sexuality*. Guilford, CT: Dushkin.

Palmer, C., & R. Thornhill (2003). Straw men and fairly tales: Evaluating reactions to *A Natural History of Rape*. *The Journal of Sex Research*, 40:249–255.

Panksepp, J. (1992). A critical role for "affective neuroscience" in resolving what is basic about basic emotions. *Psychological Review*, 9:554–560.

Panksepp, J. (1998). Attention deficit hyperactivity disorders, psychostimulants, and intolerance of childhood playfulness: A tragedy in the making. *Current Directions in Psychological Science*, 7:91–98.

Panksepp, J., J. Moskal, J. B. Panksepp, & R. Kroes (2002). Comparative approaches in evolutionary psychology: Molecular neuroscience meets the mind. *Neuroendocrinology Letters*, 23:105–115.

Park, R., E. Burgess, & R. McKenzie (1925). *The City*. Chicago: University of Chicago Press.

Parsey, R., M. Oquendo, N. Simpson, R. Ogden, R. Heertum, V. Arango, & J. Mann (2002). Effects of sex, age, and aggressive traits in man on brain serotonin 5-HT$_{iA}$ receptor binding Potential measured by PET using [C-11] WAY-100635. *Brain Research*, 954:173–182.

Parsons, L., & D. Osherson (2001). New evidence for distinct right and left brain systems for deductive versus probabilistic reasoning. *Cerebral Cortex*, 11: 954–965.

Passas, N. (1995). Continuities in the anomie tradition. In F. Adler & W. Laufer (Eds.), *The legacy of anomie theory*, pp. 91–112. New Brunswick, NJ: Transaction.

Patrick, C. (1994). Emotions and psychopathy: Startling new insights. *Psychophysiology*, 31:319–330.

Patterson, O. (1998). *Rituals of blood: Consequences of slavery in two American centuries*. Washington, DC: Civitas Counterpoint.

Patterson, O. (2000). Taking culture seriously: A framework and an Afro-American illustration. In L. Harrison & S. Huntington (Eds.), *Culture matters: How values shape human progress*, pp.202–230. New York: Basic Books.

Patton, G., B. McMorris, J. Toumbourou, S. Hemphill, S. Donath, & R. Catalano (2004). Puberty and the onset of substance use and abuse. *Pediatrics*, 114:300–306.

Paul, D. (1984). Eugenics and the left. *Journal of the History of Ideas*, 45:567–590.

Paus, T., A. Zijdenbos, K. Worsley, D. Collins, J. Blumenthal, J. Giedd, J. Rapoport, & A. Evans (1999). Structural maturation of neural pathways in children and adolescents: In vivo study. *Science*, 283:1908–1911.

Pedersen, F. (1991). Secular trends in human sex ratios: Their influence on individual and family behavior. *Human nature*, 2:271–291.

Penn, A. (2001). Early brain wiring: Activity-dependent processes. *Schizophrenia Bulletin*, 27:337–348.

Perry, B. (1997). Incubated in terror: Neurodevelopmental factors in the "cycle of violence." In J. Osofsky (Ed.). *Children in a Violent Society*, pp. 124-149. New York: Guilford Press

Perry, B. (2002). Childhood experience and the expression of genetic potential: What childhood neglect tells us about nature and nurture. *Brain and Mind*, 3:79–100.

Perry, B., & R. Pollard (1998). Homeostasis, stress, trauma, and adaptation: A neurodevelopmental view of childhood trauma. *Child and Adolescent Psychiatric Clinics of America*, 7:33–51.

Peters, J., T. Shackelford, & D. Buss (2002). Understanding domestic violence against women: Using evolutionary psychology to extend the feminist functional analysis. *Violence and Victims*, 17:255–264.

Petronis, A, I.Gootesman, T. Crow, L. DeLisi, A. Klar, F. Macciardi, M. McInnis, F. Mahon, A. Paterson, D. Skuseet, & G. Sutherlan (2000). Psychiatric epigenetics: A new focus for a new century. *Molecular Psychiatry*, 5:342-346.

Phelps, E. (2006). Emotion and cognition: Insights from studies of the human amygdala. *Annual Review of Psychology*, 57:27–53.

Pigliucci, M., C. Murren, & C. Schlichting (2006). Phenotypic plasticity and evolution by genetic assimilation. *Journal of Experimental Biology*, 209: 2362–2367.

Pillsworth, E., & M. Haselton. (2005). The evolution of coupling. *Psychological Inquiry*, 16:98–104.

Pine, D., J. Coplan, G. Wasserman, L. Miller, J. Fried, M. Davies, T. Cooper, L. Greenhill, D. Shaffer, & B. Parsons (1997). Neuroendocrine response to fenfluramine challenge to boys: Associations with aggressive behavior and adverse rearing. *Archives of General Psychiatry*, 54:839–846.

Pinker, S. (1997). *How the mind works*. New York: Norton.

Pinker, S. (2002). *The blank slate: The modern denial of human nature*. New York: Viking.

Pitchford, I. (2001). The origins of violence: Is psychopathy an adaptation? *Human Nature Review*, 1:28–38.

Pitman, R., & S. Orr (1993). Psychophysiological testing for post-traumatic stress disorder. *American Academy of Psychiatry and Law*, 21:37–52.

Plavcan, J., & C. van Schaik (1997). Intrasexual competition and body weight dimorphism in anthropoid primates. *American Journal of Physical Anthropology*, 103:37–68.

Plomin, R. (2003). General cognitive ability. In R. Plomin, J. Defries, I. Craig, & P. McGuffin (Eds.), *Behavioral genetics in the postgenomic era*, pp.183–201. Washington, DC: American Psychological Association.

Plomin, R. (2005). Finding genes in child psychology and psychiatry: When are we going to be there? *Journal of Child Psychology and Psychology*, 46:1030–1038.

Plomin, R., K. Ashbury, & J. Dunn (2001). Why are children in the same family so different? Nonshared environment a decade later. *Canadian Journal of Psychiatry*, 46:225–233.

Plomin, R & K.Asbury (2005). Nature and Nurture: Genetic and Environmental Influences on Behavior. *The Annals of the American Academy of Political and Social Science*, 600: 86-98.

Plomin, R., & C. Bergeman (1991) The nature of nurture: Genetic influences on "environmental" measures. *Behavioral and Brain Sciences*, 14:373–427.

Plomin, R., J. Defries, I. Craig, & P. McGuffin (2001). *Behavioral genetics* (4th ed.). New York: Worth Publishers.

Plomin, R., & S. Petrill (1997). Genetics and intelligence: What's new? *Intelligence*, 24:53–77.

Pollak, S. (2005). Early adversity and mechanisms of plasticity: Integrating affective neuroscience with developmental approaches to psychopathology. *Development and Psychopathology*, 17:735–752.

Pope, C., & Snyder, H. (2003). Race as a factor in juvenile arrests. Juvenile justice bulletin (NCJ 189180). Washington, DC: Office of Juvenile Justice and Delinquency Prevention.

Popenoe, D. (1993). American family decline, 1960–1990: Evidence from the 1980s. *Journal of Marriage and the Family*, 55:527–542.

Popenoe, D. (1994). The family condition of America: Cultural change and public policy. In H. Aaron, T. Mann, & T. Taylor (Eds.), *Values and public policy*, pp. 81-111. Washington, DC: The Brookings Institute.

Porter, M. (1995). The competitive advantage of the inner city. *Harvard Business Review*, 73:63–64.

Posthuma, D., E. de Geus, & D. Boomsma (2003). Genetic contributions to anatomical, behavioral, and neurophysiological indices of cognition. In R. Plomin, J. Defries, I. Craig, & P. McGuffin (Eds.), *Behavioral genetics in the postgenomic era*, pp. 141–161. Washington, DC: American Psychological Association.

Powell, K. (2006). How does the teenage brain work? *Nature*, 442:865–867.

Pratt, T., & F. Cullen (2000). The empirical status of Gottfredson and Hirschi's general theory of crime: A meta-analysis. *Criminology*, 38:931–964.

Prayer, D., G. Kasprian, E. Krampl, B. Ulm, L. Witzani, L. Prayer, & P. Brugger (2006). MRI of normal fetal brain development. *European Journal of Radiology*, 57:199–216.

Price, B., & N. Sokoloff (1995). Theories and facts about women offenders. In B. Price & N. Sokoloff (Eds.), *The Criminal justice system and women: Offenders, victims, and workers*, pp.1–10. New York: McGraw-Hill.

Quartz, S., & T. Sejnowski (1997). The neural basis of cognitive development: A constructivist manifesto. *Behavioral and Brain Sciences*, 20:537–596.

Quay, H. (1997). Inhibition and attention deficit hyperactivity disorder. *Journal of Abnormal Child Psychology*, 25:7–13.

Quinlan, R., & Quinlan, M. (2007). Evolutionary ecology of human pair bonds: Cross cultural tests of alternative hypotheses. *Cross-Cultural Research*, 41:149–169.

Quinney, R. (1975). Crime control in capitalist society: A critical philosophy of legal order. In I. Taylor, P. Walton, & J. Young (Eds.), *Critical criminology*, pp. 181–202. Boston: Routledge & Kegan Paul.

Quinsey, V. (2002). Evolutionary theory and criminal behavior. *Legal and Criminological Psychology*, 7:1–14.

Quinton, D., A. Pickles, B. Maughan, & M. Rutter (1993). Partners, peers and pathways: Assortative pairing and continuities in conduct disorder. *Development and Psychopathology*, 5:763–783.

Qvarnstrom, A., J. Brommer, & L. Gustafsson (2006). Testing the genetics underlying the co-evolution of mate choice and ornament in the wild. *Nature*, 44:84–86.

Raine, A. (1993). *The psychopathology of crime: Criminal behavior as a clinical disorder*. San Diego, CA: Academic Press.

Raine, A. (1997). Antisocial behavior and psychophysiology: A biosocial perspective and a prefrontal dysfunction hypothesis. In D. Stoff, J. Breiling, & J. Maser (Eds.), *Handbook of antisocial behavior*, pp. 289–304. New York: John Wiley.

Raine, A., J. Meloy, S. Bihrle, J. Stoddard, L. LaCasse, & M. Buchsbaum (1998). Reduced prefrontal and increased subcortical brain functioning assessed using positron emission tomography in predatory and affective murderers. *Behavioral Sciences and the Law*, 16:319–332.

Raine, A., T., Lencz, S., Bihrle, L. Lacasse, & P. Colletti (2000). Reduced prefrontal gray matter volume and reduced autonomic activity in antisocial personality disorder. *Archives of General Psychiatry*, 57:119-127.

Rakic, P. (1996). Development of the cerebral cortex in human and non-human primates. In M. Lewis, (Ed.), *Child and Adolescent Psychiatry: A Comprehensive Textbook*, pp. 9-30. New York: Williams and Wilkins.

Raleigh, M., M. McGuire, G. Brammer, D. Pollock, & A. Yuwiler (1991). Serotonergic mechanisms promote dominance acquisition in adult vervet monkeys. *Brain Research*, 559:181–190.

Raley, S., & S. Bianchi (2006). Sons, daughters, and family processes: Does gender of children matter? *Annual Review of Sociology*, 32:401–421.

Rasche, C. (1995). Minority women and domestic violence: The unique dilemmas of battered women of color. In B. Price & N. Sokoloff (Eds.), *The criminal justice system and women: Offenders, victims, and workers*, pp. 246–261. New York: McGraw-Hill.

Raz, A. (2004). Brain imaging data of ADHD. *Neuropsychiatry*, August:46–50.

Repetti, R., S. Taylor, & T. Seeman (2002). Risky families: Family social environment and the mental and physical health of offspring. *Psychological Bulletin*, 128:330–336.

Replogle, R. (1990). Justice as superstructure: How vulgar is vulgar materialism? *Polity*, 22:675–699.

Restak, R. (2001). *The secret life of the brain.* New York: co-published by Dana Press and Joseph Henry Press.

Reyna, V., & F. Farley, F. (2006). Risk and rationality in adolescent decision making: Implications for theory, practice, and public policy. *Psychological Science in the Public Interest*, 7: 1–44.

Rhee, S., & I. Waldman (2002). Genetic and environmental influences on antisocial behavior: A meta-analysis of twin and adoption studies. *Psychological Bulletin*, 128:490–529.

Richter-Levin, G. (2004). The amygdala, the hippocampus, and emotional modulation of memory. *The Neuroscientist*, 10:31–39.

Ridley, M. (1996). *The origins of virtue: Human instincts and the evolution of cooperation.* New York: Viking.

Ridley, M. (2003). *Nature via nurture: Genes, experience and what makes us human.* New York: HarperCollins.

Riggins-Caspers, K., R. Cadoret, J. Knutson, & D. Langbehn (2003). Biology–environment interaction and evocative biology–environment correlation: Contributions of harsh discipline and parental psychopathology to problem adolescent behaviors. *Behavior Genetics*, 33:205–220.

Ritzer, G. (1992). *Contemporary sociological theory.* New York: Alfred A. Knopf.

Roberts, B., K. Walton, & W. Viechbauer (2006). Patterns of mean-level change in personality traits across the life course: A meta-analysis of longitudinal studies. *Psychological Bulletin*, 132:1–25.

Robinson, M. (2004). *Why crime? An integrated systems theory of antisocial behavior.* Upper Saddle River, NJ: Prentice-Hall.

Robinson, W. (1950). Ecological correlations and the behavior of individuals. *American Sociological Review*, 15:351 357.

Rodkin, P., T. Farmer, R. Pearl, & R. Van Acker (2000). Heterogeneity of popular boys: Antisocial and prosocial configurations. *Developmental Psychology*, 36:14–24.

Rodseth, L., & S. Novak (2006). The impact of primatology on the study of human society. In J. Barkow (Ed.), *Missing the revolution: Darwinism for social scientists*, pp. 188–220. New York: Oxford University Press.

Romine, C., & C. Reynolds (2005). A model of the development of frontal lobe functioning: Findings from a meta-analysis. *Applied Neuropsychology*, 12:190–201.

Rose, S. (1999). Precis of *Lifelines:* Biology, freedom, determinism. *Behavioral and Brain Sciences*, 22:871–921.

Rose, H., & S. Rose (Eds.) (2000). *Alas, poor Darwin: Arguments against evolutionary psychology*. London: Jonathan Cape.

Rosenbaum, D., A. Lurigio, & R. Davis (1998). *The prevention of crime: Social and situational strategies*. Belmont, CA: West/Wadsworth.

Rosenfeld, R., & B. Nicodemus (2003). The transition from adolescence to adult life: Physiology of the transition phase and its evolutionary basis. *Hormone Research*, 60:74–77.

Rossi, A. (1964). Equality between the sexes: An immodest proposal. Daedalus, Spring:607–652.

Rossi, A. (1977). A biosocial perspective on parenting. *Daedalus*, 106:1–31.

Rossi, A. (1984). Gender and Parenthood. American Sociological Review, 49:1–19.

Rossi, A. (1997). The impact of family structure and social change on adolescent sexual behavior. *Children and Youth Services Review*, 19:369-400.

Rothbart, M. (2007). Temperament, development, and personality. *Current Directions in Psychological Science*, 16:207–212.

Rothbart, M., A. Ahadi, & D. Evans (2000). Temperament and personality: Origins and outcomes. *Journal of Personality and Social Psychology*, 78:122–135.

Rowe, D. (1992). Three shocks to socialization research. *Behavioral and Brain Sciences*, 14:401–402.

Rowe, D. (1994). *The limits of family influence: Genes, experience, and behavior.* New York: Guilford Press.

Rowe, D. (1996). An adaptive strategy theory of crime and delinquency. In J. Hawkins (Ed.), *Delinquency and crime: Current theories*, pp. 268–314. Cambridge: Cambridge University Press.

Rowe, D. (1997). A place at the policy table? Behavior genetics and estimates of family environmental effects on IQ. *Intelligence*, 24:133–158.

Rowe, D. (2002a). *Biology and crime.* Los Angeles: Roxbury.

Rowe, D. (2002b). On genetic variation in menarche and age at first sexual intercourse: A critique of the Belsky-Draper hypothesis. *Evolution and Human Nature*, 23:365–372.

Rowe, D., D. Almeida, & K. Jacobson (1999). School context and genetic influences on aggression in adolescence. *Psychological Science*, 10:277–280.

Rowe, D., K. Jacobson, & E. Van den Oord (1999). Genetic and environmental influences on vocabulary IQ: Parents' education level as moderator. *Child Development*, 70:1151–1162.

Rowe, D., & Osgood, D. (1984). Heredity and sociological theories of delinquency: A reconsideration. *American Sociological Review*, 49, 526–540.

Ruden, R. (1997). *The craving brain: The biobalance approach to controlling addictions*. New York: HarperCollins.

Ruffie, J. (1986). *The population alternative: A new look at competition and the species.* New York: Random House.

Runciman, C. (2001). Was Max Weber a selectionist in spite of himself? *Journal of Classical Sociology*, 1:15–32.

Runciman. W. (2005). Stone age sociology. *Journal of the Royal Anthropological Institute*, 11:129–142.

Rushton, J. (1988). The reality of racial differences: A rejoinder with additional evidence. *Personality and Individual Differences*, 9:1035–1040.

Rushton, J. (1990). Sir Francis Galton, epigenetic rules, genetic similarity theory, and human life history. *Journal of Personality*, 58:117–140.

Rushton, J. (1991). Race differences: A reply to Mealey. *Psychological Science*, 2:126.

Rushton, J. (1994). The equalitarian dogma revisited. *Intelligence*, 19:263–280.

Rushton, J. (1995). Race and crime: International data for 1989–1990. *Psychological Reports*, 76:307–312.

Rushton, J. (1997). *Race, evolution, and behavior: A life history perspective.* New Brunswick, NJ: Transaction.

Rushton, J. (2004). Placing intelligence into an evolutionary framework or how g fits into the r-K matrix of life history traits including longevity. *Intelligence*, 32:321–328.

Rushton, J., C. Littlefield, & C. Lumsden (1986). Gene-culture coeveolution of complex social behavioe: Human altruism and mate choice. *Proceedings of the National Academy of Sciences*, 83:7340-7343.

Rushton, J., D. Fulker, M. Neale, D. Nias, & H. Eysenk (1986). Altruism and aggression: The heritability of individual differences. *Journal of Personality and Social Psychology*, 50:1192–1198.

Rushton, J., & G. Whitney (2002). Cross-national variation in violent crime rates: Race, r-K theory, and income. *Population and Environment*, 23:501–511.

Rutter, M. (1996). Introduction: Concepts of antisocial behaviour, of cause, and of genetic influence. In G. Bock & J. Goode (Eds.), *Genetics of criminal and antisocial behavior*, pp. 1–20. Chichester, England: Wiley.

Rutter, M. (2007). Gene-environment interdependence. *Developmental Science*, 10:12–18.

Sagan, C. (1995). *The demon-haunted world: Science as a candle in the dark.* New York: Random House.

Sampson, R. (1985), Race and criminal violence: A demographically disaggregated analysis of urban homicide. *Crime and delinquency*, 31:47-82.

Sampson, R (1995). The community. In Wilson, J. & J. Petersilia (Eds.), *Crime*, pp. 193-216. San Franciso: ICS Press.

Sampson, R. (1999). Techniques of research neutralization. *Theoretical Criminology*, 3:438–450.

Sampson, R. (2000). Whither the sociological study of crime. *Annual Review of Sociology*, 26:711–714.

Sampson, R. (2004). Neighborhood and community: Collective efficacy and community safety. *New Economy*, 11:106–113.

Sampson, R., & L. Bean (2006). Cultural mechanisms and killing fields: A revised theory of community-level racial inequality. In R. Peterson, L. Krivo, & J. Hagan (Eds.), *The many colors of crime: Inequalities of race, ethnicity, and crime in America*, pp. 8–36. New York: New York University Press.

Sampson, R., & J. Laub (2005). A life-course view of the development of crime. *American Academy of Political and Social Sciences*, 602:12–45.

Sampson, R., & W. J. Wilson (2000). Toward a theory of race, crime, and urban inequality. In S. Cooper (Ed.), *Criminology*, pp. 149–160. Madison, WI: Coursewise.

Sanderson, S. (2001). *The evolution of human sociality: A Darwinian conflict perspective.* Lanham, MD: Rowman & Littlefield.

Sanjiv, K., & E. Thaden (2004). Examining brain connectivity in ADHD. *Psychiatric Times,* January: 40–41.

Sarfraz, H. (1997). Alienation: A theoretical overview. *Pakistan Journal of Psychological Research,* 12:45–60.

Saudino, K. (2005). Behavioral genetics and child temperament. *Journal of Developmental and Behavioral Pediatrics,* 26:214–223.

Saur, N. (1992). Forensic anthropology and the concept of race: If races don't exist, why are forensic anthropologists so good at identifying them? *Social Science and Medicine,* 34:107–111.

Savage, J., & B. Vila (2003). Human ecology, crime, and crime control: Linking individual behavior and aggregate crime. *Social Biology,* 50:77–101.

Sawhill, I., & J. Morton (2007). *Economic mobility: Is the American dream alive and well?* Washington, DC: The Economic Mobility Project/Pew Charity Trusts.

Sayers, S. (2005). Why work? Marx and human nature. *Science and Society,* 69:606–616.

Scafidi, B. (2008). *The taxpayer costs of divorce and unwed childbearing: First ever estimates for the nation and all fifty states.* New York: Institute for American Values

Scarpa, A., & A. Raine (2003). The psychophysiology of antisocial behavior: Interactions with environmental experiences. In A. Walsh & L. Ellis (Eds.), *Biosocial criminology: Challenging environmentalism's supremacy,* pp. 209–226. Hauppauge, NY: Nova Science.

Scarr, S. (1981). *Race, social class, and individual differences in* IQ. Hillsdale, NJ: Lawrence Erlbaum Associates.

Scarr, S. (1992). Developmental theories for the 1990s: Development and individual differences. *Child Development,* 63:1–19.

Scarr, S. (1993). Biological and cultural diversity: The legacy of Darwin for development. *Child Development,* 64:1333–1353.

Scarr, S. (1995). Psychology will be truly evolutionary when behavior genetics is included. *Psychological Inquiry,* 6:68–71.

Schildkraut, J., S. Murphy, R. Palmieri, E. Iversen, P. Moorman, Z. Huang, S. Halabi, B. Calingaert, A. Gusberg, A. Marks, & A. Berchuck (2007). Trinucleotide repeat polymorphisms in the androgen receptor gene and risk of ovarian cancer. *Cancer Epidemiology, Biomarkers and Prevention,* 16:473–480.

Schmallager, F. (2004). *Criminology today.* Upper Saddle River, NJ: Prentice-Hall.

Schmidt, F., & K. Hunter (2004). General mental ability in the world of work: Occupational attainment and job performance. *Journal of Personality and Social Psychology,* 86:162–173.

Schmous, W. (2003). Is Durkheim the enemy of evolutionary psychology? *Philosophy of the Social Sciences,* 33: 25–52.

Schon, R., & M. Silven (2007). Natural parenting—back to basics in infant care. *Evolutionary Psychology,* 5:102–183.

Segal, N., & K. McDonald (1998). Behavioral genetics and evolutionary psychology: Unified perspective on personality research. *Human Biology,* 70:159–184.

Segerstrale, U. (2006). Evolutionary explanation: Between science and values. In J. Barkow (Ed.), *Missing the revolution: Darwinism for social scientists,* pp. 121–143. Oxford: Oxford University Press.

Seligman, D. (1992*). A question of intelligence: The IQ debate in America.* New York: Birch Lane Press.

Senger, H. (1993). Human rights and crime. Paper presented at the 11th International Congress of Criminology, August 22–28, Budapest, Hungary.

Sergeant, J., H. Geurts, S. Huijbregts, A. Scheres, & J. Ooserlan (2003). The top and bottom of ADHD: A neuropsychological perspective. *Neuroscience and Biobehavioral Reviews,* 27:583–592.

Serran, G., & P. Firestone (2004). Intimate partner homicide: A review of the male proprietariness and the self-defense theories. *Aggression and Violent Behavior,* 9:1–15.

Sesardic, N. (2003). Evolution of human jealousy: A just-so story or a just-so criticism? *Philosophy of the Social Sciences,* 33:427–443.

Shapiro, I. (1991). Resources, capacities, and ownership: The workmanship ideal and distributive justice. *Political Theory,* 19:47–72.

Shavit, Y., & A. Rattner (1988). Age, crime, and the early lifecourse. *American Journal of Sociology,* 93:1457–1470.

Shaw, C., & H. McKay (1972). *Juvenile delinquency and urban areas.* Chicago: University of Chicago Press.

Shedler, J., & Block, J. (1990). Adolescent drug use and psychological health. *American Psychologist,* 45, 612–630.

Shelden, R., S. Tracy, & W. Brown (2001). *Youth gangs in American society* (2nd ed.). Belmont, CA: Wadsworth.

Shi, S., T. Cheng, L. Jan, & Y. Jan (2004). The immunoglobin family member dendrite arborization and synapse maturation 1 (Dasm1) controls excitatory synapse maturation. *Proceedings of the National Academy of Sciences,* 101:13246–13351.

Shively, C., D. Friedman, H. Gage, M. Bounds, C. Brown-Proctor, J. Blair, J. Henderson, M. Smith, & N. Buchheimer (2006). Behavioral depression and positron emission tomography-determined serotonin 1A receptor binding potential in cynomolgus monkeys. *Archives of General Psychiatry,* 63:396–403.

Shore, R. (1997). *Rethinking the brain: New insights into early development.* New York: Families and Work Institute.

Shover, N. (1985). *Aging criminals.* Beverly Hills, CA: Sage.

Shreeve, J. (1994). Terms of estrangement. *Discover,* 15:57–63.

Siegel, L. (1986). *Criminology.* Belmont, CA: Wadsworth.

Sigmund, K., E. Fehr, & M. Nowak (2002). The economics of fair play. *Scientific American,* 286:82–87.

Silfver, M., & H. Klaus (2007). Empathy, guilt, and gender: A comparison of two measures of guilt. *Scandinavian Journal of Psychology,* 48:239–246.

Singer, P. (2000). *A Darwinian left: Politics, evolution, and cooperation.* New Haven, CT: Yale University Press.

Sisk, C., & J. Zehr (2005). Pubertal hormones organize the adolescent brain and behavior. *Frontiers in Neuroendocrinology,* 26:163–174.

Skilling, T., V. Quinsey, & W. Craig (2001). Evidence of a taxon underlying serious antisocial behavior in boys. *Criminal Justice and Behavior,* 28:450–470.

Skinner, L., K. Carrol, & K. Berry (1995). A typology for sexually aggressive males in dating relationships. *Journal of Offender Rehabilitation,* 22:29–45.

Smetana, J., N. Campione-Barr, & A. Metzger (2006). Adolescent development in interpersonal and societal contexts. *Annual Review of Psychology,* 57:255–284.

Smith, D. (1993). Brain, environment, heredity, and personality. *Psychological Reports,* 72:3–13.

Smith, H., & R. Bohm (2008). Beyond anomie: Alienation and crime. *Critical Criminology: An International Journal,* 16:1–15.

Smith, M. (1987). Evolution and developmental psychology: Toward a sociobiology of human development. In C. Crawford, M. Smith, & D. Krebs (Eds.), *Sociobiology and psychology: Ideas, issues, and applications,* pp. 225–252. Hillsdale, NJ: Lawrence Erlbaum Associates.

Snyderman, M., & S. Rothman (1988). *The IQ controversy, the media and public policy.* New Brunswick, NJ: Transaction.

Sobel, M. (2006). What do randomized studies of housing mobility demonstrate?: Causal influence in the face of interferences. *Journal of the American Statistical Association,* 101:1398–1407.

Sober, E., & D. Wilson (1998). *Unto others: The evolutionary psychology of unselfish behavior.* Cambridge, MA: Harvard University Press.

Sokol, R., V. Delaney-Black, & B. Nordstrom (2003). Fetal alcohol spectrum disorder. *Journal of the American Medical Association,* 290:2996–2999.

Sokoloff, N., & B, Price (1995). The criminal law and women. In B. Price & N. Sokoloff (Eds.), *The criminal justice system and women: Offenders, victims, and workers,* pp.11–29. New York: McGraw-Hill.

Sowell, T. (1987). *A conflict of visions: Ideological origins of political struggles.* New York: William Morrow.

Sowell, E., P. Thompson, & A. Toga (2004). Mapping changes in the human cortex throughout the span of life. *Neuroscientist,* 10:372–392

Spear, L. (2000). Neurobehavioral changes in adolescence. *Current Directions in Psychological Science,* 9:111–114.

Spelke, E. (1999). *Object perception: The faculty of segmenting word into objects.* http://humanitas.ucsb.edu/users/steen/CogSci/Spelke.html.

Spencer, T., J. Biederman, T. Wilens, & S. Faraone (2002). Overview and neurobiology of attention-deficit/hyperactivity disorder. *Journal of Clinical Psychiatry,* 63:3–9.

Spergel, I. (1995). *The youth gang problem: A community approach.* New York: Oxford University Press.

Spiro, M. (1975). *Children of the kibbutz.* Cambridge, MA: Harvard University Press.

Spiro, M. (1980). *Gender and culture: Kibbutz women revisited.* New York: Schocken.

Stacey, J. (1993). Good riddance to 'the family': A response to David Popenoe. *Journal of Marriage and the Family,* 55:545–547.

Stallings, M., R. Corely, B. Dennhey, J. Hewwitt, K. Krauter, J. Lessem, S. Mikulich-Gilbertson, S. Rhee, A. Smolen, S. Young, & T. Crowley (2005). A genomewide search for quantitative trait loci that influence antisocial drug dependence in adolescence. *Archives of General Psychiatry,* 62:1042–1051.

Stampp, K. (1956). *The peculiar institution: Slavery in the antebellum South.* New York: Vintage.

Stark, R. (1996). Deviant places: A theory of the ecology of crime. In P. Cordella & L. Siegal (Eds.), *Readings in contemporary criminological theory,* pp.128–142. Boston: Northeastern University Press.

Steffensmeier, D,. & E. Allan (1996). Gender and crime: Toward a gendered theory of female offending. *Annual Review of Sociology,* 22:459–488.

Steffensmeier, D., H. Zhong, J. Ackerman, J. Schwartz, & S. Agha (2006). Gender gap trends for violent crimes, 1980 to 2003: A UCR-NCVS comparison. *Feminist Criminology,* 1:71–98.

Steimer, T. (2002). The biology of fear- and anxiety-related behaviors. *Dialogues in Clinical Neurosciences,* 4:231–249.

Stein, D., J. Fan, J. Fossella, & V. Russell (2007). Inattention and hyperactivity-impulsivity: Psychobiological and evolutionary underpinnings of ADHD. *CNS Spectrums, The International Journal of Neuropsychiatric Medicine,* 12:190–196

Steinberg, L. (2005). Cognitive and affective development in adolescence. *Trends in Cognitive Sciences,* 9:69–74.

Steinberg, L. (2007). Risk taking in adolescence: New perspectives from brain and behavioral research. *Current Directions in Psychological Science,* 16:55–59.

Sterzer, P., C. Stadler, A. Krebs, A. Kleinschmidt, & F. Poustka (2003). Reduced anterior cingulate activity in adolescents with antisocial conduct disorder confronted with affective pictures. *NeuroImage*, 19 (Supplement 1), 123.

Storey, A., K. Delahunty, D. McKay, C. Walsh, & S. I. Wilhelm (2006). Social and hormonal bases for individual differences in the parental behaviour of birds and mammals. *Canadian Journal of Experimental Psychology*, 60, 237–245.

Sullivan, C., & M. Maxfield (2003). Examining paradigmatic development in criminology and criminal justice: A content analysis of research methods syllabi in doctoral programs. *Journal of Criminal Justice Education*, 14:269–285.

Summers, C. (2002). Social interaction over time, implications for stress responsiveness. *Integrative and Comparative Biology*, 42:591–599.

Sutherland, E. (1939). *Principles of criminology*. Philadelphia: J. B. Lippincott.

Sutherland, E., & D. Cressey (1974). Criminology (9th ed.). Philadelphia: J. B. Lippincott.

Sutton, S., & R. Davidson (1997). Prefrontal brain asymmetry: A biological substrate of the behavioral approach and inhibition systems. *Psychological Science*, 8:204–210.

Sutton, W., & E. Linn (1976). *Where the money was: Memoirs of a bank robber*. New York: Viking.

Swaab, D. (2004). Sexual differentiation of the human brain: Relevance for gender identity, transsexualism and sexual orientation. *Gynecological Endocrinology*, 19:301–312.

Swaab, D. (2007). Sexual differentiation of the brain and behavior. *Clinical Endocrinology and Metabolism*, 21:431–444.

Symons, D. (1987). If we're all Darwinians, what's the fuss about? In C. Crawford, M. Smith, & D. Krebs (Eds.), *Sociobiology and psychology: Ideas, issues, and applications*, pp.121–146. Hillsdale, NJ: Lawrence Erlbaum Associates.

Tancredi, L. (2005). *Hardwired behavior: What neuroscience reveals about morality*. Cambridge: Cambridge University Press.

Tang H., T. Quertermous, B. Rodriguez, S. Kardia, X. Zhu, A. Brown, J. Pankow, M. Province, S. Hunt, E. Boerwinkle, N. Schork, & N. Risch (2005). Genetic structure, self-identified race/ethnicity, and confounding in case-control association studies. *American Journal of Human Genetics*, 76:268–75.

Taylor, I. (1999). Crime and social criticism. *Social Justice*, 26:150–168.

Taylor, J. (1992). *Paved with good intentions: The failure of race relations in contemporary America*. New York: Carroll & Graff.

Taylor, I., P. Walton, & J. Young (1973). *The new criminology*. New York: Harper & Row.

Taylor, L., B. Zuckerman, V. Harik, & B. Groves (1994). Witnessing violence by young children and their mothers. *Journal of Developmental and Behavioral Pediatrics*, 15:120–123.

Taylor, S. (2006). Tend and befriend: Biobehavioral bases of affiliation under stress. *Current Directions in Psychological Science*, 15:273–277.

Taylor, S., L. Klien, B. Lewis, T. Gruenwald, R. Gurung, & J. Updegraff (2000). Biobehavioral responses to stress in females: Tend-and-befriend, not fight-or-flight. *Psychological Review*, 107:411–429.

Teicher, M., Y. Ito, C. Glod, F. Schiffer, & H. Gelbard (1997). Early abuse, limbic system dysfunction, and borderline personality disorder. In J. Osofsky (Ed.), *Children in a violent society*, pp. 177–207. New York: Guilford Press.

Terranova, A., A. Morris, & P. Boxer (2008). Fear reactivity and effortful control in overt and relational bullying: A six-month longitudinal study. *Aggressive Behavior*, 34:104–115.

Thernstrom, S., & A. Thernstrom (1997). *America in black and white: One nation indivisible.* New York: Simon & Schuster.

Thornberry, T., D. Huizinga, & R. Loeber (2004). The causes and correlates studies: Findings and policy implication. *Juvenile Justice*, 9:3–19.

Thornberry, T., E. Wei, M. Stouthamer-Loeber, & J. Van Dyke (2000). Teenage fatherhood and Delinquent behavior. *Juvenile Justice Bulletin.* Washington, DC: Office of Justice and Delinquency Prevention.

Thornhill, R., & C. Palmer (2000). *A natural history of rape: Biological bases of sexual coercion.* Cambridge, MA: MIT Press.

Tittle, C. (2000). Theoretical developments in criminology. *National Institute of Justice 2000, Vol. 1: The nature of crime: Continuity and change.* Washington, DC: National Institute of Justice.

TJBHE (2008). The widening racial scoring gap on the SAT college admission test. *The Journal of Blacks in Higher Education,* January, 2008. http://www.jbhe.com/features/college_admissions-test.html.

Toga, A., P. Thompson, & E. Sowell (2006). Mapping brain maturation. *Trends in Neuroscience,* 29:148–159.

Tooby, J. (1999). The view from the president's table: The most testable concept in biology. *Human Behavior and Evolution Society Newsletter,* 8:1–6.

Tooby, J., & L. Cosmides (1992). The psychological foundation of culture. In J. Barkow, L. Cosmides, & J. Tooby (Eds.), The adapted mind: Evolutionary psychology and the generation of culture, pp. 19–136. New York: Oxford University Press.

Tralau, J. (2005). The effaced self in the utopia of the young Karl Marx. *European Journal of Political Theory,* 4:193–412.

Trevarthen, C. (1992). Emotions of human infants and mothers and development of the brain. *Behavioral and Brain Sciences,* 15:524–525.

Trivers, R. (1971). The evolution of reciprocal altruism. *Quarterly Review of Biology,* 46:35–57.

Trivers, R. (1991). Deceit and self-deception: The relationship between communication and consciousness. In M. Robinson & L. Tiger (Eds.), *Man and beast revisited* (pp. 175-191). Washington, DC: Smithsonian Institution Press.

Trivers, R. (2002). *Natural selection and social theory.* Oxford: Oxford University Press.

Trudge, C. (1999). Who's afraid of genetic determinism? *Biologist,* 46:96.

Tucker, K. (2002). *Classical social theory: A contemporary approach.* Oxford: Blackwell.

Turkheimer, E. (2000). Three laws of behavior genetics and what they mean. *Current Directions in Psychological Science,* 9:160–164.

Turkheimer, E. & I. Gottesman (1991). Is $H^2 = 0$ a null hypothesis anymore? *Behavioral and Brain Sciences,* 14:410-411.

Turkheimer, E., A. Haley, M. Waldron, B. D'Onofrio, & I. Gottesman (2003). Socioeconomic status modifies heritability of IQ in young children. *Psychological Science,* 14:623–628.

Turkheimer, E. & M. Waldron (2000). Nonshared environment: A theoretical, methodological, and quantitative review. *Psychological Bulletin,* 126:78-108.

Turner, S. (2007). Social theory as cognitive neuroscience. *European Journal of Social Theory,* 10:357–374.

Turner, H., D. Finkelhor, & R. Ormrod (2006). The effects of lifetime victimization on the mental health of children and adolescents. *Social Science and Medicine,* 62:13–27.

Udry, R. . (1994). The nature of gender. *Demography,* 31:561–573.

Udry, R. (1995). Sociology and biology: What biology do sociologists need to know? *Social Forces,* 73:1267–1278.

Udry, R. (2000). Biological limits of gender construction. *American Sociological Review*, 65:443–457.

Udry, R. (2003). *The National Longitudinal Study of Adolescent Health (Add Health)*. Chapel Hill, NC: Carolina Population Center, University of North Carolina at Chapel Hill.

United States Census Bureau (2000). *The Population of the United States*. Washington, DC: U.S. Census Bureau.

van As, A., G. Fieggen, & P. Tobias (2007). Sever abuse of infants—an evolutionary price for human development? *South African Journal of Children's Health*, 1:54–57.

van Bokhoven, I., W. Matthys, S. van Goozen, & H. van Engeland (2005). Prediction of adolescent outcome in children with disruptive behaviour disorders. *European Child and Adolescent Psychiatry*, 14:153–163.

van Bokhoven, I., S. Van Goozen & H. van Engeland, B., Schaal, L. Arseneault, J.Séguin, D. Nagin, F. Vitaro, & R. Tremblay (2005). Salivary cortisol and aggression in a population-based longitudinal study of adolescent males, *Journal of Neural Transmission* 112:1083–1096.

van Bokhoven, I., S. van Goozen, H. Engeland, B. Schaal, L. Arseneault, J. Seguin, J. Assaad, D. Nagin, F. Vitaro, & R. Tremblay (2006). Salivary testosterone and aggression, delinquency, and social dominance, in a population-based longitudinal study of adolescent males. *Hormones and Behavior*, 50:118–125.

van den Berghe, P. (1987). Incest taboos and avoidance: Some African applications. In C. Crawford, M. Smith, & D. Krebs (Eds.), *Sociobiology and psychology: Ideas, issues, and applications*, pp. 353–371. Hillsdale, NJ: Lawrence Erlbaum Associates.

van den Berghe, P. (1990). Why most sociologists don't (and won't) think evolutionarily. *Sociological Forum*, 5:173–185.

van Goozen, S., G. Fairchild, H. Snoek, & G. Harold (2007). The evidence for a neurobiological model of childhood antisocial behavior. *Psychological Bulletin*, 133:149–82.

van Honk, J., J. Peper, & D. Schutter (2005). Testosterone reduces unconscious fear but not consciously experienced anxiety: Implications for the disorders of fear and anxiety. *Biological Psychiatry*, 58:218–225.

Van Hooff, J. (1990). Intergroup competition in animals and man. . In J. Van der Dennen & V. Falger (Eds.), *Sociobiology and conflict: Evolutionary perspectives on competition, cooperation, violence and warfare*, pp. 23–54. London: Chapman and Hall.

van Voorhees, E., & A. Scarpa (2004). The effects of child maltreatment on the hypothalamic-pituitary-adrenal axis. *Trauma, Violence, and Abuse*, 5:333–352.

Vandell, D. L. (2000). Parents, peer groups, and other socializing influences. *Developmental Psychology*, 36: 699-710.

Vaughn, M. (2008). Substance abuse and crime: Biosocial foundations. In A. Walsh & K. Beaver (Eds.), *Biosocial Criminology: New directions in theory and research*, pp. 176–189. New York: Routledge.

Vedder, R. & L. Gallaway (1993). Declining black unemployment. *Society*, 30:57-63.

Viding, W., J. Blair, T. Moffitt, & R. Plomin (2005). Evidence for substantial genetic risk for psychopathy in 7-year-olds. *Journal of Child Psychology and Psychiatry*, 46:592–597.

Vila, B. (1994). A general paradigm for understanding criminal behavior: Extending evolutionary ecological theory. *Criminology*, 32:311–358.

Vila, B. (1997). Human nature and crime control: Improving the feasibility of nurturant strategies. *Politics and the Life Sciences*, 16:3–21.

Virkkunen, M., D. Goldman, & M. Linnoila (1996). Serotonin in alcoholic violent offenders. In Bock, G. & J. Goode (Eds.), *Genetics of criminal and antisocial behavior*, pp. 168-177. Chichester, England: Wiley.

Vold, G. (1958). *Theoretical Criminology*. Newark: University of Delaware Press.

Vold, G., & T. Bernard (1986). *Theoretical criminology*. New York: Oxford University Press.

Vold, G., T. Bernard, & J. Snipes (1998). *Theoretical criminology*. New York: Oxford University Press.

Walinsky, A. (1997). The crisis of public order. In M. Fisch (Ed.), *Criminology 97/98*, pp. 8–15. Guilford, CT: Dushkin.

Walker, E. (2002). Adolescent neurodevelopment and psychopathology. *Current Directions in Psychological Science*. 11:24–28.

Wallace, W. (1990). Rationality, human nature, and society in Weber's theory. *Theory and Society*, 19:199–223.

Walsh, A. (1990). Illegitimacy, child abuse and neglect, and cognitive development. *Journal of Genetic Psychology*, 151:279–285.

Walsh, A. (1995a). *Biosociology: An emerging paradigm*. New York: Praeger.

Walsh, A. (1995b). Parental attachment, drug use, and facultative sexual strategies. *Social Biology*, 42:95–107.

Walsh, A. (1995c). Genetic and cytogenetic intersex anomalies: Can they help us to understand gender differences in deviant behavior? *International Journal of Offender Therapy and Comparative Criminology*, 39:151–164.

Walsh, A. (1997). Methodological individualism and vertical integration in the social sciences. *Behavior and Philosophy*, 25:121–136.

Walsh, A. (2000a). Evolutionary psychology and the origins of justice. *Justice Quarterly*, 17:841–864.

Walsh, A. (2000b). Behavior genetics and anomie/strain theory. *Criminology*, 38:1075–1107.

Walsh, A. (2000c). Human reproductive strategies and life history theory. In J. Bancroft (Ed.), *The role of theory in sex research*, pp. 17–29. Bloomington: Indiana University Press.

Walsh, A. (2002). *Biosocial criminology: Introduction and integration*. Cincinnati, OH: Anderson.

Walsh, A. (2003). Intelligence and antisocial behavior. In A. Walsh & L. Ellis (Eds.), *Biosocial criminology: Challenging environmentalism's supremacy*, pp. 105–124. Huntington, NY: Nova Science.

Walsh, A. (2004). *Race and crime: A biosocial analysis*. Hauppauge, NY: Nova Science.

Walsh, A. (2005). African Americans and serial killing in the media: The myth and the reality. *Homicide Studies*, 9:271–291.

Walsh, A. (2006). Evolutionary psychology and criminal behavior. In J. Barkow (Ed.), *Missing the revolution: Darwinism for social scientists*, pp. 225–268. Oxford: Oxford University Press.

Walsh, A. (2009). Criminal behavior from heritability to epigenetics: How genetics clarifies the role of the environment. In A. Walsh & K. Beaver, *Biosocial criminology: New directions in theory and research*, pp. 29–49. New York: Routledge.

Walsh, A., & L. Ellis (2004). Ideology: Criminology's Achilles' heel? *Quarterly Journal of Ideology*, 27:1–25.

Walsh, A., & L. Ellis (2007). *Criminology: An interdisciplinary approach*. Thousand Oaks, CA: Sage.

Walsh, A., & C. Hemmens (2008). *Law, justice, and society: A sociolegal introduction*. New York: Oxford University Press.

Walsh, A., & H.-H. Wu (2008). Differentiating antisocial personality disorder, psychopathy, and sociopathy: Evolutionary, genetic, neurological, and sociological considerations. *Criminal Justice Studies*, 21:135–152.

Ward, B. (1999). Sex machines and prisoners of love: Male rhythm and blues, sexual politics and the black freedom struggle. In P. Ling & S. Monteith (Eds.), *Gender in the civil rights movement*, pp. 41–67. New York: Garland.

Ward, T., & R. Siegert (2002). Rape and evolutionary psychology: A critique of Thornhill and Palmer's theory. *Aggression and Violent Behavior*, 7:145–168.

Ward, D., & C. Tittle (1994). IQ and delinquency: A test of two competing explanations. *Journal of Quantitative Criminology*, 10:189–212.

Warr, M. (2002). *Companions in crime: The social aspects of criminal conduct*. New York: Cambridge University Press.

Waschbusch, D., W. Pelham, Jr., J. Jennings, A. Greiner, R. Tarter, & H. Moss (2002). Reactive aggression in boys with disruptive behavior disorders: Behavior, physiology, and affect. *Journal of Abnormal Child Psychology*, 30:641–656.

Watters, E. (2006). DNA is not destiny. *Discover: Science, Technology and the Future*. November.

Weaver, I., N. Cervoni, F. Champagne, A. D'Alessio, S. Sharma, J. Seckl, S. Dymov, M. Szyf, & M. Meaney (2004). Epigenetic programming by maternal behavior. *Nature Neuroscience*, 7:847–854.

Weber, Max (1978). *Economy and Society: An outline of interpretative sociology* (Vol. 2; Edited by G. Roth & C. Wittich). Berkeley: University of California Press.

Weeks, M., M. Singer, M. Grier, & J. Schensul (1996). Gender relations: Sexuality and AIDS risk among African American and Latina women. In C. Sargent & C. Brettell (Eds.), *Gender and health: An international perspective*, pp. 338–370. Upper Saddle River, NJ: Prentice-Hall.

Weinberger, D., B. Elvevag, & J. Giedd (2005) *The adolescent brain: A work in progress*. Washington, DC: The National Campaign to Prevent Teen Pregnancy.

Weinhold, B. (2006). Epigenetics: The science of change. *Environmental Health Perspectives*, 114:161–167.

Wellford, C. (1997). Controlling crime and achieving justice: The American Society of Criminology 1996 Presidential Address. *Criminology*, 35:1–11.

West, D., & D. Farrington (1977). *The delinquent way of life*. New York: Crane Russak.

Western, B. (2003). *Incarceration, employment, and public policy*. New Jersey Institute for Social Justice. http://www.njisj.org/reports/western_report.html.

Wexler, B. (2006). *Brain and culture: Neurobiology, ideology, and social change*. Cambridge, MA: MIT Press.

White, A. (2004). *Substance use and the adolescent brain: An overview with the focus on alcohol*. Durham, NC: Duke University Medical Center.

White, J., D. McMullin, K. Swartout, S. Sechrist, & A. Gollehon (2008). Violence in intimate relationships. A conceptual and empirical examination of sexual and physical aggression. *Children and Youth Services Review*, 30:338–351.

Widom, C., & L. Brzustowicz (2006). MAOA and the "cycle of violence": Childhood abuse and neglect: MAOA genotype and the risk for violent and antisocial behavior. *Biological Psychiatry*, 60:684–689.

Wiebe, R. (2004). Psychopathy and sexual coercion: A Darwinian analysis. *Counseling and Clinical Psychology Journal*, 1:23–41.

Wiebe, R. (2004). Expanding the model of human nature underlying self-control theory: Implications of the constructs of self-control and opportunity. *The Australian and New Zealand Journal of Criminology*, 37:64–84.

Wiebe, R (2009). Psychopathy. In A. Walsh & K. Beaver (Eds.), *Biosocial criminology: New directions in theory and research*, pp. 224–242. New York: Routledge.

Weiss, S., P. Wilson, & D. Morrison (2004). Maternal tactile stimulation and neurodevelopment of low birth weight infants. *Infancy*, 5:85–107.

Wikstrom, P., & R. Loeber. (2000). Do disadvantaged neighborhoods cause well-adjusted children to become adolescent delinquents? A study of male juvenile serious offending, individual risk and protective factors and neighborhood context. *Criminology*, 38:1109–1142.

Wilkinson, G. (1990). Food sharing in vampire bats. *Scientific American*, 262:64–70.

Wilkinson, W. (2005). Capitalism and human nature. *CATO Policy Report*, 27:1–15.

Williams, L., M. Barton, A. Kemp., B. Liddell, A. Peduto, E. Gordon, & R. Bryant (2005). Distinct amygdala-autonomic arousal profiles in response to fear signals in healthy males and females. *NeuroImage*, 28:618–626.

Williams, J., & E. Taylor (2006). The evolution of hyperactivity, impulsivity and cognitive diversity. *Journal of the Royal Society Interface*, 3:399–413.

Willoughby, M. (2003). Developmental course of ADHD symptomology during the transition from childhood to adolescence: A review with recommendations. *Journal of Child Psychology and Psychiatry*, 43:609–621.

Wilson, E. O. (1975). *Sociobiology: The new synthesis*. Cambridge, MA: Harvard University Press.

Wilson, E. O. (1978). *On human nature*. Cambridge, MA: Harvard University Press.

Wilson, E. O. (1990). Biology and the social sciences. *Zygon*, 25:245–262.

Wilson, E. O. (1998). *Consilience: The unity of knowledge*. New York: Alfred A. Knopf.

Wilson, G. (1983). *Love and instinct*. New York: Quill.

Wilson, G., I. Sakura-Lemessy, & J. West (1999). Reaching the top: Racial differences in mobility paths in upper-tier occupations. *Work and Occupations*, 26:165–186.

Wilson, J. Q. (2000). *The marriage problem: How our culture has weakened families*. New York: HarperCollins.

Wilson, J., & R. Herrnstein (1985). *Crime and human nature*. New York: Simon & Schuster.

Wilson, M., & M. Daly (1985). Competitiveness, risk taking and violence: The young male syndrome. *Ethology and Sociobiology*, 6:59–73.

Wilson, M., & M. Daly (1997). Life expectancy, economic inequality, homicide and reproductive timing in Chicago neighborhoods. *British Medical Journal*, 314:1271–1274.

Wilson, W. J. (1987). *The truly disadvantaged*. Chicago: University of Chicago Press.

Wismer Fries, A., T. Ziegler, J. Kurian , S. Jacoris, & S. Pollak (2005). Early experience in humans is associated with changes in neuropeptides critical for regulating social behavior. *Proceedings of the National Academy of Sciences*, 102:17237–17240

Wolfgang, M., T. Thornberry, & R. Figlio (1987). *From boy to man, from delinquency to crime*. Chicago: University of Chicago Press.

Wong, A., I. Gottesman, & A. Petronis (2005). Phenotypic differences in genetically identical organisms: The epigenetic perspective. *Human Molecular Genetics*, 14:11–18.

Wood, P., W. Gove, J. Wilson, & J. Cochran (1997). Nonsocial reinforcement and habitual criminal conduct: An extension of learning theory. *Criminology*, 35:335–366.

Woodger, J. (1948). *Biological principles*. London: Routledge & Kegan Paul.

Wrangham, R., & D. Peterson (1996). *Demonic males: Apes and the origins of human violence*. Boston: Houghton Mifflin.

Wrangham, R., & M. Wilson (2004). Collective violence: Comparisons between youths and chimpanzees. *Annals of the New York Academy of Sciences*, 1036:233–256.

Wright, J. (2009). Inconvenient truths: Science, race and crime. In A. Walsh & K. Beaver, *Biosocial criminology: New directions in theory and research*, pp. 137–153. New York: Routledge.

Wright, J., & K. Beaver (2005). Do parents matter in creating self-control in their children? A genetically informed test of Gottfredson and Hirschi's theory of low self-control. *Criminology*, 43:1169–1202.

Wright, J., K. Beaver, M. Delisi, & M. Vaughn (2008). Evidence of negligible parenting influence on self-control, delinquent peers, and delinquency in a sample of twins. *Justice Quarterly*, 25:544–569.

Wright, J., D. Boisvert, K. Dietrich, & M. Ris (2009). The ghost in the machine and criminal behavior: Criminology for the 21st century. In A. Walsh & K. Beaver (Eds.), *Biosocial criminology: New directions in theory and research*, pp. 73–89. New York: Routledge.

Wright, J., K. Dietrich, M. Ris, R. Hornung, S. Wessel, & B. Lanphear (2008). Association of prenatal and childhood blood lead concentrations with criminal arrests in early childhood. *PLoS Medicine*, 5:732–740.

Wright, R. (1994). *The moral animal: Evolutionary psychology and everyday life*. New York: Pantheon Books.

Wright, R. (1995). The biology of violence. *New Yorker*, 71 (March, 13):68–78.

Wright, R. A., & J. Miller (1998). Taboo until today? The coverage of biological arguments in criminology textbooks, 1961 to 1970 and 1987 to 1996. *Journal of Criminal Justice*, 26:1–19.

Wright, W. (1999). *Born that way: Genes, behavior, personality*: New York: Knopf.

Yacubian, J., T. Sommer, K. Schroeder, J. Glascher, R. Kalisch, B. Leuenberger, D. F. Braus, & C. Buchel (2007). Gene-gene interaction associated with neural reward Sensitivity. *Proceedings of the National Academy of Sciences*, 104:8125–8130.

Yang, Y., A. Raine, T. Lencz, S. Bihrle, L. LaCasse, & P. Colletti (2005). Volume reduction in prefrontal gray matter in unsuccessful criminal psychopaths. *Biological Psychiatry*, 57:1103–1108.

Yoav, G., O. Man, S. Pääbo, & D. Lancet (2003). Human specific loss of olfactory receptor genes. *Proceedings of the National Academy of Sciences*, 100:3324–3327.

Young, J. (ed.) (1994). *The exclusive society: Social exclusion, crime and difference in late modernity*. Thousand Oaks, CA: Sage.

Young, J. (2003). Merton with energy, Katz with structure: The sociology of vindictiveness and the criminology of transgression. *Theoretical Criminology*, 7:389–414.

Young, V., & A. Sulton (1996). Exclude: The current status of African-American scholars in the field of criminology and criminal justice. In A. Sulton (Ed.), *African-American perspectives on crime causation, criminal justice administration and crime prevention*, pp. 1–16. Boston: Butterworth-Heinemann.

Young, L., & Z. Wang (2004). The neurobiology of pair bonding. *Nature Neuroscience*, 7:1048–1054.

Zahn, M. (1999). Thoughts on the future of criminology. The American Society of Criminology Presidential Address. *Criminology*, 37:1–15.

Zeifman, D., & C. Hazan (1997). Attachment: The bond in pair-bonds. In J. Simpson & D. Kenrick (Eds.), *Evolutionary social psychology*, pp. 237–263. Mahwah, NJ: Lawrence Erlbaum Associates.

Zelazniewicz, A. (2007). Evolution of human intelligence: Hypotheses for the causes. *European Anthropological Association Summer School eBook*, 1:241–248.

Zhang, B. (1994). *Marxism and human sociobiology: The perspective of economic reforms in China*. Albany: State University of New York Press.

Zuckerman, M. (1990). The psychophysiology of sensation-seeking. *Journal of Personality*, 58:314-345.

Zuckerman, M. (2007). *Sensation seeking and risky behavior*. Washington, DC: American Psychological Association.

Index

Printed in the United States
by Baker & Taylor Publisher Services